Lecture Notes in Control and Information Sciences

Edited by M. Thoma and A. Wyner

137

S. L. Shah, G. Dumont (Eds.)

Adaptive Control Strategies for Industrial Use

Proceedings of a Workshop
Kananaskis, Canada, 1988

Springer-Verlag
Berlin Heidelberg New York
London Paris Tokyo Hong Kong

Editors
Dr. Sirish L. Shah
Dept. of Chemical Engineering
University of Alberta
Edmonton
Canada T6G 2G6

Dr. Guy Dumont
Dept. of Electrical Engineering
University of British Columbia
2356 Main Hall
Vancouver
Canada V6T 1W5

ISBN 3-540-51869-x Springer-Verlag Berlin Heidelberg New York
ISBN 0-387-51869-x Springer-Verlag New York Berlin Heidelberg

Printing: Mercedes-Druck, Berlin
Binding: B. Helm, Berlin
2161/3020-543210 Printed on acid-free paper.

PREFACE

This book contains proceedings of the workshop on Adaptive Control Strategies for Industrial Use, held at Lodge Kananaskis, Alberta, Canada during June of 1988. Over 70 participants, 30 from industry and the remaining from academia, from 10 countries came together. 20 of the 26 papers presented at this workshop are published in this volume.

The objective of this workshop was to bring together engineers from industry and scientists from universities to focus attention on new developments and practical enhancements for using adaptive control in industry. The workshop was held over a two and one-half day period and provided a forum for a tutorial introduction, through survey-type plenary sessions, to the state of the art in adaptive process control. Attention was also focussed through technical sessions, vendor demonstrations and panel discussions on the process control needs of industry and the mechanisms for transfer of current adaptive control technology between academia and industry.

Two panel discussions were held during the workshop. The first panel discussion was titled "Process Control Needs of the Industrial Community". Panelists from the petroleum, petrochemical, mining and pulp and paper sectors of industry participated in the discussion. The general consensus was that while PID control is suitable for a majority of loops, adaptive control is appropriate for many difficult loops in industry. In addition, lack of suitably trained personnel was identified as the cause of slow technology transfer. The potential benefits of adaptive feedforward control were stressed in this as well as the second panel discussion. The second panel discussion titled "Is a General-Purpose Adaptive Controller Feasible?" was held with several leading academics as panelists. The answer to this question was a unanimous: "No" in the sense that a universal adaptive controller was not only impossible but also undesirable. The discussion then focussed attention on "What important features should the next generation of adaptive controllers have?" The answer to this question led to a long discussion with the main conclusion that a necessary requirement for a practical adaptive controller is an intelligent supervisory system which would oversee the integrated performance and tuning of the estimator, the controller and the appropriate signal conditioning and filtering. The workshop ended on this positive note.

There are a number of people to whom we owe many thanks for making this a successful meeting. Firstly, we would like to acknowledge the administrative and secretarial help we received through the Department of Chemical Engineering, University of Alberta, and the Pulp and Paper Centre, University of British Columbia, as well as the Conference Grants Committee at the University of Alberta for partial financial support of this meeting. Finally, it is a pleasure to acknowledge the help of our graduate students, who looked after the program preparation, registration details, and audio-visual requirements at the workshop.

May, 1989 Sirish L. Shah and Guy Dumont

TABLE OF CONTENTS

RECENT DEVELOPMENTS IN ADAPTIVE CONTROL

IMPLEMENTATION ISSUES IN ADAPTIVE CONTROL

APPLICATIONS OF ADAPTIVE CONTROL

AI AND INDUSTRIAL ADAPTIVE CONTROLLERS

Recent Developments in Adaptive Control

SELF-TUNING MULTISTEP OPTIMISATION CONTROLLERS

David W. Clarke
Department of Engineering Science, Oxford University
Oxford OX1 3PJ, England

Abstract

The widespread demand for increased product quality and the growing use of high-throughput energy-intensive plant means that alternatives to multi-loop PID controllers must be investigated. One significant approach, compatible with current powerful microcomputers, is multistep or long-range predictive control. This uses a process model and an optimization algorithm to determine the best set of future controls for achieving a desired closed-loop perfomance. The design of these controllers depends on the choices of model, optimization method, and performance index, but they have been applied successfully to a wide range of processes, including multi input/output, constrained, dead-time, and multi-modal plant.

The paper discusses the basic philosophy of long-range predictive control, the criteria for selecting an appropriate model structure, the choice of cost-function and its unconstrained and constrained optimization. In practice the process model must be determined experimentally, leading to a self-tuned or adaptive design, and modifications to the standard recursive least-squares parameter estimator are described. Of particular importance is the correct conditioning and filtering of data, especially when handling unmodelled dynamics. The initialisation and industrial application of self-tuned predictive control is outlined.

1 Introduction

Consider a spray-drying tower. A slurry is forced through a nozzle near the top of the tower so that droplets fall through a counter-current flow of heated air to settle as dried powder at the base. A 'quench' damper modulates the flow of heated air entering the tower and an exhaust damper controls the rate of air extraction at the top. The air flow through the tower is to be regulated: too high a flow entrains the particles whereas if it is too low the drying action is insufficient. Tower pressure must be set below atmospheric mainly for safety reasons: the operators might wish to open an inspection hatch. There is significant one-way interaction between the variables, as tower pressure is affected by both flow dampers, and quite strong nonlinearity: equally-spaced increments in the exhaust damper cause changes in the pressure with gains varying by 5 to 1. The relatively fast yet simple dynamics indicate a sample interval of less than 2 seconds.

Preheated oil, mixed with recycled gas, is fed into a hydrotreater reactor where the sulphur and nitrogen are converted to hydrogen sulphide and ammonia, and unsaturated hydrocarbons are saturated. The exothermic reactions are controlled by the addition of cold quench gas between each of the reactor beds using valves which must not be more than 60% open during normal operation. The control objective, for this highly interactive process with complex dynamics, is to maintain 'weighted average bed temperature' at a set-point whilst minimizing energy costs and ensuring that variables stay within prespecified limits. Measured disturbances, usable as feedforward signals, are the variations in feed flow-rate and recycled gas temperature. With a time to steady-state of 90 minutes and a sample interval of 3 minutes there are 30 samples over the plant's rise-time.

A high-speed compliant link is found to have detectable flexure modes ranging from 18Hz up to over 1kHz. It is controlled by a direct-drive DC motor and the tip (end-effector) position is sensed by a light untorqued rigid link. The angular positions of hub and tip are transmitted via shaft encoders to multiple microcomputers for feedback control. Variations of end-mass, such as when picking up a load, modify the modal frequencies. The required path for the tip is predetermined so that *future* values of reference are known. The link is designed for fast slewing which inevitably involves torque saturation of the motor and requires a sample rate of at least 60Hz.

The above are examples of typical high-performance control problems for which classical approaches are unsuitable without a great deal of effort in design and tuning. Yet they are all cases for which long-range predictive control (LRPC) has been successfully applied to real plant: MIMO Generalized Predictive Control (GPC) using a DEC LSI11 for the spray-dryer (Lambert,E., 1987); constrained Dynamic Matrix Control (DMC) using an IBM PC-AT for the reactor (Cutler and Hawkins, 1987); SISO GPC using twin Motorola 68020s for the flexible link (Lambert,M., 1987). This paper develops the basic ideas of LRPC, discusses what choices the designer has in achieving different performance objectives, and shows how the methods can be used in practice. It concentrates on methods such as GPC which can have a self-tuned or adaptive mode and for which there are theorems to demonstrate the stabilization of unstable, nonminimum-phase plant.

A long-range predictive controller is a combination of the following basic components:

- A *model* $\mathcal{M}(\theta)$ with which future plant ouputs $y(t+j)$ can be predicted at time t based on assumptions about present and future controls. A good choice of model structure is crucial for an effective LRPC design.
- Knowledge of the time-behaviour of the future set-point $w(t+j)$. If known in detail, as in robotics or some batch process applications, these are called *preprogrammed* set-points; otherwise a simple choice is to make the future set-point have the known current value.
- A cost- or objective-function $J(\mathbf{e}, \mathbf{u})$ where $\mathbf{e}$ is the vector of future system-

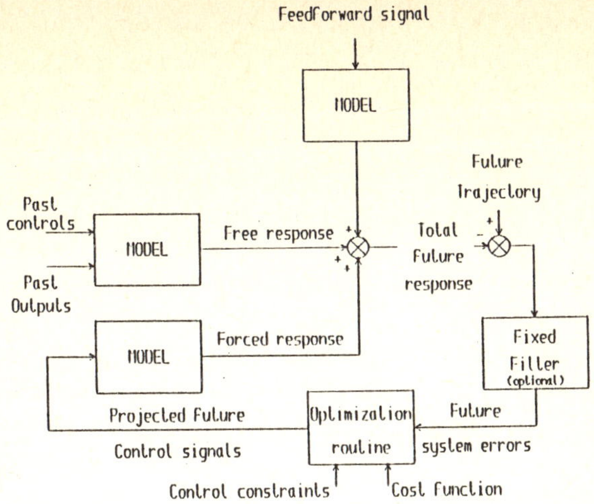

Figure 1: The structure of an LRPC loop

errors and $\mathbf{u}$ is the vector of appropriate controls. Important choices here include the *range* of future errors to include in the cost (e.g. one possibility is over the last half of the rise-time) and in the *number* of future control actions to consider as contributing to the future response.

- An optimization routine which minimises J, either *unconditionally* (assuming there are no constraints) or subject to known constraints such as control amplitude limits. The unconditional minimum can be derived analytically and hence involves the minimum number of computations.
- (Optionally) various filters and transfer functions added to the design to broaden the range of performance objectives (e.g. closed-loop pole-placement) or to improve its robustness against practical inevitabilities such as unmodelled dynamics and disturbances.

The general structure of a long-range predictive controller is shown in Fig.1. Clearly there are a large range of possibilities for each of the components described above, so that very many distinct designs can be produced. Practical considerations, however, restrict the freedom of choice: these points are discussed in the following sections.

One of the most important questions is whether to choose full-value, offset, or incremental signals as a basis of the design. Suppose internal calculations take place using variables $\tilde{u}$, $\tilde{y}$ representing the plant's input and output, then we can select:

Full-value : $\tilde{u}(t) \equiv u(t)$, so that if $\tilde{u}$ is the result of the control computations its value is directly transmitted to the plant. Similarly $\tilde{y}(t) \equiv y(t)$ is used for the measured variable.

Offset : $\tilde{u}(t) \equiv u(t) - u_0$, where u_0 is some predetermined mean-value which is added to the algorithm output to obtain the transmitted control. Similarly $\tilde{y}(t) \equiv y(t) - y_0$; here the mean-value can be obtained either *ab initio* from a steady-state reading or recursively by low-pass filtering.

Incremental : $\tilde{u}(t) \equiv u(t) - u(t-1) \equiv \Delta u(t)$, where Δ is the backward-differencing operator. Hence a design with $\tilde{u}(t)$ produces a control $u(t) = u(t-1) + \tilde{u}(t)$. This is seen automatically to append an integrator in the forward-path – clearly desirable in order to have zero offset against constant load-disturbances. The internal variable $\tilde{y}(t)$ is also given by the differenced data $y(t) - y(t-1)$, being zero when the measured variable is constant.

There are many reasons why the incremental form is to be preferred. In self-tuning or adaptive control the estimated model is generally a local-linearization about the current operating point, but if the offset approach is used with constant $[u_0, y_0]$ these might not correspond to the correct values if the plant or its operating point change. In the control calculation of LRPC future values of $\tilde{u}$ are considered: one reasonable assumption is that there is a *control horizon* beyond which the control becomes constant and hence the corresponding increments are zeros. This means that there are significantly fewer variables involved in the optimization, leading to faster computations. This will be explored in more detail later.

There is one valid objection to the simple use of incremental data is that high frequencies (where there will be noise and effects of unmodelled dynamics) are emphasised. This is overcome by appropriate filtering (see later); indeed the offset approach using a computed u_0 obtained by the low-pass filter:

$$\hat{u}_0(t) = \beta \hat{u}_0(t-1) + (1-\beta)u(t),$$

gives an internal variable $\tilde{u}$ with overall *high-pass* filtering:

$$\tilde{u}(t) = \frac{\beta\Delta}{1-\beta q^{-1}} u(t),$$

where q^{-1} is the backward-shift operator. This filter blocks low frequencies (from the Δ operator) and has unit gain at high frequencies. Hence an incremental formulation in conjunction with suitable filtering has similar behaviour to the offset approach.

2 Choice of process model and output prediction

The purpose of a model is to predict the output response $\tilde{y}(t+j)$ based on past known inputs and measured outputs $\{\tilde{u}(t-i); \tilde{y}(t-i)\}$, and depending on the additional effect of current and future controls. (Obviously if there are constraints on the plant's control actuation such as motor-torque limits these should be taken into account as accurately as possible so that $\tilde{u}$ reflects the *actual* value affecting the plant rather than the possibly clipped demand of the LRPC algorithm). A model can be thought of as having two aspects: its *structure* $\mathcal{M}(.)$ and its actual *parameter* set $\{\theta\}$.

The derivation and computer implementation of an LRPC algorithm depends on the assumed structure $\mathcal{M}$. A 'good' design gives satisfactory answers to the following questions:

- Can $\mathcal{M}$ represent a very general class of plants? For example, can it deal with dead-time, unstable, lightly-damped, high-order systems simply by changes in parameter values? If so, then an LRPC code would not need *ad hoc* modifications when applied to different problem areas.
- Is the number of parameters minimal with $\mathcal{M}$ still giving adequate predictions? For then the computational burden could be minimised.
- Can prior knowledge be easily incorporated? This is particularly relevent to adaptive algorithms: for example if in robot manipulation the only unknown is the load mass a simple algorithm could be used to determine its value, and bounds on likely loads could be prespecified as an error-check.
- Is there a realistic assumption about load-disturbances? Some algorithms attempt to model these directly (which might be a fruitless exercise); most approaches make only implicit (and often unacknowledged) assumptions here. In practice the *minimal* assumption is that there is a constant load-disturbance (corresponding say to steady-state heat-loss or to constant load-torque) implying that even with a zero control signal the measured output would be non-zero. This is a further reason for insisting on offset or incremental models.

For any given application, associated with the structure are the particular parameters which need to be determined. This can be achieved by detailed mathematical modelling or simulation, but more often direct experiments on the plant are required. If the plant dynamics are reasonably time and set-point independent, a prior exercise can provide once-and-for-all parameters. If they vary with the set-point a series of related experiments could provide sets of parameters suitable for 'gain-scheduling'. In general, however, variations can (in principle: more difficult in practice) be handled by an adaptive algorithm which tracks changes as reflected in the plant's I/O behaviour. Hence acquiring good parameters involves answering the following questions:

- Do I need to perform specific open-loop experiments on the plant or is 'normal operating data' (such as with a closed-loop test) acceptable?

- Must I inject a predetermined test-signal such as a step, or can other signals be used?
- Should I/O data be prefiltered to accentuate the model fit over particular frequency ranges?
- What is the effect of noise, nonlinearities, and unmodelled dynamics on the quality of the model and the subsequent closed-loop LRPC behaviour?
- Will my estimator be able to track time-variations?

Details of parameter estimation are given later; here we simply note that parameter estimation is simpler if $\mathcal{M}$ involves a minimal number of parameters.

The simplest general-purpose model is the impulse response or weighting sequence from which the output $y(t)$ is derived by the convolution sum:

$$y(t) = \sum_{i=1}^{\infty} h_i u(t-i).$$

The only assumption here is superposition: the plant is linear with arbitrary dynamics having parameters $\{h_i\}$ being points on its unit-pulse response. In principle, however, there are an infinite number of parameters, so for implementations there must be *truncation* after some point N, assuming $[h_i = 0, i > N]$. This model can be written in operator form as:

$$y(t) = H(q^{-1})u(t),$$

where $H(q^{-1})$ is the FIR polynomial:

$$H(q^{-1}) = h_1 q^{-1} + h_2 q^{-2} + \cdots + h_N q^{-N}.$$

The problem with FIR models is that they require a very large number of parameters to represent stiff dynamic systems accurately. The sample interval h must be smaller than the smallest time-constant of interest and the model 'length' must be such that Nh exceeds the plant's settling-time. A typical choice of settling-time for overdamped dynamics is 5 times the largest time-constant: hence with only a 1:10 range of time-constants at least 50 parameters may be necessary.

A closely related plant representation is its step-response. Instead of taking the input to be a series of pulses it is considered as a set of 'moves' or increments. Superposition then provides the output:

$$y(t) = s_1 \Delta u(t-1) + s_2 \Delta u(t-2) + \cdots + s_i \Delta u(t-i) + \cdots,$$

where the parameters $\{s_i\}$ are points on the unit-step response. Again s_i must be truncated at the point N where the response has settled, and previous moves are assumed to provide an 'initial condition' y_0, giving:

$$y(t) = y_0 + \sum_{i=1}^{N} s_i \Delta u(t-i).$$

Similar considerations about the numbers of parameters apply; indeed $\{h\}$ and $\{s\}$ are related by the iterations:

$$s_0 = 0; s_i = s_{i-1} + h_i, \quad i = 1, 2, \dots N;$$

$$h_i = s_i - s_{i-1} = \Delta s_i, \quad i = 1, 2, \dots N.$$

Put simply, the step-response is the integral of the pulse-response.

An alternative with a long history in self-tuning control is the difference-equation:

$$\begin{aligned} y(t) + a_1 y(t-1) \;&+\; a_2 y(t-2) + \cdots a_{na} y(t-na) = \\ b_1 u(t-1) \;&+\; b_2 u(t-2) + \cdots b_{nb} u(t-nb). \end{aligned}$$

This is often called the DARMA (Deterministic AutoRegressive and Moving-Average) model, having the operator form:

$$A(q^{-1})y(t) = B(q^{-1})u(t),$$

where A and B are polynomials of degree na and nb in the backward-shift operator. All the above model forms can be related by:

$$\Delta S(q^{-1}) = H(q^{-1}) = A^{-1}(q^{-1})B(q^{-1}),$$

though note here that an n'th order DARMA model can give an exact representation of a stiff n'th order plant as it does not need truncation. In particular it can emulate *unstable* processes which do not admit pulse- or step-response models. It can also handle deadtime by appropriate changes to the order of the $B(q^{-1})$ polynomial: k samples of deadtime increases nb by k and the leading k parameters become zero. Hence in adaptive applications where the deadtime might vary one possibility is to use a relatively high order B and accept that some leading or trailing coefficients might become insignificant. It is important here to ensure that the controller design (such as the choice of horizons) is insensitive to these changes: LRPC, unlike some approaches such as minimum variance, is acceptably robust.

A SISO state-space model, giving access to extensive theoretical and algorithmic results, is of the form:

$$\begin{aligned} \mathbf{x}(t+1) &= \mathbf{A}\mathbf{x}(t) + \mathbf{b}u(t) \\ y(t) &= \mathbf{c}^T\mathbf{x}(t). \end{aligned}$$

It is interesting to note that if a unit-pulse sequence {1,0,0,...} is injected into this model the response is given by:

$$\{h\} = \{0, \mathbf{c}^T\mathbf{b}, \mathbf{c}^T\mathbf{A}\mathbf{b}, \dots\} \Longrightarrow h_i \equiv \mathbf{c}^T\mathbf{A}^{i-1}\mathbf{b}.$$

It is possible to convert directly from a DARMA model into state-space using an observable canonical form with $n = \max(na, nb)$ and:

$$\mathbf{A} = \begin{bmatrix} -a_1 & 1 & 0 & 0 & \cdots & 0 \\ -a_2 & 0 & 1 & 0 & \cdots & 0 \\ \vdots & & & & & \\ -a_n & 0 & 0 & 0 & \cdots & 1 \end{bmatrix},$$

$$\begin{aligned} \mathbf{b} &= [b_1, b_2, \ldots, b_n]^T, \\ \mathbf{c} &= [1, 0, 0, \ldots, 0]^T. \end{aligned}$$

At some stage in an LRPC design some estimate of the states (which in general are not directly accessible) is required, for which an observer of the form:

$$\hat{\mathbf{x}}(t) = \mathbf{A}\hat{\mathbf{x}}(t-1) + \mathbf{b}u(t-1) + \mathbf{k}(y(t) - \mathbf{c}^T\hat{\mathbf{x}}(t-1))$$

can be used. It is possible, however, to define a *nonminimal* state comprising simply of past inputs and outputs:

$$\mathbf{x}(t) \equiv [y(t), y(t-1), \ldots, y(t-na); u(t-1), u(t-2), \ldots, u(t-nb)]^T.$$

In this case all elements of the state are accessible; moreover a state-feedback of the form $u(t) = \mathbf{k}^T\mathbf{x}(t)$ gives a controller which can be immediately interpreted as a transfer-function.

There are many advocates for each of the model structures. A tentative (and possibly prejudiced) assessment is:

Impulse-response : Easy to formulate the corresponding LRPC designs. Standard PRBS/ crosscorrelation is usable for parameter estimation. Needs many parameters for a good fit with stiff or lightly-damped dynamics, so adaptation might be poor. Cannot be applied to unstable plant (unless stabilized by inner-loop feedback). Some highly complex process dynamics might be handled well: consider for example a plant with parallel paths having significantly different dynamics. Significant truncation problem: how big should N be?

Step-response : As above: initial parameter estimation (reaction curve) even easier *provided* that there are no load-disturbances during the test. Its incremental model formulation is more 'natural'.

DARMA : Minimal parameterization. Can deal with dead-time, lightly-damped and unstable dynamics. No truncation problem. Need to choose two (na, nb) rather than one (N) model orders (in practice choose na equal to the number of 'difficult' poles and nb large enough to deal with the expected range of dead-time). Must use an algorithm (Section 4) for parameter estimation, though there are simple results for obtaining second-order dead-time models from reaction curves.

State-space : Very general and can cope with prior stuctural information. The multivariable case is a simple extension. Parameter estimation is difficult (use Extended Kalman Filtering): for a general model note that $\mathbf{A}$ has n^2 parameters yet only n distinct values (e.g. eigenvalues) can be determined. Hence canonical forms such as phase-variable or observable should be used. In practice many approaches estimate a DARMA model and convert to state-space.

2.1 Multi-loop and multivariable models

In principle the extension to the multivariable case is straightforward. Suppose that the plant has nu inputs and ny outputs. Then its dynamics can be considered as an $ny \times nu$ matrix of transfer-functions with each output $y_l(t)$ being affected by all the inputs $[u_k; k = 1 \cdots nu]$. For example with a pulse-response model:

$$y_l(t) = \sum_{k=1}^{nu} \sum_{i=1}^{N_{kl}} h_{ikl} u_k(t-i).$$

Hence the problem of parameterization is even more critical for high-speed or adaptive control.

It is important to note that the number of inputs and outputs need not be the same. For example if there are auxiliary measurements they can be treated as *feedforward* signals or measurable load-disturbances which can be taken into account in the prediction and subsequent optimization, hence providing improved disturbance rejection. Similarly if there are excess controls suitable costings or constraints can be adopted which can lead to a useful solution. Perhaps the most important consideration is whether to treat the problem as essentially multiloop or fully multivariable:

Multiloop : The control design is an extension to normal SISO practice where *previous* controls for other loops are treated as load-disturbances. Hence data from the other loops take part in the *predictions*.

Multivariable : The plant is centrally controlled as an $nu \times ny$ problem for which at time t all controls are computed together, taking into account, say, the effect on loop l of *future* controls in loop k. Hence other loops take part in both the predictions and the final control computations.

In principle the multivariable approach should give the best performance but in practice the margin of superiority can be small, particularly as it involves more parameters. The telling argument for the multiloop approach is that it allows for a *distributed* implementation, which is current process control practice. For example it permits sequential loop-closing whereas a multivariable approach usually needs to have all loops closed simultaneously.

Given the above considerations, a MIMO DARMA plant can be given a multi-loop structure of the form:

$$A_l(q^{-1})y_l(t) = \sum_{k=1}^{nu} B_{kl}(q^{-1})u_k(t), \quad l = 1 \dots ny.$$

For a full multivariable design we use a vector difference equation which can be written:

$$(\mathbf{A}_0 + \mathbf{A}_1 q^{-1} + \cdots + \mathbf{A}_{na} q^{-na})\mathbf{y}(t) = (\mathbf{B}_1 q^{-1} + \cdots + \mathbf{B}_{nb} q^{-nb})\mathbf{u}(t),$$

where $\mathbf{u}(t)$ and $\mathbf{y}(t)$ are the I/O vectors and $\mathbf{A}_i$, $\mathbf{B}_i$ are appropriately dimensioned matrices. In much of the self-tuning literature for MIMO plant prior knowledge of its *interactor matrix* is required as well as assumptions of nonsingularity of the leading $\mathbf{B}_1$ matrix: this is not the case for LRPC designs.

2.2 Long-range prediction

The future plant output behaviour $y(t+j)$ is assumed to be the sum of two effects:

1. $p(t+j)$, being the free response to the past known control signals, and:
2. $p'(t+j)$, the additional response due to the signals $\tilde{\mathbf{u}}$, which are to be determined by the optimization algorithm.

In some algorithms where there is an explicit model for the disturbances these can also be predicted in, say, a minimum-variance sense. To avoid undue complexity we restrict attention to deterministic predictors, noting how disturbance effects will be handled in practice. In particular we are interested in predictors which *preserve mean levels*, as this is a prerequisite for offset-free closed-loop performance. This is tested by considering whether the mean prediction $\bar{p}$ equals the output mean $\bar{y}$ under steady-state conditions with constant controls, output and load-disturbance.

Note that LRPC makes only a restricted assumption about plant linearity. The signal $p(t+j)$ can be generated by a *nonlinear* model and the design is still acceptable provided that the future controls are 'small' enough to admit local linearization. Note further that the 'free' response could be based on assuming either zero future controls or a constant future control (i.e. zero control increments), depending on the choice of model.

As a simple example consider the FIR model written at time $(t+j)$:

$$y(t+j) = \sum_{i=1}^{N} h_i u(t+j-i),$$

for which we have:

$$\begin{aligned} p(t+j) &= \sum_{i=j+1}^{N} h_i u(t+j-i), \text{ and:} \\ p'(t+j) &= \sum_{i=1}^{j} h_i u(t+j-i). \end{aligned}$$

Clearly this model does not satisfy the offset-free requirement, so some *estimate* of the load-disturbance must be provided and added to $p(t+j)$. This is also the case with the standard DMC predictor, for which:

$$\begin{aligned} p(t+j) &= y_0 + \sum_{i=j+1}^{N} s_i \Delta u(t+j-i), \text{ and:} \\ p'(t+j) &= \sum_{i=1}^{j} s_i \Delta u(t+j-i). \end{aligned}$$

Here y_0 can either be a fixed value determined during initialisation, or more effectively it can be back-calculated using the current measured value:

$$y_0 = y(t) - \sum_{i=1}^{N} s_i \Delta u(t-i).$$

Note that for this model the free response assumes constant future controls: $u(t+j) = u(t-1)$.

A DARMA model also suffers from the above problem unless it is written in *incremental* form, obtained from multiplying by the differencing operator Δ:

$$\begin{aligned} A(q^{-1})\Delta y(t) &= B(q^{-1})\Delta u(t), \text{ or:} \\ A(q^{-1})\tilde{y}(t+j) &= B(q^{-1})\tilde{u}(t+j). \end{aligned}$$

This form was inspired by the Controlled AutoRegressive and *Integrated* Moving Average (CARIMA) model which also represents load disturbances acting on the plant as filtered Brownian Motion or random steps.

A j-step predictor can be obtained by recursively solving a set of Diophantine equations, but it is much simpler computationally just to iterate this equation making the following substitutions:

$$\begin{aligned} \tilde{u}(t+j-i) &= \Delta u(t+j-i), && \text{for } j < i \\ \tilde{u}(t+j-i) &= 0, && \text{for } j \geq i \\ \tilde{y}(t+j-i) &= \Delta y(t+j-i), && \text{for } j \leq i \\ \tilde{y}(t+j-i) &= \tilde{y}(t+j-i), && \text{for } j > i. \end{aligned}$$

This means using known past data for u and y, assuming future zero control increments Δu, and taking y predictions as data for subsequent predictions. This is shown in pseudo-code by Table 1, where DU and DY are arrays containing past values of Δu and Δy, DPY contains prediction increments $\tilde{y}$, and PY contains the full-valued predictions $p(t+j)$.

Note again that means are preserved for if the plant is constant with all of Δu and Δy being zero then the predictions $p(t+j)$ will all equal the current measured value. To obtain the forced components $p'(t+j)$ it is seen that, because

Table 1: Predictions with a DARMA (CARIMA) model

```
for j = 1 ... horizon
{   s = 0;
for  i = 1 ... na
s = s − a_i × {if j > i then DPY(j − i) else DY(i − j + 1)};
for i = 1 ... nb
s = s + b_i × {if j ≥ i then 0 else DU(i − j)};
DPY(j) = s;
PY(j) = DPY(j)+{if j = 1 then Y else DPY(j − 1)} };
```

of the incremental form of the model, these must equal those of the step-response formulation:

$$p'(t+j) = \sum_{i=1}^{j} s_i \Delta u(t+j-i), \quad j = 1 \dots N_2.$$

Collecting N_2 output predictions into a vector $\mathbf{y}_p$ it is seen that they satisfy the key equation:

$$\mathbf{y}_p = \mathbf{G}\tilde{\mathbf{u}} + \mathbf{p}, \quad \text{where:}$$

$$\begin{aligned} \mathbf{y}_p &= [y(t+1), y(t+2), \dots, y(t+N_2)]^T, \\ \tilde{\mathbf{u}} &= [\Delta u(t), \Delta u(t+1), \dots, \Delta u(t+N_2-1)]^T, \\ \mathbf{p} &= [p(t+1), p(t+2), \dots, p(t+N_2)]^T, \end{aligned}$$

and the $N_2 \times N_2$ matrix $\mathbf{G}$ is of banded triangular form:

$$\mathbf{G} = \begin{bmatrix} g_1 & 0 & \cdots & 0 \\ g_2 & g_1 & \cdots & 0 \\ \vdots & \vdots & \cdots & \vdots \\ g_{N_2} & g_{N_2-1} & \cdots & g_1 \end{bmatrix}.$$

The above equations correspond to the case where predictions over the whole range from 1 to the *upper output horizon* N_2 are required, depending on N_2 control increments $\Delta u(t+j)$. In practice only a subset of each should be involved: it is found that this gives better control behaviour with provable stabilization properties. In general these subsets correspond to:

Lower output horizon : in many cases the output cannot deviate much from the current $y(t)$ for the first few samples or, for nonminimum-phase plant, it might even deviate in the 'wrong' direction. Hence predictions could be started after some initial point N_1 for which there is a significant positive-going output change.

Control horizon : suppose it is assumed that only a restricted number of future 'moves' are to be made, i.e. that after NU increments future values of Δu are to be zero. This reduces the dimension of $\tilde{\mathbf{u}}$ from N_2 to NU, which is significant for calculations with typical values of N_2 being 10...50 and of NU being 1...5.

With these assumptions the matrix $\mathbf{G}$ becomes $(N_2 - N_1 + 1) \times NU$:

$$\mathbf{G} = \begin{bmatrix} g_{N_1} & g_{N_1-1} & \cdots & 0 \\ g_{N_1+1} & g_{N_1} & \cdots & 0 \\ \vdots & \vdots & \cdots & \vdots \\ g_{N_2} & g_{N_2-1} & \cdots & g_{N_2-NU+1} \end{bmatrix}.$$

The properties of this matrix, such as its singular values, gives valuable insight into the choices of horizon in LRPC.

For the deterministic case or where the future load disturbance is indeed constant, and the plant model is correct, the above equations will give an *exact* prediction of the future plant output. In practice the model will be correct neither in its structure nor in its parameters, and the disturbances will be random. Hence there will be a prediction discrepancy reflecting these sources of error which would probably grow as the horizon extends. Data filters can be used to enhance the robustness of the predictions against certain types of uncertainty: these will be discussed later as part of the model tuning problem.

3 Predictive control design

A sample instant t we have available the set-point sequence $\{w(t+j)\}$ and the predictions $y_p(t+j)$. We note that one component of the predictions is affected by control decisions yet to be taken. Hence we can choose to assert a future control sequence so as to minimise a cost-function:

$$J(N_1, N_2, NU, \lambda) \equiv \sum_{j=N_1}^{N_2} (y_p(t+j) - w(t+j))^2 + \lambda \sum_{j=1}^{NU} \Delta u^2(t+j-1).$$

In terms of the vectors defined previously this can be written:

$$J = (\mathbf{G}\tilde{\mathbf{u}} + \mathbf{p} - \mathbf{w})^T (\mathbf{G}\tilde{\mathbf{u}} + \mathbf{p} - \mathbf{w}) + \lambda \tilde{\mathbf{u}}^T \tilde{\mathbf{u}}.$$

The values of N_1, N_2, NU and λ are *design constants* which are adopted to achieve the control objectives. Details of the choices is found below: here we provide an informal discussion.

For minimising the cost J there are NU 'degrees of freedom'. If the value of λ is zero then J_{min} corresponds to a minimisation of the sum-of-squares of

$NY \equiv (N_2 - N_1 + 1)$ system errors over the chosen horizons. If NY equals NU and the solution is feasible then the minimum value of J is zero: the output will achieve the set-point exactly over the chosen horizons. Provided that NY at least equals the internal plant state dimension n, this implies that the error and all relevent derivatives are zero by time $t + N_2$ and that the set-point is attained in a minimum-ripple sense. However if NY is smaller than n and N_1 is also small, the control is attempting to drive the plant too fast to the set-point: this might induce large controls, intersample ripple, and in the limit, instability. Note that the classical minimum-variance self-tuner has $NY = 1$ and N_1 the assumed dead-time k of the plant: this is well-known to be an unstable controller for nonminimum-phase dynamics.

Consider now another limiting case where N_2 is very large (say well beyond the plant's settling time) and NU is just 1. The algorithm choses that value of *constant* future control which minimises the system errors over a long range. Suppose the set-point follows a step: the control also follows a step of magnitude sufficient to drive the output to the set-point. This is known as *mean-level* control for which the open and closed-loop modes are the same (for stable plant): it is found to be highly robust.

What is the rôle of λ? Traditional LQ design suffers from the problem that λ is the main 'tuning-knob'. In the GPC approach the horizons are used for coarse setting of control objectives with λ simply available for fine trimming. Increasing its value trades system error minimisation against control signal variance. Hence a non-zero λ is most useful for 'high-performance' applications where the larger values of NU give more energetic control signals. It has a subsidiary function of improving the numerical properties of the optimization algorithm.

Given the choices of the design constants the predictive control approach is to use a *receding-horizon* strategy in which for every sample instant t:

1. From the current measured value $y(t)$ and previous I/O data, the prediction sequence $p(t + j)$ is computed and the set-point sequence $w(t + j)$ evaluated.

2. The future-control-sequence vector $\tilde{\mathrm{u}}$ is calculated to minimise the cost-function J.

3. The *first* element $\tilde{u}_1$ is extracted to provide the current control $u(t) = u(t - 1) + \tilde{u}_1$ which, after suitable clipping by known actuator limits, is transmitted to the plant.

4. All sequences, etc., are shifted in preparation for the next sample.

3.1 Unconstrained and constrained optimization

If there are no amplitude or rate limits on the actuation signal, minimisation of J can be performed analytically to give the optimising solution:

$$\tilde{\mathbf{u}}^* = (\mathbf{G}^T\mathbf{G} + \lambda\mathbf{I})^{-1}\mathbf{G}^T(\mathbf{w} - \mathbf{p}).$$

Note that if NU equals NY and $\mathbf{G}$ is invertible, then for zero control weight λ this can be written as:

$$\tilde{\mathbf{u}}^* = \mathbf{G}^{-1}(\mathbf{w} - \mathbf{p}),$$

and substituting into the key prediction equation:

$$\mathbf{y}_p = \mathbf{G}\tilde{\mathbf{u}} + \mathbf{p} = \mathbf{G}\mathbf{G}^{-1}(\mathbf{w} - \mathbf{p}) + \mathbf{p} = \mathbf{w}.$$

This confirms the above discussion that the set-point is attained exactly starting at time $(t + N_1)$, and shows that this is possible only if the horizons are chosen for invertibility of $\mathbf{G}$. Note that if there is a nonzero deadtime then b_1 (and possibly others) is zero, implying that g_1 (and others) are zero. Hence a clearly necessary condition for invertibility with $NY = NU$ is that N_1 exceeds the plant deadtime.

In the general case a solution depends on the invertibility of the $NU \times NU$ matrix $(\mathbf{G}^T\mathbf{G} + \lambda\mathbf{I})$: the rank of this matrix must equal NU. As $\mathbf{G}^T\mathbf{G}$ is at least positive semi-definite then invertibility is guaranteed for small nonzero λ. In most cases, however, even a zero λ can be used provided NY exceeds NU. For example, with $NU = 1$ the solution is:

$$\Delta u^*(t) = \Big(\sum_{j=N_1}^{N_2} [g_j(w(t+j) - p(t+j))]\Big)/\Big(\sum_{j=N_1}^{N_2} g_j^2\Big).$$

This is feasible provided at least *one* of the g_j is nonzero: i.e. the plant must have a nonzero step-response!

It is useful to consider the computational burden associated with the optimising solution. Note that only the first element of $\tilde{\mathbf{u}}^*$ is required and that in the nonadaptive case the inversion needs be performed just once. Hence the $1 \times NU$ row-vector:

$$\bar{\mathbf{g}}^T \equiv [1, 0, \ldots, 0](\mathbf{G}^T\mathbf{G} + \lambda\mathbf{I})^{-1}\mathbf{G}^T$$

can be precomputed so that the only on-line calculation is the inner-product:

$$\Delta u(t) = \sum_{i=1}^{NY} \bar{g}_i[w(t + N_1 - 1 + i) - p(t + N_1 - 1 + i)].$$

For the adaptive case the matrix computations must be performed each time there is a significant change in the estimated plant model. The time-consuming aspect of inversion is clearly minimised by having a small NU: fortunately this is often the best choice in process control because of the consequential smooth and robust performance. It is interesting to see that a multiprocessing solution is possible with separate tasks devoted to parameter estimation, to the matrix computations for deriving $\bar{\mathbf{g}}$, and to fast inner-product feedback control.

All practical control problems are affected by input constraints such as torque saturation in DC motor drives or stroke limits in flow valves. The common approach is to design on a linear basis and to make *ad hoc* adjustments for dealing with saturation (for example anti-windup measures in PID regulators). One nice feature of LRPC is that constraints can be handled within the same general framework simply by changing the optimisation algorithm. Note that by using a Quadratic Programming algorithm (as obtained from many texts on numerical optimisation) we can deal with state constraints (e.g. limits on auxiliary plant outputs such as intermediate temperatures) as well as input limits. However a full QP solution can be very time-consuming and may only be suitable for process applications with sample intervals of the order of minutes. The discussion below concentrates on a particular method for high-speed work.

Suppose that the rate-of-change of the control signal is limited by, say:

$$\alpha_j \leq \Delta u(t+j-1) \leq \beta_j, \text{ for } j = 1, \ldots, NU.$$

The full unconstrained optimising solution provides a sequence $\{\tilde{u}^*(t+j-1)\}$ for which a particular control is *feasible* when it lies within its constraints. If all NU are feasible then no difficulty arises; similarly if only the *first* $\tilde{u}^*(t)$ lies outside a limit then the best solution is simply to clip it:

$$\Delta u(t) = sat(\tilde{u}^*(t), \alpha_1, \beta_1).$$

The problem arises when one or more *future* controls is infeasible; for simplicity suppose that the constrained optimum has one element on the boundary, say:

$$\Delta u(t+j-1) = \alpha_j.$$

We can use a Lagrange multiplier to find the constrained optimum by considering the function:

$$L \equiv J + 2\mu_j[\Delta u(t+j-1) - \alpha_j].$$

The constrained optimum is found when the derivative of L vanishes:

$$\mathbf{G}^T(\mathbf{G}\tilde{\mathbf{u}} + \mathbf{p} - \mathbf{w}) + \lambda\mathbf{I}\tilde{\mathbf{u}} + \mu_j\mathbf{e}_j = \mathbf{0},$$

where $\mathbf{e}_j$ is a vector containing 1 in element j. Hence the constrained control sequence is given by:

$$\tilde{\mathbf{u}} = \tilde{\mathbf{u}}^* + (\mathbf{G}^T\mathbf{G} + \lambda\mathbf{I})^{-1}\mu_j\mathbf{e}_j.$$

The unknown Lagrange multiplier μ_j is obtained by observing the jth equation giving:

$$\alpha_j = \Delta u^*(t+j-1) + h_{jj}\mu_j,$$

where h_{jj} is element (j,j) of the matrix $(\mathbf{G}^T\mathbf{G} + \lambda\mathbf{I})^{-1}$.

The procedure is as follows. First the unconstrained control sequence is calculated; if control increment j is infeasible then μ_j is evaluated and used to obtain the constrained optimal control, assuming that the constrained optimum lies on the constraints violated by $\tilde{\mathbf{u}}^*$. If more than one future control is infeasible we need a vector of Lagrange mutipliers, but the principles behind the solution are the same.

Simulation and practical experience with a high-speed compliant mechanical link shows that for certain plant the method can give a much improved response to a large step change in set-point. Such plant are characterized by sign changes in control as found with type-2 dynamics where initial acceleration is followed by deceleration. For more common process plant the improvement is not nearly so marked and in particular when NU is 1 (as often used with process control) the procedure is unnecessary.

3.2 Selecting the controller design constants

Suppose that the 'real' plant has p poles and a deadtime of k samples. A minimal state-space model would have dimension $n' = p+k$ and be, except for pathological cases, controllable and observable (C+O). However as we are considering an incremental formulation the plant dimension is increased by 1 ($n = n'+1$) to reflect the implicit augmentation by a integrator. Consider now the control problem at time t. The plant can be driven from any state $\mathbf{x}(t)$ to any other state in n samples. Hence a control horizon NU of at least n is required for this high-performance application.

In practice neither p nor k would be known and indeed many processes are governed by PDEs for which there is no exact finite state-space model. We must consider cases of overmodelled (model n_m greater than the plant n) or undermodelled plant. In the overmodelled case there will be cancelling poles and zeros in the model and hence it is no longer C+O, though in practice it will be stabilisable and detectable (S+D). When there are unmodelled dynamics it is important to ensure that nevertheless the model gives a good fit within the frequency band of the closed-loop. This can be achieved by the filters described below.

Given this preliminary discussion it is worth presenting choices of constants for which there are proven stable controls:

State-dead-beat : A stable state-dead-beat controller is produced if the plant model is C+O and $N_1 = n, N_2 \geq 2n-1, NU = n, \lambda = 0$.

Pole-placement : For the same conditions, a pole-placement controller with closed-loop poles at the roots of $P(q^{-1})$ arises from GPC control of the *auxiliary* output $\psi(t) = P(q^{-1})y(t)$.

Finite-horizon : A stable finite-horizon control arises with a S+D model of state dimension n_m with $N_1 \geq n_m, NU = N_1, NY \geq n_m, \lambda = \epsilon > 0$.

Infinite-horizon LQ : The closed loop is stable for a S+D model for $NU, N_2 \rightarrow \infty$, $NU = N_2 - n_m + 1$, $\lambda = 0$, or for $NU = N_2 \rightarrow \infty, \lambda > 0$.

Mean-level : A stable plant with an S+D model is closed-loop stable with a mean-level control having GPC constants $NU = N_1 = 1, N_2 \rightarrow \infty, \lambda = 0$.

These results are conservative. For example the mean-level controller has been found to give stable control even for some types of unstable plant. Note that there is no minimum-phase restriction in any of the above cases.

The applicability of these control objectives depends on the plant and on the desired performance. For most process plant mean-level control is adequate: it has the simplest choice of constants, the computation is minimal, and it tends to be highly robust. More vigorous action is found by increasing NU: even going from 1 to 2 has a marked effect. It is useful to consider NU as corresponding to the number of process poles that should be cancelled. For example if there are two unstable open-loop poles, NU should be at least 2. A summary of choices is:

1. $N_1 = n_m$: there is no point in costing outputs which, because of the deadtime k, are not affected by the current control.

2. N_2 large: up to the plant's rise-time so that possible negative-going nonminimum-phase behaviour does not dominate the cost minimisation.

3. $NU \leq n_m$: make NU as small as possible consistent with suitably vigorous control action.

4. $\lambda = 0$: one problem with this constant is that its effect is scale-dependent (e.g. 0.01 might have little effect in one case but be significant in others) and difficult to choose. The best policy is to start from zero and make minor adjustment to 'trim' the closed-loop performance. A very small value such as 10^{-10} is useful in aiding numerical robustness.

4 Tuning the model parameters

Given the chosen model structure $\mathcal{M}$ an LRPC design needs a 'good' set of corresponding parameters, which can be obtained either from a physico-chemical model or, more usually, from direct plant experimentation. Sometimes the experiments can be very simple: for DMC a step-response is all that is required. However it is generally very difficult to eliminate all sources of disturbance during a step. Moreover models such as the DARMA form require algorithmic treatment of the data for computing their parameters. Hence it is better to use a Recursive Prediction Error (RPE) method for estimation, as RPE can cope with both disturbances and different model structures. It can also handle the adaptive problem in which the plant parameters change with time or with set-point. The idea behind RPE is to use the model and the *current* parameter estimates with *old* data to predict the most recent output measurement. The error is then the discrepancy between the prediction $\hat{y}(t|t-1)$ and the outcome $y(t)$:

$$\epsilon(t) = y(t) - \hat{y}(t|t-1).$$

The parameter estimates are then updated using the recursion:

$$\hat{\theta}(t) = \hat{\theta}(t-1) + j(t)a(t)\mathbf{M}(t)\mathbf{x}(t)\epsilon(t),$$

- $\mathbf{x}(t)\epsilon(t)$ corresponds to the gradient-vector pointing in the direction of steepest descent down the local error surface;
- $\mathbf{M}(t)$ is a matrix which rotates this vector towards the global best-fit direction;
- $a(t)$ is an acceleration factor;
- $j(t)$ is a 'jacketting' switch, having the value 1 if the current data is 'good' and 0 if it is 'bad'.

One popular *off-line* method for estimating parameters to data is Least-Squares, in which the fitting criterion minimises the sum-of-squares of the prediction errors over all the data. To place this in RPE form the recursions:

$$\begin{aligned} \mathbf{k}(t) &= \frac{\mathbf{P}(t-1)\mathbf{x}(t)}{\beta + \mathbf{x}^T\mathbf{P}(t-1)\mathbf{x}(t)} \\ \mathbf{P}(t) &= (\mathbf{I} - \mathbf{k}(t)\mathbf{x}^T(t))\mathbf{P}(t-1)/\beta \\ \hat{\theta}(t) &= \hat{\theta}(t-1) + \mathbf{k}(t)\epsilon(t) \end{aligned}$$

are used. Here β is a 'forgetting-factor' such that $1/(1-\beta)$ (sometimes called the *asymptotic sample length* or ASL) reflects the amount of past data which contributes significantly to the current estimates. Hence $\beta = 1$ can be used for self-tuning where it is felt that the plant parameters are indeed fixed: the estimates tend to converge to a fixed set despite any noise. However if there are likely to be time-variations in the plant dynamics (or even if it is felt that its structure differs from the model), the ASL can be set to correspond to the expected speed of change. The above is called Recursive Least Squares (RLS): as one of the most important algorithms in self-tuning, much effort has been devoted to increasing its reliability in applications whilst retaining its power for rapid convergence. Updating of the matrix $\mathbf{P}$ is best with a factorized (e.g. UDU) approach; automatic rescaling of variables can improve conditioning; 'variable forgetting factors' can handle 'blow-up' when data lacks excitation; 'dead-bands' eliminate small perturbations in data which may have mainly nonlinear, unmodelled, or noise components.

To use RLS with, say, a DARMA model we simply make the correspondencies:

$$\begin{aligned} \theta^T &= [a_1, a_2, \ldots; b_1, b_2, \ldots] \\ \mathbf{x}^T &= [-y(t-1), -y(t-2), \ldots; u(t-1), u(t-2), \ldots]. \end{aligned}$$

This choice uses 'full-value' signals which is generally inappropriate and it is preferable to employ differenced data as then unwanted dc components are blocked. Note that this pretreatment of data implicitly assumes a noise model and its consequence is discussed below.

4.1 Practical data filtering

Suppose the model adopted for the plant is of the general linear form:

$$y(t) = H(q^{-1})u(t) + N(q^{-1})e(t),$$

where H and N are infinite polynomials with the term in $e(t)$ providing a general disturbance. With no loss of generality we can assume that $n_0 = 1$. The one-step-ahead predictor is required for recursive estimation and is derived by:

$$\begin{aligned}
y(t+1) &= Hu(t+1) + Ne(t+1) \\
&= Hu(t+1) + e(t+1) + q(N-1)e(t). \\
\text{But: } e(t) &= N^{-1}[y(t) - Hu(t)], \\
\text{Hence: } y(t+1) &= H(1 - 1 + N^{-1})u(t+1) + (1 - N^{-1})y(t+1) + e(t+1), \\
\text{so: } \hat{y}(t+1|t) &= N^{-1}[Hu(t+1) + (N-1)y(t+1)].
\end{aligned}$$

This is causal as both h_0 and $n_0 - 1$ are both zero. If $e(t)$ is a sequence of IID random variables the predictor is also minimum variance; otherwise it is simply a 'good' heuristic filter. Note again that the prediction process involves some implicit disturbance assumption. Suppose that the *true* plant model is given by:

$$y(t) = H_0(q^{-1})u(t) + \nu_0(t), \text{ where:}$$

$$\phi_\nu = \lambda_0 |N_0(e^{j\omega})|^2$$

is the disturbance spectrum. Then the prediction error is given by:

$$\begin{aligned}
\epsilon(t) &= N^{-1}[y(t) - Hu(t)] \\
&= N^{-1}[(H_0 - H)u(t) + \nu_0(t)].
\end{aligned}$$

Hence the spectrum of the prediction errors is:

$$\phi_\epsilon = \frac{|\tilde{H}|^2 \phi_u + \phi_\nu}{|N|^2},$$

where $\tilde{H}$ is the modelling error of the control path. Now RLS attempts to minimise the sum-of-squares of the prediction errors using a cost $V(\theta) \equiv \sum_{t=1}^{N} \epsilon^2$ and if the number of data-points N tends to infinity this is equivalent to minimising:

$$V(\theta) = \frac{1}{2\pi} \int_{-\pi}^{\pi} \phi_\epsilon(\omega)\, d\omega.$$

If the signal-to-noise ratio is large then the estimation is equivalent to minimising the integral of $|\frac{\tilde{H}}{N}|^2 \phi_u$ over all relevent frequencies up to the Nyquist frequency. If the model is correctly parameterised then $\tilde{H}$ has a minimum value of zero, but if not the minimum value corresponds to a fit which is *frequency-weighted* by both ϕ_u and N. For example if $\phi_u/|N|^2$ is low-pass (e.g. if the test signal has mainly low-frequency components such as a step) then it would be expected that the low-frequency error of the model (e.g. its gain) would be reduced.

Consider then the use of RLS to estimate the parameters of a DARMA plant:

$$A_0(q^{-1})y(t) = B_0(q^{-1})u(t).$$

Often to eliminate the effects of *dc* offsets the data is differenced, giving a model:

$$A(q^{-1})\Delta y(t) = B(q^{-1})\Delta u(t) + \epsilon(t),$$

where ϵ is its prediction-error. Converting to the general linear form we obtain:

$$y(t) = A^{-1}(q^{-1})B(q^{-1})u(t) + [A(q^{-1})\Delta]^{-1}\epsilon(t),$$

so that the 'noise-polynomial' N becomes $[A\Delta]^{-1}$ and consequently the frequency weighting on the model error in RLS is $|A\Delta|^2\phi_u$. It is clearly seen that the weighting greatly accentuates the fit at high frequencies at the expense of a serious loss of fit at low frequencies. This causes a significant problem when there is unmodelled dynamics: RLS 'latches' on to high- rather than low-frequency poles. In effect the model's step response matches well in its initial stages (transient) but not in the steady-state. GPC, however, relies on a reasonable steady-state (model gain) accuracy for long-range predictions, particularly for mean-level applications.

The solution to this problem is simple. Instead of just differencing the data, a band-pass filter is used so that the data seen by RLS is of the form:

$$y'(t) = \frac{\Delta}{T(q^{-1})}y(t); \quad u'(t) = \frac{\Delta}{T(q^{-1})}u(t),$$

where T is an appropriate filtering polynomial chosen to match the desired closed-loop bandwidth of the process. The degree of T should be at least that of A. This has been found to be highly effective in applications: in particular GPC with a compliant arm having many high-frequency modes gave poor results until T was adopted. Its use is strongly recommended.

Note: the above describes the use of T-filtering for providing good estimates of the plant dynamics within a given bandwidth. $T(q^{-1})$ has another rôle as an *observer* polynomial which changes the disturbance-rejection modes of GPC whilst maintaining the set-point response. Proper design of T provides enhanced *robustness* of the closed-loop against modelling errors, which is beneficial for both the adaptive and nonadaptive cases.

4.2 Initialising a self-tuner

For a fixed-parameter GPC algorithm there are several choices to be made:

1. the sample interval h;
2. the process time-delay k;
3. the orders na and nb of the dynamics polynomials $A(q^{-1})$, $B(q^{-1})$;
4. values of the parameters in $A(q^{-1})$ and $B(q^{-1})$;
5. the filtering polynomial $T(q^{-1})$, interpreted for robustness enhancement;
6. the control design constants NU, N_1, N_2 and λ.

For use in self-tuning or adaptive control there are also:

1. the *range* of likely dead-times $k_{min} \ldots k_{max}$;
2. the forgetting-factor β or other design constants for the RPE algorithm.

If the plant behaviour is well-known, say from an analytical model, these are all quite easy to choose. A more searching question is, "Can I approach an *arbitrary* plant and, from simple experiments, determine the LRPC design constants in a uniform and automatic way?". Experience has given a positive answer: much of the set-up can be derived from a knowledge of the settling-time and from a 'step-up/ step-down' test on the manipulated variable. Space precludes a full discussion here, but one procedure is as follows:

1. Set the control computer to sample the plant relatively rapidly and take a series of measurements (for a plant under say loose PI control) to determine the noise level.
2. Inject a step and trace the output until it 'settles' to derive an approximate gain.
3. During the step-down, mark the dead-time, the 20% and 60% points on the response and use these to give an initial *continuous-time* second-order model with dead-time.
4. Choose a sample interval so that there are 10...20 points over the rise-time and use the corresponding Z-transform model to initialise the A and B polynomials. The T polynomial can be chosen to give roll-off within the Nyquist frequency.

The LRPC loop can now be closed, starting perhaps with a low-gain ($NU = 1$) setting. If required, further refinement of the model can be achieved using RLS with, say, extended A and B polynomials.

Table 2: Some applications of LRPC

Type of process	Reported by	Controller	
Absorbtion Plants	Ydstie (1982)	SISO	EHAC
	Chesna & Ydstie (1985)	SISO	EHAC
	Ydstie *et al* (1985)	SISO	EHAC
	Hiram & Kershenbaum (1985)	MISO	EHAC
Engine test-bed	Tacey (Ricardo Ltd.) (1987)	SISO	GPC
	Hesketh & Sandoz (1987)	MIMO	LQ
Distillation columns	Richalet *et al* (1978)	SISO/MIMO	IDCOM
	Luecke *et al* (1985)	SISO	DMC
Heating systems	Chesna & Ydstie (1985)	SISO	EHAC
	de Keyser & van Cauwenberghe (1985)	SISO	EPSAC
	Jota (1987)	SISO/MIMO	GPC
Robot manipulators	Lambert, M. (1987)	SISO	GPC
	Lelic & Wellstead (1987)	SISO/MIMO	GPP
Chemical reactors	Cutler & Ramaker (1980)	MIMO	DMC
	Ergrand (1980)	?	IDCOM
	Martin & Van Horn (1982)	MIMO	IDCOM
	Cutler & Hawkins (1987)	MIMO	DMC
Extruder	M'Saad *et al* (1987)	SISO	LQ,GPC
Tracking system	Favier (1987)	MIMO	GPC
Furnace	Richalet *et al* (1978)	MIMO	IDCOM
	Cutler & Ramaker (1980)	SISO	DMC
	de Keyser & van Cauwenberghe (1981)	SISO	EPSAC
Steam boiler	Richalet *et al* (1978)	MIMO	IDCOM
Dryers	Lambert,E. (1987)	SISO	GPC
	Lambert,E. (1987)	MIMO	GPC
Cement mill	Al-Assaf (1988)	SISO	GPC
Dredging ship	de Keyser & van Cauwenberghe (1985)	SISO	EPSAC

5 Some practical applications

Table 2 provides a list of some LRPC applications on real processes (i.e. not just on simulated plant models). Reported experience has been good: the control set-up appears to be straightforward yet the closed-loop performance has been consistently better than PID. This is particularly the case when the plant dynamics is complex (dead-time, low-damped poles, interacting) and where measured disturbances can be used for feedforward. With dead-time processes such as rolling mills and paper machines LRPC gives greater design flexibilty compared with classical Smith predictors. In particular the merging of constrained optimization (often done by LP for the static case) with dynamic prediction promises to provide excellent control based on ecomomic criteria.

6 Conclusions

Long-range predictive control has matured into a highly effective method for dealing with a whole range of industrial problems. The basic principles are simple to understand and combine well with modern computers and numerical optimisation algorithms. Concrete practical trials have demonstrated that it is easy to use and gives good performance. This paper has described some of the considerations behind the development of LRPC, showing that in the GPC algorithm attention is focussed on key issues such as model structure, costings and control horizons for provable stability, robust adaptation with appropriate data filtering, and on simple ways to set-up a self-tuned algorithm.

7 Acknowledgements

I would like to thank all members of the Self-tuning Control Group at Oxford for their help in the research described in this paper. In particular Coorous Mohtadi made many significant theoretical developments, Terence Tsang produced the constrained algorithm, whilst David Peel, Eugene and Martin Lambert performed the industrial trials. The work is part of a programme sponsored by the UK SERC investigating the application of Generalized Predictive Control.

8 Bibliography

Al-Assaf,Y. (1988). Self-tuning control: theory and applications. D.Phil Thesis, Oxford University.

Åström ,K.J. and Wittenmark,B. (1973). On self-tuning regulators. *Automatica,* **9**, 185-199.

Bruijn,P.M., Bootsma,L.J. and Verbruggen,H.B. (1980). Predictive control using impulse response models. *IFAC Symposium on Digital Computer Applications to Process Control*, Dusseldorf, FRG.

Bruijn,P.M. and Verbruggen,H.B. (1984). Model algorithmic control using impulse response models. *Journal A*, **25**, 69-74.

Chesna,S.A. and Ydstie,B.E. (1985). Self-tuning and adaptive control of chemical processes. *Proc. Workshop on Adaptive Control*, Yale, USA.

Clarke,D.W. (1984). Self-tuning control of nonminimum-phase systems. *Automatica* , **20**, 501-517.

Clarke,D.W. and Gawthrop,P.J. (1975). Self-tuning controller. *Proc. IEE*, **122**, 929-934.

Clarke,D.W. and Zhang,L. (1987). Long-range predictive control using weighting-sequence models. *Proc. IEE*, **134**, Pt.D, 187-195.

Clarke,D.W., Kanjilal,P.P. and Mohtadi,C. (1985). A generalized LQG approach to self-tuning control. *Int.J.Control*, **41**, 1509-1544.

Clarke,D.W., Mohtadi,C. and Tuffs,P.S. (1987). Generalized predictive control. Parts 1 and 2. *Automatica*, **23**, 137-160.

Cutler,C.R. and Hawkins,R.B. (1987). Constrained multivariable control of a hydrocracker reactor. *Proc. ACC*, Minneapolis, USA.

Cutler,C.R. and Ramaker,B.L. (1980). Dynamic matrix control - a computer control algorithm. *Proc. JACC*, San Francisco, USA.

Dawkins,J. and Briggs,P.A.N. (1965). A method for using weighting functions as system description in optimal control. *Proc. IFAC Symposium*, Teddington, UK.

de Keyser,R.M.C. and van Cauwenberghe,A.R. (1981). Self-tuning predictive control. *Journal A*, **22**, 167-174.

de Keyser,R.M.C. and van Cauwenberghe,A.R. (1982). Applications of self-tuning predictive control. *Journal A*, **23**, 1-10.

de Keyser,R.M.C. and van Cauwenberghe,A.R. (1982). Typical application possibilites for self-tuning predictive control. *Proc. IFAC Symposium on Identification and System Parameter Estimation*, Washington, DC, USA.

de Keyser,R.M.C. and van Cauwenberghe,A.R. Extended prediction adaptive control. *Proc. IFAC Symposium on Identification and System Parameter Estimation*, York, UK.

Ergrand,J.C. (1980). Applications of multivariable control control in a refinery and implementation. *Proc. JACC*, San Francisco, USA.

Favier,G. (1987). Self-tuning long-range predictive controllers. *Proc. IFAC World Congress*, Munich, FRG.

Foigel,J.K. and Richalet,J. (1979). Self-adapting IDCOM. *Proc. IFAC Symposium on Identification and System Parameter Estimation*, Darmstadt, FRG.

Garcia,C.E. and Morari,M. (1982). Internal Model Control. 1.A unifying review and some new results. *I & EC Process Design and Development*, **21**, pp.308-323.

Jota,F.G. (1987). The application of self-tuning control technique to a multivariable process. D.Phil Thesis, Oxford University.

Karny,M., Halouskova,A., Bohm,J., Kulhavy,R. and Nedoma,P. (1985). Design of linear quadratic adaptive control: theory and algorithms. *Kybernetica supplement.*

Lambert,E. (1987). The industrial application of long range prediction: D.Phil Thesis, Oxford University.

Lambert,M. (1987). Adaptive control of flexible systems: D.Phil Thesis, Oxford University.

Lawson,C.L. and Hanson,R.J. (1974). *Solving least-squares problems.* Prentice-Hall.

Lelic,M.A. and Wellstead,P.E. (1987). A generalized pole-placement self-tuning controller. Part II - an application to robot manipulator control. *Int.J.Control*, **46**, 569-602.

Little,D.L. and Edgar,T.F. (1986). Predictive control using constrained optimal control. *Proc. ACC.*

Lueke,R.H., Lewis,J.C., Lin,H.Y. and Yoon,W.K. (1985). Dynamic matrix control of a batch distillation column. *Proc. ACC*, Boston, USA.

Martin,G.D. (1981). Long-range predictive control. *AIChE Journal*, **27**, 748-753.

Martin,G.D. and van Horn,L.D. (1982). IDCOM hierarchical multivariable control application. *Proc. Conf. on Industrial Control Systems*, Hamilton, Canada.

Maurath,P.R., Seborg,D.E. and Mellichamp,D.A. (1985). Predictive controller design by principal components analysis. *Proc. ACC*, Boston, USA.

Mehra,R.K., Kessel,W.C., Rault,A., Richalet,J. and Papon,J. (1978). Model algorithmic control using IDCOM for the F100 jet engine multivariable control design problem. in *Alternatives for linear multivariable control*, ed. Sain,M.K., Peczkowski,J.L. and Melsa,J.L.

Menga,G. and Mosca,E. (1979). MUSMAR: a new class of multivariable adaptive regulators. *Proc. IFAC Symposium on Identification and System Parameter Estimation*, Darmstadt, FRG.

Mohtadi,C. (1986). Studies in advanced self-tuning algorithms: D.Phil thesis, Oxford University.

Mohtadi,C. and Clarke,D.W. (1986). Generalized predictive control, LQ, or pole-placement: a unified approach. *Proc. CDC*, Athens, Greece.

Montague,G.A., Morris,A.J., Wright,A.R., Aynsley,M. and Ward,A.C. (1986). Online estimation and adaptive control of penicillin fermentation. *Proc.IEE*, **133**, Pt.D, 240-246.

Morshedi,A.M., Cutler,C.R. and Skrovanek,T.A. (1985). Optimal solution of dynamic matrix control with linear programming techniques (LDMC). *Proc. ACC*, Boston, USA.

M'saad,M., Duque,M. and Landau,I.D. (1987). Thermal process robust adaptive control. *Proc. IFAC World Congress*, Munich, FRG.

Peterka,V. (1984). Predictor-based self-tuning control. *Automatica*, **20**, 39-50.

Reid,J.G., Chaffin,D.E. and Silverthorn,J.T. (1981). Output predictive algorithmic control: precision tracking with application to terrain following. *AIAA J.Guidance and Control*, **4**, 502-509.

Reid,J.G., Mehra,R.K. and Kirkwood,E.,Jr. (1979). Robustness properties of output predictive dead-beat control: SISO case. *Proc. CDC*, Ft Lauderdale, USA.

Richalet,J., Rault,A., Testud,J.L. and Papon,J. (1978). Model predictive heuristic control: applications to industrial processes. *Automatica*, **14**, 413-428

Rohrs,C.E., Valavani,L., Athans,M. and Stein,G. (1982). Robustness of adaptive control algorithms in the presence of unmodelled dynamics. *Proc. CDC*.

Rouhani,R. and Mehra,R.K. (1982). Model algorithmic control (MAC); basic theoretical properties. *Automatica*, **18**, 401-414.

Shah,S.L., Mohtadi,C. and Clarke,D.W. (1987). Multivariable adaptive control without a prior knowledge of the delay matrix. *Systems and Control Letters*, **9**, 295-306.

Tacey,E. (1987). The thought processors. *The Engineer*, May.

Tuffs,P.S. and Clarke,D.W. (1985). FAUST: a software package for self-tuning control. *Proc. IEE Conf. Control '85*, Cambridge, UK.

Tuffs,P.S. and Clarke,D.W. (1985). Self-tuning control of offset: a unified approach. *Proc.IEE*, **132**, Pt.D, 100-110.

van Cauwenberghe,A.R. and de Keyser,R.M.C. (1985). Self-adaptive long-range predictive control. *Proc. ACC*, Boston, USA.

Ydstie,B.E. (1984). Extended horizon adaptive control. *Proc. IFAC 9th World Congress*, Budapest, Hungary.

Wellstead,P.E., Prager,D. and Zanker,P. (1979). Pole assignment self-tuning regulator. *Proc.IEE*, **126**, 781-787.

INFORMATION AND INTEGRATED CONTROL

Graham C. Goodwin and Mario E. Salgado
Department of Electrical Engineering & Computer Science
University of Newcastle NSW 2308
Australia

ABSTRACT

This paper will reflect on certain practical issues which arise in the application of adaptive control. A key problem that we wish to address is how to bring ideas of robust control into harmony with the notion of adaptation. In particular, we will argue that a central issue is the role that the available information plays in specifying the design procedure. We advocate that there should be a consistency between the kind of information used in robust control design and that used in adaptive control design. We will discuss how this might be achieved and we coin a new term, namely 'integrated control' for an approach to control design in which robust considerations and adaptation are part of a unified point of view. Our objective is to give physical insight into the problem rather than to give a rigorous mathematical treatment.

1. INTRODUCTION

As is well known (Aström, 1987) the history of adaptive control can be traced back at least 4 decades. Early work was made difficult by the lack of suitable hardware on which the algorithms could be implemented and a lack of a suitable supporting design theory.

It was not until the late 1970's that the dynamics of adaptive control was even partially understood. However, this initial theory was applicable only to very idealized circumstances (Goodwin et.al 1980, Morse 1980, Narendra, 1980) and hence was not particularly helpful as a practical design aid. Indeed, during the early 1980's it was shown (Rohrs et.al. 1985) that the existing theory of adaptive control was excessively sensitive to the assumptions. For example, a globally convergent algorithm could be turned into a globally unstable algorithm by making an infinitesimal change in the system's complexity.

In recent years, this difficulty has been partially resolved since it has been shown that the sensitivity of the theory to the assumptions can be reduced by adding various modifications to the basic algorithms. Such modifications include better parameterizations (e.g. δ rather than shift operator models (Middleton et.al. 1988)), frequency domain focusing (e.g. by use of band pass filters in the estimator) and time domain focusing (e.g. by deadzones and/or normalization techniques (Praly, 1983)). Coincidentally, some of these modifications were already in use in a number of the practical realizations of adaptive control (Aström, 1987). The end result of the above developments has been the consolidation of a view of adaptive control based on a combination of a robust parameter estimator together with a certainty equivalence control system design procedure. These strategies are aimed at achieving good performance in the face of certain types of model uncertainty. Within this framework, the role of the parameter estimator in adaptive control is to provide a mechanism for dealing with structured uncertainty. In addition, certain safeguards are included to deal with the unstructured uncertainty (Middleton et.al., 1988).

An alternative method for dealing with model uncertainty in control system design is the approach of robust control (Horowitz, 1963). In this framework one aims to design a fixed (usually linear) control law which allows to trade off performance versus robustness to uncertainties and, at a minimum, to preserve stability in the worst case. To achieve this result, robust control design requires information to be available on the nature of the model uncertainty.

It therefore seems that adaptive control and robust control are tackling the same underlying problem but are based on different information structures. However, the origins of these differences are historical in nature rather than conceptual. Obviously, the information structures have to be made compatible if one is to have a rational basis for comparing the two approaches. Further, once the information structures have been rationalized, then it should be possible to integrate the two approaches into a unified design philosophy. Since this new approach combines robust and adaptive control but is not identical to either, we coin a new term, namely, 'integrated control' for the resulting methodology. The practical implementation of this approach will be different from the majority of existing commercial adaptive controllers. This paper gives some suggestions how such a controller might be realized.

2. INFORMATION STRUCTURES

A key issue in the design of any robust control system is the way in which the available information about the system is described. The usual way of presenting this information is in the form of a nominal model plus associated uncertainty.

A critical choice is the structure of the nominal model since this is pivotal in many respects including the fact that it summarises a vast amount of prior knowledge about the nature of the system and the fact that it underpins all further information processing and data extrapolation.

There are many ways of describing the nominal model including state space, frequency response, impulse response, transfer functions, etc. Also, within each of these choices there exist other structural issues such as order, parameterization, etc. These different choices are relevant to both the modelling issue and to the ultimate control system design procedure.

There are also many ways of specifying the uncertainty about the model. Mostly some form of bound is used either on the parameters, the impulse response or the frequency response. Obviously the tighter that we can specify the uncertainty, then the more useful is the model. For example, a simple specification of upper and lower bounds for each parameter is less informative than a detailed picture of the shape of the uncertainty region in the parameter space. This is illustrated in Figure 2.1 where θ^* denotes the nominal parameters and the shaded region gives the complete uncertainty description.

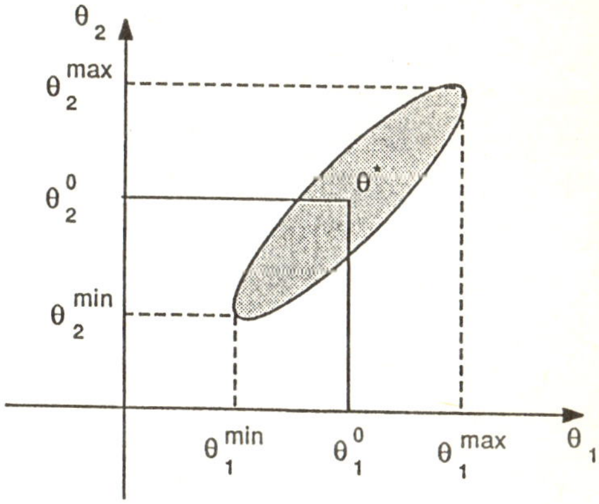

Figure 2.1 Illustration of parametric uncertainty

Another illustration of different ways to describe uncertainty is the specification of bounds for the frequency response, then a more informative description is obtained if one also specifies some inherent smoothness of the response between adjacent frequencies (Goodwin and Salgado, 1988).

Some care is necessary in specifying the nature of the uncertainty. For example, in some cases, parameter uncertainty may dominate as in a rocket that has a known minimum and maximum weight. However, it is more common that a measure of unstructured uncertainty will also be required since the nominal model will have lower complexity than the real system.

The above description of uncertainty forms the basis of work in robust control. Also, in recent times, it has been the starting point for adaptive control. However, the parameter estimation procedures in common use in current adaptive controllers yields an estimated nominal model without updating the associated measure of model uncertainty. The adaptive control design is then based on the estimated nominal model alone which is inconsistent with the robust control philosophy.

The approaches of robust and adaptive control can be integrated provided the estimation module of the adaptive controller yields information on both the estimated nominal model and the updated uncertainty.

3. CONNECTIONS BETWEEN INFORMATION AND ESTIMATION

One way of looking at estimation is as a mapping from the available prior information (including the nominal model structure, the known a-priori bounds on the parameters, the form of the unstructured uncertainty, prior information about the environment surrounding the system, plus the observed data from the system) into an a-posteriori summary of the nature of the system (including nominal parameter estimates, confidence regions for the parameters, updated data about the model uncertainty and the system's environment). There are different levels of complexity that can be used in giving this a-posteriori summary. Well known examples of how this information can be presented from the field of adaptive control, include

- a classification of the system's step response in terms of a finite set of possible patterns (Foxboro's Exact)

- specification of the critical gain and critical frequency on the system's frequency response (SATT ECA)

- nominal parameter estimates and their covariance (BB and ASEA's Novatune).

These models have been found to be well suited to associated design strategies. However, they are traditional approaches to adaptive control since they do not readily match corresponding strategies for robust control design.

We will attempt to describe below an alternative viewpoint of estimation which is consistent with the information needed for robust control design. To set the scene, let us consider an abstract situation. Say L denotes the unknown true system and let $M(\theta)$ denote a parameterized nominal model. We assume that the unstructured component of the model can be quantified, e.g. by saying that there exists some parameter value $\theta = \theta_0$, such that the residual unstructured model error $L-M(\theta_0)$ has certain known quantitive properties. The estimator should now map this a-priori information plus the measured data into some a posteriori summary, e.g. a nominal estimated model $M(\hat{\theta})$ and some measure of the unmodelled error, i.e. $L-M(\hat{\theta})$.

It is common in statistical estimation to assume that the model, M, has the same structure as the true system, L, and that the only source of errors is random noise in the data. In this case, one can develop unbiased estimates for θ_0 (Goodwin and Payne, 1977). Also, the estimation error between $\hat{\theta}$ and θ_0 is typically an inverse function of the data length (at least asymptotically). However, a more common situation in practice is that the model has a lower complexity then that of the system. In this case, it is likely that systematic errors (commonly called bias) will dominate. Moreover, these errors will be roughly independent of data length (Ljung, 1987).

The bias error mentioned above is not always a negative aspect. Firstly, it is known that sometimes large parameter errors are not reflected in the performance of the model. For example, the frequency response of a linear system over a particular bandwidth can vary insignificantly for certain types of large parameter variations. Secondly, the bias can be helpful in that the systematic error is deliberately adjusted by the parameter estimator so as to give the best fit to the observed data. However, the difficulty is that the bias is data dependent and thus a model fitted to one set of data may perform badly under different conditions. It is therefore important that the relationship between the data characteristics, the bias distribution and the ultimate model performance be quantified. To our knowledge this has not been previously achieved, although Ljung (1987) has given a qualitative analysis of the factors influencing bias distribution.

In the next section we will briefly describe a technique for achieving the objective outlined above.

4. EVALUATION OF BIAS DISTRIBUTION

For illustrative purposes, let us assume that the system under study is linear, single-input, single-output and has one extra measurement available for feedforward. Thus the system may appear as in Figure 4.1.

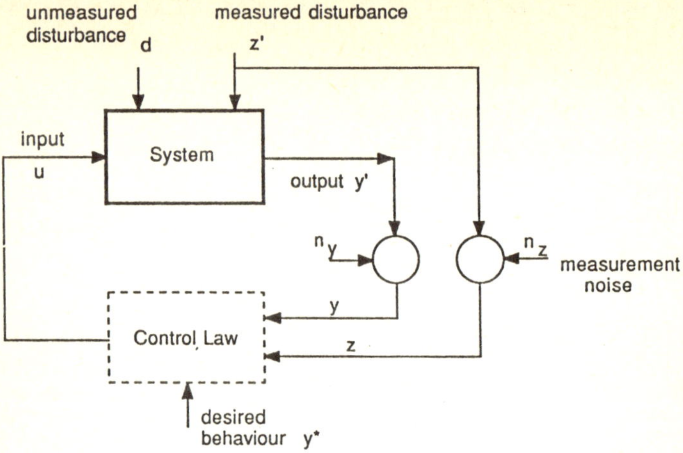

Figure 4.1 A simple class of systems

A suitable model for this system is

$$y' = G_0 u + G_1 z' + H_1 u + H_2 z' + H_3 d \tag{4.1}$$

where G_0, G_1, H_1, H_2, H_3 are linear transfer functions; and G_0, G_1 describe the modelled part of the response and H_1, H_2, H_3 represent the unmodelled part of the response.

Of course to specify the form of the estimator we need to decide on a representation of the various transfer functions of interest. There are many possibilities (e.g. frequency domain, step response, etc). However, to illustrate the idea we choose to use a rational transfer function

$$G_0 = \frac{B}{A}, \quad G_1 = \frac{F}{A} \tag{4.2}$$

Then (4.1) can be rewritten as

$$Ay = Bu + Fz + \eta \tag{4.3}$$

where η denotes the modelling error given by:

$$\eta = An_y + AH_1 u - (AH_2 + F)n_z + AH_2 z + AH_3 d \tag{4.4}$$

A natural question that arises in conjunction with the model (4.2) is how to deal with pure time delays. It has been argued elsewhere that time delays are readily included by use of rational approximations (Salgado et. al 1988).

The model format given in (4.3) is in general unsuitable for estimation purposes since the error η will dominate the response. For example, it will almost always be the case that there will be non-negligible d.c. offsets between input and output. Hence, it is usually desirable to band-pass filter all signals to eliminate both high and low frequency components. A generally accepted rule of thumb is that the upper cut-off frequency should be about 1/5th of the sampling frequency whilst the lower cut-off frequency should be about $\frac{1}{100}$ th of the sampling frequency. Note that the upper cut-off is determined by the need to eliminate high frequency errors whilst the lower cut-off is determined by the need to eliminate off-sets and by finite word length considerations. (Middleton and Goodwin, 1989).

Applying this band-pass filter to (4.3) gives

$$Ay_f = Bu_f + Fz_f + \eta_f \tag{4.5}$$

where y_f, u_f, z_f, η_f are the band-pass filtered versions of y, u, z, η respectively.

We next introduce a further filter $\frac{1}{E}$ where E is a stable polynomial which has the same degree as A. The purpose of this filter is to compensate for the term A which appears on the right hand side of (4.4). Hence (4.5) now becomes

$$\frac{A}{E} y_f = \frac{B}{E} u_f + \frac{F}{E} z_f + \frac{1}{E} \eta_f \tag{4.6}$$

To specify the model of (4.6) completely we need to choose an appropriate operator and parameterization. Since we are dealing with sampled signals, some form of discrete time model is called for. It is tradition to model discrete time systems by using the backward shift operator q^{-1}. However, it has recently been pointed out (Middleton and Goodwin, 1989) that this is inferior to use the delta operator, δ, (or some related form) where

$$\delta = C\left(\frac{q-1}{\Delta}\right) \tag{4.7}$$

where q is the forward shift operator

Δ is the sampling period

C is a constant.

The advantages of the δ operator over the shift operator, include (Middleton and Goodwin, 1988)

- convergence of all discrete results to the underlying continuous counterpart as $\Delta \to 0$ for $C=1$
- superior numerical properties
- ability to use continuous intuition for discrete design.

The factor C in (4.7) is a frequency scaling factor and by choosing this proportional to Δ, then the dynamics of differing systems can be brought to the same normalized range.

A further issue of interest is the detailed parameterization of A, B, E and F. We shall assume that it is known that the system has k pure integrators. (Note that we can lump into pure integrations any stable or unstable pole having magnitude less than about $\frac{1}{20\Delta}$ since these all contribute 90° phase shift in the frequency band of interest). Hence we parameterize the model as follows:

$$A(\delta) = \delta^k[a_n\delta^n + a_{n-1}\delta^{n-1} + \ldots + 1] \tag{4.8}$$

$$B(\delta) = b_m\delta^m + b_{m-1}\delta^{m-1} + \ldots b_0 \tag{4.9}$$

$$F(\delta) = f_\ell\delta^\ell + f_{\ell-1}\delta^{\ell-1} + \ldots f_0 \tag{4.10}$$

$$E(\delta) = e_{n+k}\delta^{n+k} + \ldots + e_{k+1}\delta^{k+1} + \delta^k + e_{k-1}\delta^{k-1} + \ldots + e_0 \tag{4.11}$$

Note that we have fixed the lowest power of δ in A to be 1 rather than the highest power. Similarly, we have fixed e_k to be 1. The reasons for this choice are

- b_0 becomes the d.c. gain (velocity constant, etc.) for the system
- if we overestimate the order of A, then the redundant coefficients converge to zero. (However, if we normalize from the left then all coefficients have to go to ∞ to accommodate the overmodelling).

Substituting (4.8) to (4.11) into (4.6) allows us to express y_f as a linear regression as follows

$$y_f' = \phi^T\theta_0 + \eta_f \tag{4.12}$$

where

$$y_f' = \left[1 - \frac{e_{k-1}\,\delta^{k-1} + \ldots + e_0}{E}\right] y_f \tag{4.13}$$

$$\phi^T = \left[\frac{\delta^{n+k}}{E} y_f, \ldots, \frac{\delta^{k+1}}{E} y_f, \frac{\delta^m}{E} u_f \ldots \frac{1}{E} u_f, \frac{\delta^\ell}{E} z_f, \ldots, \frac{1}{E} z_f\right] \tag{4.14}$$

$$\theta_0^T = (e_{n+k} - a_n, \ldots, e_{k+1} - a_1, b_m, \ldots, b_0, f_\ell, \ldots, f_0) \tag{4.15}$$

The filtered derivatives in the above regression can be readily calculated in state space form.

The model (4.12) is now in a form suitable for parameter estimation. Again there are many alternatives for the precise form of the parameter estimator. However, the most common choice is some variant of ordinary least squares. In this case, the estimated parameter is simply given by

$$\hat{\theta} = \left[\sum_{t=1}^{N} \phi_t \phi_t^T\right]^{-1} \left[\sum_{t=1}^{N} \phi_t (y_f')_t\right] \tag{4.16}$$

The above expression describes block least squares over N data points. This can be converted into various alternative forms including recursive least squares, exponential weighted least squares, etc.

Using (4.12) we immediately obtain

$$\tilde{\theta} = \left[\sum_{t=1}^{N} \phi_t \phi_t^T\right]^{-1} \left[\sum_{t-1}^{N} \phi_t (\eta_f)_t\right] \tag{4.17}$$

where $\tilde{\theta} = \hat{\theta} - \theta_0$.

We see from (4.17) that in order to compute a bound on $\tilde{\theta}$, we need to say something about the correlation between the regression vector, ϕ_t, and the error sequence $(\eta_f)_t$. This, in turn, requires suitable assumptions to be made about the various terms in (4.4). However, provided this is done then it is possible to translate this into information about $\tilde{\theta}$ via (4.17) (Goodwin and Salgado, 1988).

Even if information about $\tilde{\theta}$ is available this falls short of giving the information required to design a robust controller.

Again to be specific, let us assume that the design will be carried out in the frequency domain. In this case, we need some measure of the difference between the model frequency response, $\hat{M}(j\omega)$, evaluated at $\hat{\theta}$, and the true system frequency response, $L(j\omega)$. This error can be evaluated as follows:

$$S(\omega) = |L(j\omega)-\hat{M}(j\omega)|^2 = |L(j\omega)-M(j\omega)+M(j\omega)-\hat{M}(j\omega)|^2 \tag{4.18}$$

where $M(j\omega)$ is the frequency response of the nominal model (having the same structure as the estimated model).

The term $L(j\omega)-M(j\omega)$ in (4.18) describes the inherent lack of modelability of the system with respect to the given model structure. The term, $M(j\omega)-\hat{M}(j\omega)$, is a function of the error between θ_0 (the nominal parameter vector) and $\hat{\theta}$ the estimated value. This latter term can be approximated by a Taylor's series expansion as follows:

$$\hat{M}(j\omega)-M(j\omega) \simeq \left.\frac{\partial M(j\omega)}{\partial \theta}^T\right|_{\theta=\hat{\theta}} \cdot \tilde{\theta} \tag{4.19}$$

Equation (4.19) has a nice physical interpretation because it reflects the interaction of the parameter error and the smoothness of the chosen model with respect to parameter changes. Also, the error given (4.19) is a function of the particular nature of the experimental data used for the estimation.

The final result of using (4.18) is a frequency bound on the modelling error. The result could be described graphically as shown in Figure 4.2.

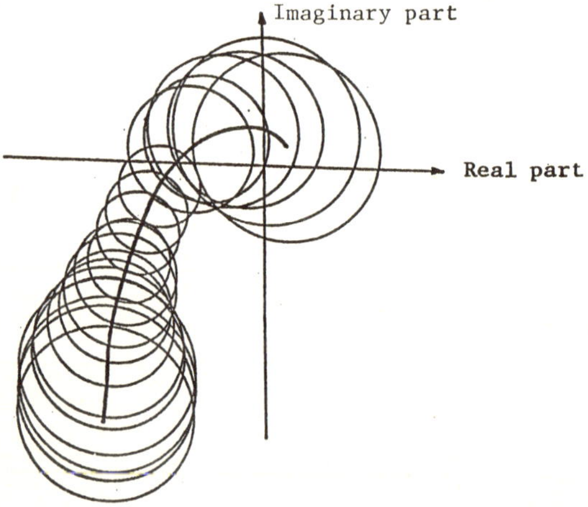

Figure 4.2 Estimated frequency response with uncertainty bound.

A key point about this diagram is that the error distribution can be changed by concentrating the input signal in critical frequency zones.

We illustrate the above ideas by two simple examples in the next section.

5. ILLUSTRATIVE EXAMPLES

Consider the class of system shown in Figure 5.1 where $\Delta G(s)$ denotes unmodelled dynamics.

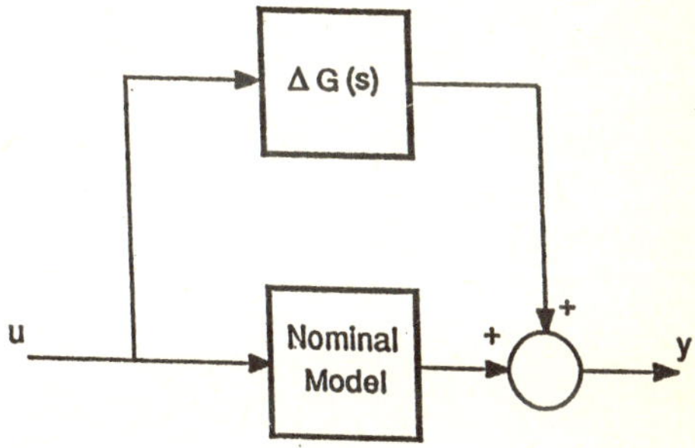

Figure 5.1 Class of systems with unmodelled dynamics.

We assume that the unmodelled dynamics are described by an impulse response which is bounded by an exponential decay with maximum time constant $\frac{1}{\beta}$.

Case (i). We first consider a simple case when the nominal model is a pure gain a_0. Say we obtain an estimate for a_0 using ordinary least squares when the input is a single sinewave of frequency ω_0. We are then interested in the prediction performance of the model at some other frequency ω. Expression (4.18) now becomes a function of both ω and ω_0. This function cannot be computed since it depends on the unknown unmodelled dynamics. However, bounds can be found on $S(\omega)$ (Goodwin, Mayne and Salgado, 1988) which can be computed using prior available information. One of these bounds is plotted against ω in Figure 5.2 for different values of ω_0 and for β normalized to 10.

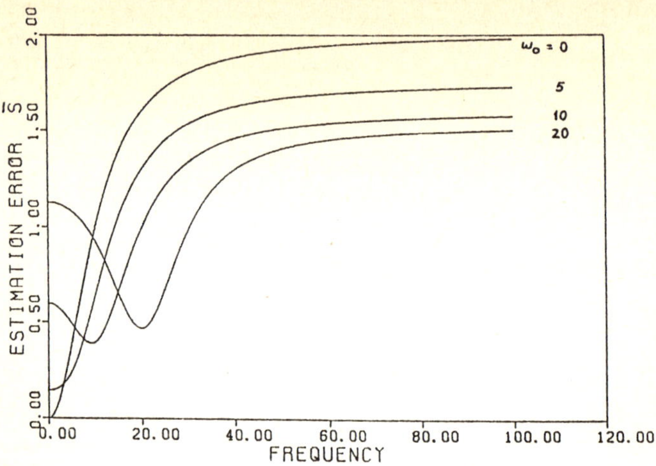

Figure 5.2. Frequency domain bound for example 1.

We make the following observations from the figure

- the minimum prediction error is zero only when $\omega_0=0$ This is consistent with intuition since it is only at d.c. that perfect modelling is possible

- the minimum prediction error occurs approximately for $\omega=\omega_0$ in all cases

- when ω_0 is greater than β, the unmodelled dynamics dominates the least squares fitting and hence the model fitted at ω_0 is really only useful for $\omega=\omega_0$

- when ω_0 is less than β, the phase shift produced by the unmodelled dynamics is relatively small and hence the model performs well at all low frequencies. Indeed the minimum prediction error occurs at a frequency lower than ω_0

Case (ii). We next consider a more complex nominal model having two parameters, i.e. $(\frac{a_1}{s} + a_0)$. Unlike case (i), this model allows perfect fitting at the single frequency, ω_0. Hence we would expect the prediction error to be zero for $\omega = \omega_0$ for all ω_0. The bound $\overline{S}(\omega)$ for the function $S(\omega)$ of equation

(4.18) is plotted against ω in Figure 5.3 for different values of ω_0 and β normalized to 10.

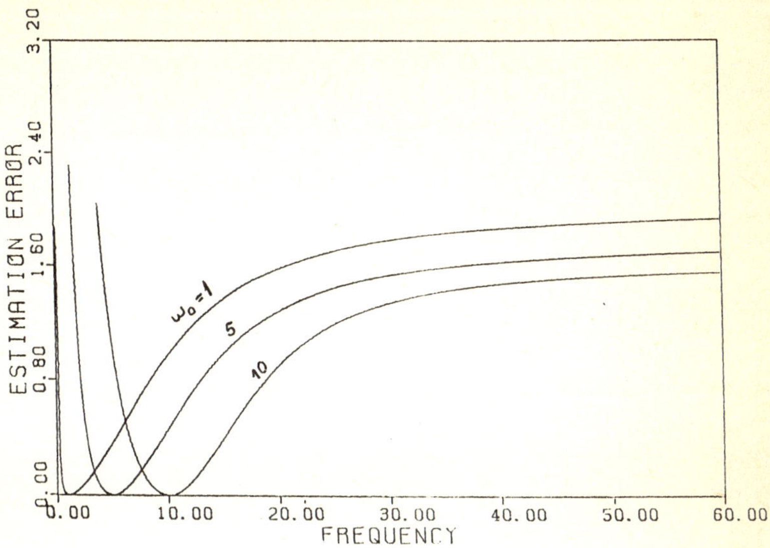

Figure 5.3. Frequency domain bound for Example 2.

From the figure we make the following observations

- the minimum prediction error is always zero and occurs for $\omega=\omega_0$
- for frequencies approaching zero, the prediction error approaches infinity. This is because any error in estimating a_1 is magnified at d.c.

Figure 5.4 shows a particular result expressed in the form suggested in Figure 4.2. The uncertainty region is shown around the frequency response of an estimated model. We see again that the error is zero for $\omega=\omega_0$ as discussed above.

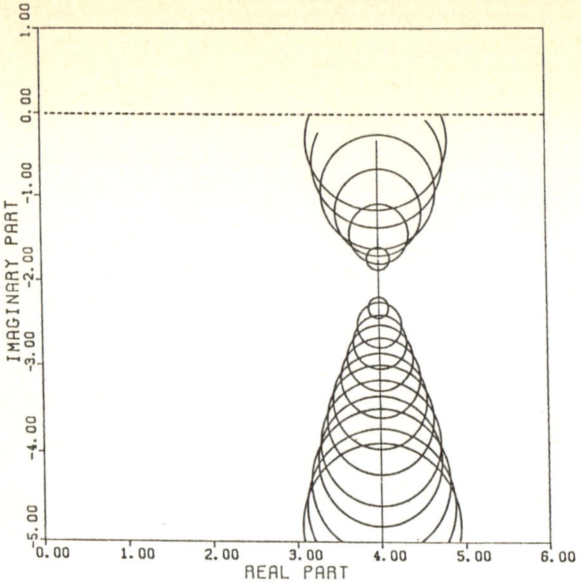

Figure 5.4. Frequency response with uncertainty bound for example 2.

6. INTEGRATED CONTROL

Given the approach to estimation outlined above, then we see that the estimator supplies information consistent with that required for robust control system design. Thus an adaptive controller can be constructed whose performance is at least as good as a fixed robust design. All that is necessary is for the information supplied by the estimator to be compared with the information available a-priori and to base the design on the most informative description.

We have not discussed the details of the robust design itself. However, this is a topic which has been extensively treated elsewhere (see, for example Horowitz 1963). The key point that we have tried to make in this paper is that adaptation should be seen as a higher level function which feeds information about the plant to the design algorithms. Since this differs from the traditional view of adaptive control, we will coin the term 'integrated control' for this design methodology.

7. CONCLUSION

This paper has discussed the role of information in control design and has indicated how consistency between robust control and adaptive control can be achieved. Using this approach, it has been argued that adaptive control and robust control can be viewed as part of a more general design methodology which we term integrated control. This paper has outlined the philosophical basis of this approach. We have built a practical realization of the strategy which has been implemented on a PC with 386 processor. The strategy has been found to work well in practice and to offer insights not available from traditional approaches to adaptive control.

REFERENCES

Aström, K.J. (1987). Adaptive control. *Proc.IEEE*, Vol.75, No.2, pp.185-217.

Goodwin, G.C., D.Q. Mayne and M.E. Salgado (1988). Uncertainty, Information and Estimation, *Technical Report EE8855*, Department of Electrical Engineering and Computer Science, University of Newcastle, Australia.

Goodwin, G.C., R.L. Payne (1977). Dynamic System Identification, Academic Press, New York.

Goodwin, G.C., P. Ramadge and P. Caines. (1980). Discrete-time multi- variable adaptive control. *IEEE Transactions on Automatic Control*, Vol.AC-25 No.3, pp.449-456.

Goodwin, C.G., M.E. Salgado (1988). A new approach to time-series analysis using restricted complexity model. Technical Report, University of Newcastle, NSW.

Horowitz, J.M. (1963). Synthesis of feedback systems. Academic Press, New York.

Ljung, L. (1987). System identification : theory for the user. Prentice Hall.

Middleton, R.H., G.C. Goodwin, D.J. Hill and D.Q. Mayne (1988). Design issues in adaptive control. *IEEE Transactions on Automatic Control*, Vol.AC-33, No.1, pp.50-58.

Middleton, R.H., and G.C. Goodwin (1989). Digital control and estimation. A unified approach. Prentice Hall, to appear.

Morse, A.S. (1980). Global stability of parameter-adaptive control systems. *IEEE Transactions on Automatic Control*, Vol.AC-25, No.3, pp.433-439.

Narandra, K.S., Y.H. Lin and L.S. Valavani (1980). Stable adaptive controller design, Part II: Proof of stability, *IEEE Transactions on Automatic Control*, Vol.AC-25, No.3, pp.440-448.

Praly, L. (1983) Robustness of indirect adaptive control based on pole placement design. Presented at the IFAC Workshop on Adaptive Control, San Francisco, CA.

Rohrs, C., L. Valavani, M. Athans and G. Stein (1985). Robustness of adaptive control algorithms in the presence of unmodelled dynamics, *IEEE Transactions on Automatic Control*, AC-30, No.9, pp.881-889.

Salgado, M.E., C.E. de Souza and G.C. Goodwin (1988). Issues in time delay modelling. IFAC Symposium on Identification, Beijing.

MULTIVARIABLE SELF-TUNING CONTROL BASED ON LAGUERRE SERIES REPRESENTATION

C.C. Zervos† and G.A. Dumont‡

† Electrical Engineering Department, University of British Columbia, Canada
‡ Pulp and Paper Research Institute of Canada and Dept. of Electrical Engineering
Pulp and Paper Centre, 2385 East Mall, University of British Columbia,
Vancouver, B.C., V6T 1W5, Canada

Abstract

In recent years the development of multivariable self-tuning control algorithms has attracted an extensive interest and there is a vast literature on various adaptive schemes. The most challenging problem in the generalization of the single-input single-output adaptive control algorithms to the multivariable case has been the representation of time-delays of MIMO systems via a delay matrix (Mohtadi, Shah, Clarke, 1987). In this paper we present a way to overcome this problem by extending the SISO discrete adaptive control algorithms based ont he orthonormal set of Laguerre functions to the MIMO case.

1 Introduction

The development of adaptive control schemes based on orthonormal series representation has shown advantages such as, reduced a-priori knowledge, easy handling of unknown and varying time delays and non-minimum phase plants, and increased robustness in presence of unmodelled dynamices. Up to now only single-input, single-output plants (SISO) have been considered. However it is straightforward to extend the adaptive controllers based on Laguerre functions previously developed (Dumont and Zervos, 1986; Zervos and Dumont, 1988a; Zervos and Dumont 1988b) to the multivariable case.

The most challenging problem in this area has been the representation of time-delays of MIMO systems via a delay matrix (Mohtadi, Shah, Clarke and 1987). The delay matrix of a MIMO system is the direct generalization of the time-delay term of a SISO system. This matrix characterizes the infinite zeros of the MIMO process but does not have a unique structure (Elliott and Wolovich, 1984). One particular lower triangular form of the delay matrix as defined by Wolovich and Falb (1976) is known as the interactor matrix. Many early results in MIMO adaptive control considered special classes of this delay matrix; *e.g.*, a diagonal matrix while others required prior knowledge of the triangular interactor matrix. However, the use of orthonormal functions in adaptive control can be extended to the MIMO case so that the requirement for knowing the system delay matrix is completely avoided. As a result, all the problems associated with the definition of the interactor matrix are completely eliminated. This is considered to be a definite advantage in the formulation of a multivariable adaptive scheme because the simplicity of the design renders the algorithm more flexible and easier to use.

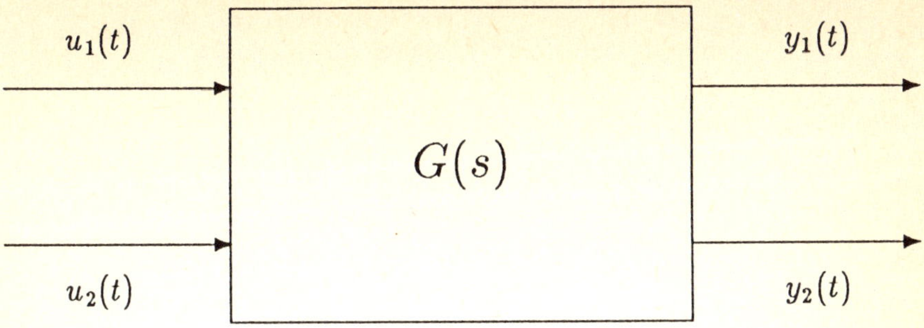

Fig 1– Multivariable 2 × 2 input-output linear system.

2 Deterministic Multivariable Self-Tuning Control

The development of the deterministic multivariable self-tuning controller is a direct extension of the SISO equivalent controller described in details in (Dumont and Zervos, 1986; Zervos and Dumont, 1988a; Zervos and Dumont 1988b). What follows is an illustrative mathematical derivation and two simulation examples that show how the principles outlined for the SISO case can be used in modelling and controlling a multivariable linear system. Let us consider, for simplicity, the 2-input, 2-output linear plant $G(s)$ shown in Figure 1, typical of a paper machine headbox, where, $u_1(t)$, $u_2(t)$ are the plant inputs and $y_1(t)$, $y_2(t)$ are the plant outputs. This system can be modelled by the Laguerre ladder network shown in Figure 2. The same principles would however apply for any $(N \times N)$ configuration.

The following equations can describe the model equations in state-space form. Two different sets of Laguerre orthonormal functions have been employed, one with time-scale p_1 and the other one with time-scale p_2. For the sake of simplicity let us also assume that both sets have the same number N of functions. By discretizing each block in Figure 2 using the network compensation method (Jury, 1958) the first set can be described by the discrete state-space description,

$$\underline{L}_1(t+1) = A_1 \underline{L}_1(t) + \underline{b}_1 u_1(t) \tag{1}$$

and the second set by,

$$\underline{L}_2(t+1) = A_2 \underline{L}_2(t) + \underline{b}_2 u_2(t) \tag{2}$$

where,

$$\underline{L}_1 = [l_{12} \;\; l_{12} \cdots l_{1N}]^T$$

$$\underline{L}_2 = [l_{21} \;\; l_{22} \cdots l_{2N}]^T$$

If we define the combined state vector to be,

$$\underline{L}^T = [\underline{L}_1^T \quad \underline{L}_2^T]$$

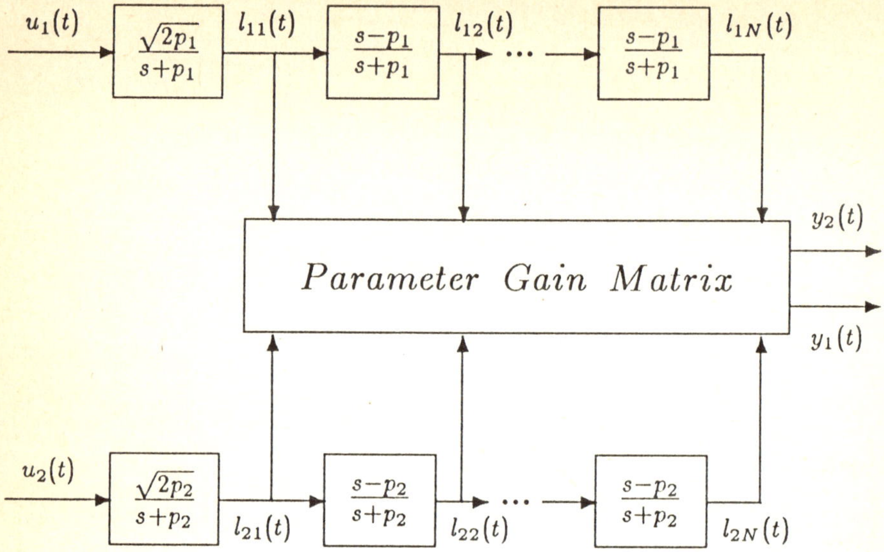

Fig 2– Laguerre Ladder Network for a Multivariable 2×2 input-output linear system.

and the output vector to be,

$$\underline{Y} = [y_1 \quad y_2]^T$$

then we can write,

$$\underline{Y}(t) = C\underline{L}(t) \tag{3}$$

Where the parameter gain matrix C is of the form,

$$C = \begin{vmatrix} \underline{C}_{11}^T & \underline{C}_{12}^T \\ \underline{C}_{21}^T & \underline{C}_{22}^T \end{vmatrix} = \begin{vmatrix} c_{11}^1 & c_{12}^1 & \cdots & c_{1N}^1 & c_{21}^1 & c_{22}^1 & \cdots & c_{2N}^1 \\ c_{11}^2 & c_{12}^2 & \cdots & c_{1N}^2 & c_{21}^2 & c_{22}^2 & \cdots & c_{2N}^2 \end{vmatrix} \tag{4}$$

where we used the notation,

$$\underline{C}_{ij}^T = [\ c_{j1}^i \quad c_{j2}^i \quad \cdots \quad c_{jN}^i\ ]$$

and i, j are both indices.

For the structure of the lower triangular matrices A_1 and A_2, and the input vectors $\underline{b}_1$ and $\underline{b}_2$ see the Appendix. The parameter gain matrix C can be estimated on-line by a multivariable least-squares method. Predictive expressions for the d-steps ahead, state vectors and outputs can be derived in an analogous way to that of the SISO case.

Assuming,

$$u_1(t) = u_1(t+1) = \cdots = u_1(t+d_1-1)$$

$$u_2(t) = u_2(t+1) = \cdots = u_2(t+d_2-1)$$

then by recursive substitution the following expressions can be derived,

$$\underline{L}_1(t+d_1) = A_1^{d_1}\underline{L}_1(t) + (A_1^{d_1-1} + \cdots + I)\underline{b}_1 u_1(t) \tag{5}$$

$$\underline{L}_2(t+d_2) = A_2^{d_2}\underline{L}_2(t) + (A_2^{d_2-1} + \cdots + I)\underline{b}_2 u_2(t) \tag{6}$$

where d_1 and d_2 are the prediction horizons. Then the required predictive output expression for $y_1(t)$ is,

$$\begin{aligned} y_1(t+d_1) &= \underline{C}_{11}^T(A_1^{d_1}-I)\underline{L}_1(t)+\underline{C}_{12}^T(A_2^{d_1}-I)\underline{L}_2(t) \\ &\quad +\underline{C}_{11}^T(A_1^{d_1-1}+\cdots+I)\underline{b}_1u_1(t) \\ &\quad +\underline{C}_{12}^T(A_2^{d_1-1}+\cdots+I)\underline{b}_2u_2(t)+y_1(t)) \end{aligned} \tag{7}$$

If we define,

$$\begin{aligned} \underline{P}_{11}^T &= \underline{C}_{11}^T(A_1^{d_1}-I) \\ \underline{P}_{12}^T &= \underline{C}_{12}^T(A_2^{d_1}-I) \\ B_{11} &= \underline{C}_{11}^T(A_1^{d_1-1}+\cdots+I)\underline{b}_1 \\ B_{12} &= \underline{C}_{12}^T(A_2^{d_1-1}+\cdots+I)\underline{b}_2 \end{aligned}$$

Then equation (7) becomes,

$$y_1(t+d_1)=\underline{P}_{11}^T\underline{L}_1(t)+\underline{P}_{12}^T\underline{L}_2(t)+B_{11}u_1(t)+B_{12}u_2(t)+y_1(t)) \tag{8}$$

In the same way we can derive the predictive expression for the output $y_2(t)$ to be,

$$y_2(t+d_2)=\underline{P}_{21}^T\underline{L}_1(t)+\underline{P}_{22}^T\underline{L}_2(t)+B_{21}u_1(t)+B_{22}u_2(t)+y_2(t)) \tag{9}$$

where,

$$\begin{aligned} \underline{P}_{21}^T &= \underline{C}_{21}^T(A_1^{d_2}-I) \\ \underline{P}_{22}^T &= \underline{C}_{22}^T(A_2^{d_2}-I) \\ B_{21} &= \underline{C}_{21}^T(A_1^{d_2-1}+\cdots+I)\underline{b}_1 \\ B_{22} &= \underline{C}_{22}^T(A_2^{d_2-1}+\cdots+I)\underline{b}_2 \end{aligned}$$

Now if we define two reference trajectories y_{r1} and y_{r2}, in the same way as in (Richalet, et al, 1978) and equate them to the r.h.s of equations (8), (9) respectively, we get a system of 2 equations with 2 unknowns, $u_1(t)$, $u_2(t)$.

$$\left\{\begin{array}{l} B_{11}u_1(t)+B_{12}u_2(t)=y_{r1}-y_1(t)-\underline{P}_{11}^T\underline{L}_1(t)-\underline{P}_{12}^T\underline{L}_2(t) \\ B_{21}u_1(t)+B_{22}u_2(t)=y_{r2}-y_2(t)-\underline{P}_{21}^T\underline{L}_1(t)-\underline{P}_{22}^T\underline{L}_2(t) \end{array}\right\} \tag{10}$$

Defining,

$$\begin{aligned} D_1 &= y_{r1}-y_1(t)-\underline{P}_{11}^T\underline{L}_1(t)-\underline{P}_{12}^T\underline{L}_2(t) \\ D_2 &= y_{r2}-y_2(t)-\underline{P}_{21}^T\underline{L}_1(t)-\underline{P}_{22}^T\underline{L}_2(t) \end{aligned}$$

then the above system (equation (10)) can be solved for the unknowns $u_1(t)$, $u_2(t)$ to give in determinant form,

$$u_1(t)=\frac{\begin{vmatrix} D_1 & B_{12} \\ D_2 & B_{22} \end{vmatrix}}{\begin{vmatrix} B_{11} & B_{12} \\ B_{21} & B_{22} \end{vmatrix}}, \quad u_2(t)=\frac{\begin{vmatrix} B_{11} & D_1 \\ B_{21} & D_2 \end{vmatrix}}{\begin{vmatrix} B_{11} & B_{12} \\ B_{21} & B_{22} \end{vmatrix}} \tag{11}$$

And these are the required control inputs to form the deterministic multivariable self-tuning scheme. Note that this requires that, $B_{11}B_{22}-B_{21}B_{12}\neq 0$.

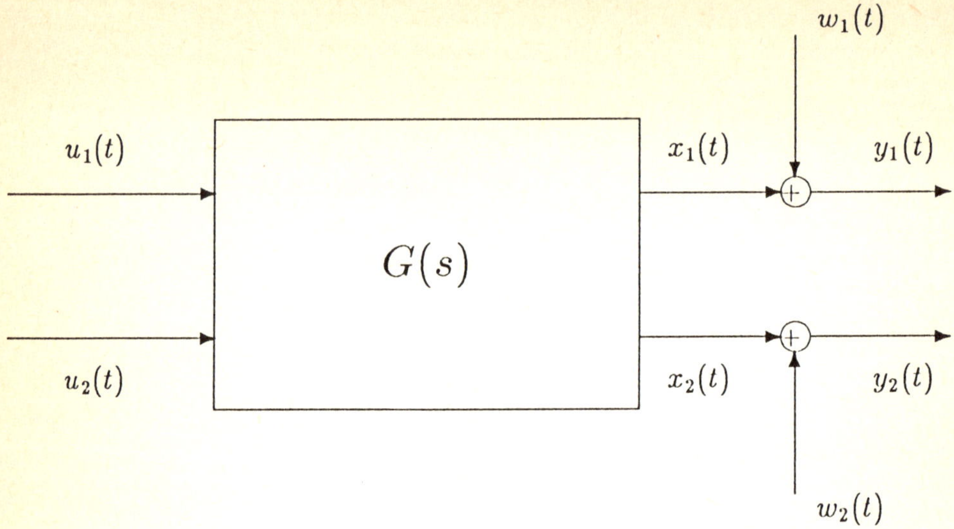

Fig 3– Stochastic multivariable 2 × 2 linear system.

3 Stochastic Multivariable Self-Tuning Control

The development of a stochastic multivariable self-tuning scheme based on the Laguerre orthonormal functions can easily be derived as an extension of the deterministic scheme by including a representation of the output process noise. Let us consider again for simplicity a 2×2 linear stochastic multivariable system as shown in Figure 3, where $w_1(t)$ and $w_2(t)$ are stochastic bounded noise sequences.

The deterministic part of the plant can be written as,

$$\underline{L}_1(t+1) = A_1 \underline{L}_1(t) + \underline{b}_1 u_1(t) \tag{12}$$

$$\underline{L}_2(t+1) = A_2 \underline{L}_2(t) + \underline{b}_2 u_2(t) \tag{13}$$

$$\begin{bmatrix} x_1 \\ x_2 \end{bmatrix} = \begin{bmatrix} \underline{C}_{11}^T & \underline{C}_{12}^T \\ \underline{C}_{21}^T & \underline{C}_{22}^T \end{bmatrix} \begin{bmatrix} \underline{L}_1 \\ \underline{L}_2 \end{bmatrix} \tag{14}$$

The stochastic part of the plant can be expressed in an analogous way by the equations,

$$\underline{N}_1(t+1) = A'_1 \underline{N}_1(t) + \underline{b}'_1 e_1(t) \tag{15}$$

$$w_1(t) = \underline{C}_1'^T \underline{N}_1(t) + e_1(t) \tag{16}$$

and,

$$\underline{N}_2(t+1) = A'_2 \underline{N}_2(t) + \underline{b}'_2 e_2(t) \tag{17}$$

$$w_2(t) = \underline{C}_2'^T \underline{N}_2(t) + e_2(t) \tag{18}$$

where, $\underline{N}_i = [n_{i1} \quad n_{i2} \ldots n_{iN'_i}]$ with $i = 1, 2$, is respectively the noise state-vector for each network representation. The noise sequences $e_1(t)$ and $e_2(t)$ are normal white noises

$N(0,1)$. Predictive expressions can then be derived for the outputs $y_1(t)$ and $y_2(t)$. Using equations (15), (17), the prediction for the output $y_1(t)$ can be written as,

$$\begin{aligned} y_1(t+d_1) &= \underline{C}_{11}^T(A_1^{d_1}-I)\underline{L}_1(t)+\underline{C}_{12}^T(A_2^{d_1}-I)\underline{L}_2(t) \\ &+\underline{C}_{11}^T(A_1^{d_1-1}+\cdots+I)\underline{b}_1u_1(t)+\underline{C}_{12}^T(A_2^{d_1-1}+\cdots+I)\underline{b}_2u_2(t) \\ &+\underline{C}_1'^T(A_1'^{d_1}-I)\underline{N}_1(t)+\underline{C}_1'^T(A_1'^{d_1-1}\underline{b}_1'(t)e_1(t)+\cdots+\underline{b}_1'(t)e_1(t+d_1+1)) \\ &+e_1(t+d_1)-e_1(t)+y_1(t) \end{aligned} \tag{19}$$

If we define,

$$\begin{aligned} \underline{P}_{11}^T &= \underline{C}_{11}^T(A_1^{d_1}-I) \\ \underline{P}_{12}^T &= \underline{C}_{12}^T(A_2^{d_1}-I) \\ B_{11} &= \underline{C}_{11}^T(A_1^{d_1-1}+\cdots+I)\underline{b}_1 \\ B_{12} &= \underline{C}_{12}^T(A_2^{d_1-1}+\cdots+I)\underline{b}_2 \\ \underline{Q}_1^T &= \underline{C}_1'^T(A_1'^{d_1}-I) \\ \underline{Q}_1'^T &= \underline{C}_1'^T(A_1'^{d_1-1}\underline{b}_1'(t)-1) \end{aligned}$$

Then the best prediction of $y_1(t+d_1 \mid t)$ is given by,

$$\begin{aligned} y_1(t+d_1 \mid t) &= \underline{P}_{11}^T\underline{L}_1(t)+\underline{P}_{12}^T\underline{L}_2(t)+B_{11}u_1(t)+B_{12}u_2(t) \\ &+\underline{Q}_1^T\underline{N}_1(t)+\underline{Q}_1'^T\hat{e}_1(t)+y_1(t) \end{aligned} \tag{20}$$

where $\hat{e}_1(t)$ is the proxying of the estimated noise $e_1(t)$ and can be calculated in the same way as it was described in (Zervos and Dumont, 1988a). The predictive expression for $y_2(t)$ is obtained, in a similar fashion:

$$\begin{aligned} y_2(t+d_2 \mid t) &= \underline{P}_{21}^T\underline{L}_1(t)+\underline{P}_{22}^T\underline{L}_2(t)+B_{21}u_1(t)+B_{22}u_2(t) \\ &+\underline{Q}_2^T\underline{N}_2(t)+\underline{Q}_2'^T\hat{e}_2(t)+y_2(t) \end{aligned} \tag{21}$$

where,

$$\begin{aligned} \underline{P}_{21}^T &= \underline{C}_{21}^T(A_1^{d_2}-I) \\ \underline{P}_{22}^T &= \underline{C}_{22}^T(A_2^{d_2}-I) \\ B_{21} &= \underline{C}_{21}^T(A_1^{d_2-1}+\cdots+I)\underline{b}_1 \\ B_{22} &= \underline{C}_{22}^T(A_2^{d_2-1}+\cdots+I)\underline{b}_2 \\ \underline{Q}_2^T &= \underline{C}_2'^T(A_2'^{d_2}-I) \\ \underline{Q}_2'^T &= \underline{C}_2'^T(A_2'^{d_1-1}\underline{b}_2'(t)-1) \end{aligned}$$

If two reference trajectories are defined, y_{r1} and y_{r2}, and equate them to the r.h.s. parts of equations (20) and (21) respectively, a system of 2 equations with 2 unknowns, $u_1(t)$ and $u_2(t)$, is formed. To solve for the required control inputs, the method described in Section 2 can be adopted.

4 Simulation examples

For illustrative purposes two examples follow that use the deterministic multivariable self-tuning scheme, described earlier, in a stochastic environment. The first one controls a simple second-order multivariable 2×2 linear plant. The second discusses how the same self-tuning scheme can be used to control the headbox of a paper machine. So many researchers have simulated this process in the past using various self-tuning regulators that it has become a benchmark for testing multivariable adaptive control schemes.

Example 1: Consider the 2×2 second-order linear stochastic multivariable plant,

$$\begin{aligned} y_1(t) &= 0.2y_1(t-1) + 0.1y_1(t-2) + u_1(t-1) \\ &\quad + 0.8u_1(t-2) + u_2(t-2) + w_1(t) \\ y_2(t) &= 0.2y_2(t-1) + 0.1y_2(t-2) + u_1(t-1) \\ &\quad + 0.1u_1(t-2) + u_2(t-1) + w_2(t) \end{aligned} \tag{22}$$

where the covariance matrix W of the disturbances is given by,

$$W = \begin{bmatrix} 0.2 & 0.35 \\ 0.35 & 0.6 \end{bmatrix} \tag{23}$$

The deterministic multivariable Laguerre self-tuner that was applied to the above plant used, $N_1 = N_2 = 8$, the Laguerre time-scales $p_1 = 0.1$, $p_2 = 0.7$, with prediction horizons $d_1 = 2$, $d_2 = 2$, and the driver block filter time constants $\alpha_1 = \alpha_2 = 0.01$. All the initial parameter estimates were zero. The initial covariance matrices for the multivariable RLS algorithm were $1000 \times I$, and all the forgetting factors were 1. The simulation run was performed over 1400 points starting from zero reference level for both outputs. The setpoint for the first output was an upward staircase-like trajectory with step jump of +3 every 200 points. The setpoint of the second output was a downward staircase-like trajectory with step jump of -2 every 350 points. Figure 4 shows the two plant outputs following their respective reference trajectories. Figure 5 shows the behaviour of the multivariable adaptive controller outputs for the same simulation run.

Example 2: The headbox is a vital element of a paper machine. Its purpose is to change the turbulent flow in the approach piping system to a sheet flow out of the headbox. The operation of the headbox has a significant effect on the characteristics of the produced paper. The headbox with its associated flow system is a complicated hydro-dynamical device. The control problem is to maintain constant jet velocity and to have a good dynamic behaviour when changing the headbox level by manipulating the stockflow and the airflow into the headbox. In the present study it is assumed that the air cushion of the headbox is pressurized using a compressor and that the airflow is manipulated either through the massflow into the headbox or through a valve at the outlet ($q(t)$ in the Figure 6).

The control of a paper machine headbox is a good example of an interactive multivariable control system. The mathematical model equations derived from physical principles show the system to be nonlinear, second order and, stiff. However, for small perturbations around a steady state operating point the process can be considered linear (Åström, 1972). Following the principles and the headbox prototype model specifications outlined

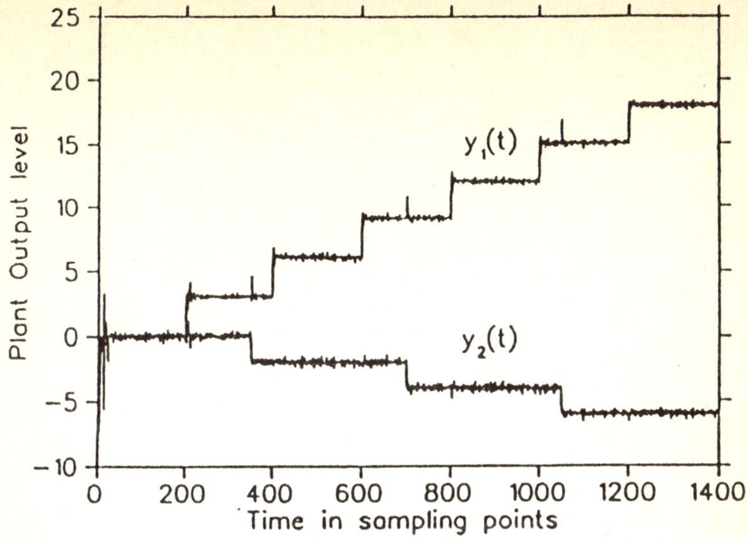

Fig 4– Example 1: Explicit multivariable self-tuning control of a 2×2 plant. System outputs tracking two respective reference trajectories.

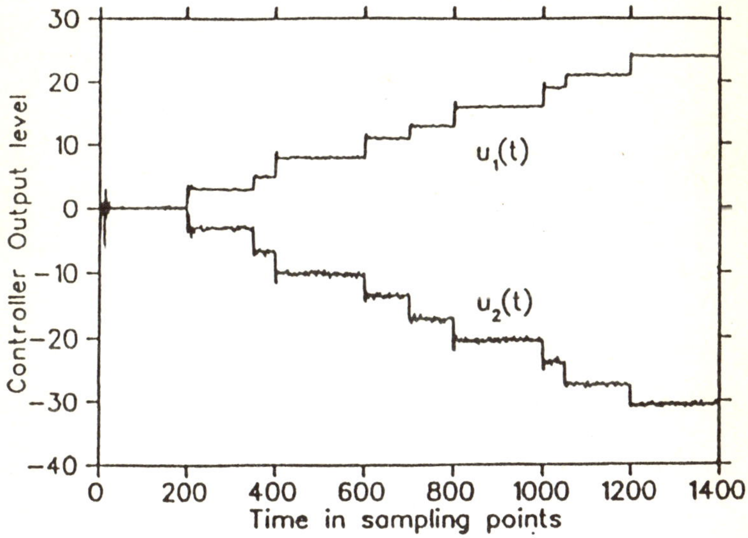

Fig 5– Example 1: Explicit multivariable self-tuning control of a 2×2 plant. Controller outputs.

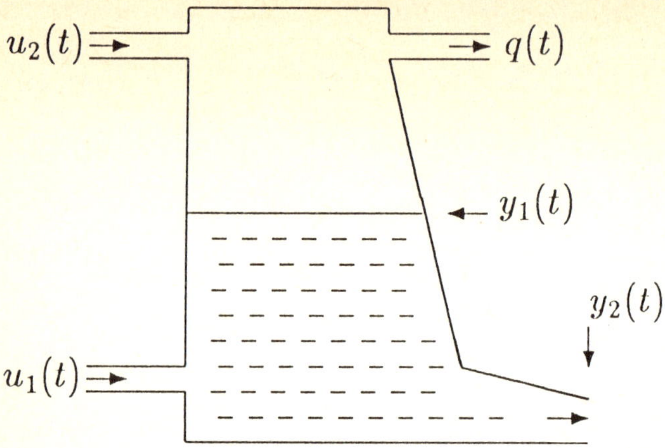

Fig 6– Example 2: Headbox of a paper-machine.

in Åström (1972) and Natarajan (1988) the continuous state-space linearized model can be derived in the form,

$$\underline{\dot{x}}(t) = \begin{bmatrix} -0.0115 & -0.1411 \\ -0.0373 & -0.5270 \end{bmatrix} \underline{x}(t) + \begin{bmatrix} 0.1 & 0.0 \\ 0.324 & 0.2 \end{bmatrix} \underline{u}(t) \tag{24}$$

$$\underline{y}(t) = \begin{bmatrix} 1.0 & 0.0 \\ 1.0 & 12.2412 \end{bmatrix} \underline{x}(t) \tag{25}$$

where,

$$\begin{array}{ll} y_1(t) & \text{is the stock level} \\ y_2(t) & \text{is the total head pressure} \\ x_1(t) & \text{is the stock level} \\ x_2(t) & \text{is the air pad density} \\ u_1(t) & \text{is the stock volume flow rate} \\ u_2(t) & \text{is the air mass flow rate} \end{array} \tag{26}$$

and the operating point was chosen to be,

$$\begin{array}{rcl} y_1(t) & = & 0.5m \\ y_2(t) & = & 4.3339\ m\ H_2O \\ x_1(t) & = & 0.5m \\ x_2(t) & = & 1.62kg/m^3 \\ u_1(t) & = & 1.0m^3/sec \\ u_2(t) & = & 0.245kg/sec \end{array} \tag{27}$$

This model was transformed to discrete form using a sampling interval of 1 *sec*. The following result was obtained,

$$\underline{x}(t+1) = \begin{bmatrix} 0.9908 & -0.1091 \\ -0.0288 & 0.5922 \end{bmatrix} \underline{x}(t) + \begin{bmatrix} 0.0803 & -0.0119 \\ 0.2505 & 0.1556 \end{bmatrix} \underline{u}(t)$$

$$\underline{y}(t) = \begin{bmatrix} 1.0 & 0.0 \\ 1.0 & 12.2412 \end{bmatrix} \underline{x}(t)$$

Eliminating $\underline{x}(t)$ and adding stochastic disturbances the following model is obtained,

$$\begin{aligned} y_1(t) &= 1.583y_1(t-1) - 0.5836y_1(t-2) \\ &\quad + 0.0802u_1(t-1) - 0.0749u_1(t-2) \\ &\quad - 0.0119u_2(t-1) - 0.0099u_2(t-2) + w_1(t) \end{aligned} \tag{28}$$

$$\begin{aligned} y_2(t) &= 1.583y_2(t-1) - 0.5836y_2(t-2) \\ &\quad + 3.1465u_1(t-1) - 3.1411u_1(t-2) \\ &\quad + 1.8927u_2(t-1) - 1.81u_2(t-2) + w_2(t) \end{aligned} \tag{29}$$

where the inputs, state variables and outputs now represent changes on the operating point conditions. To get a value of the covariance matrix of the noise $\underline{w}(t)$ the following assumptions were made in the simulations:

(a) A white noise perturbation on either input of the headbox system was taken to have standard error of 10% of their respective values, and (b) Measurement white noises with standard errors of 10% on the level and 10% on the total head pressure value were respectively assumed. Using the above assumptions on the noise disturbances we can solve the linear stochastic difference headbox state-space equations for the steady-state output noise covariance matrix by means of a discrete Lyapunov equation (Åström, 1970). The solution gives,

$$W = \mathbf{E}[\underline{w}(t)\underline{w}^T(t)] = \begin{bmatrix} 0.011 & 0.043 \\ 0.043 & 0.3127 \end{bmatrix} \tag{30}$$

The deterministic multivariable Laguerre self-tuning scheme was applied to the derived stochastic linearized headbox model. The following parameters were used : Number of Laguerre functions for each Laguerre network $N_1 = N_2 = 8$, Laguerre time-scales $p_1 = 0.06$ and $p_2 = 0.6$ respectively, prediction horizons $d_1 = d_2 = 2$. The time constants of the output reference trajectories were $\alpha_1 = \alpha_2 = 0.7$. All the initial parameters were zero. For the first 200 seconds the system runs on its normal operating point (see Figure 7) in order to allow for the estimation to obtain a model of the system. To test the system, at 200 seconds the stock level setpoint asked for a positive change of $0.3m$ for the next 200 seconds while the total head pressure setpoint was kept constant at its normal operating value. At the 400th second the stock level setpoint returned back to its normal initial operating value. At the 600th second the total head pressure setpoint asked for a positive change of 20% of its operating value for the next 200 seconds while the stock level setpoint was kept to its normal operating value. Finally at the 800th second the total head pressure setpoint returned back to its normal operating value. The output responses in Figure 7 are indicative of the good regulation obtained on both outputs.

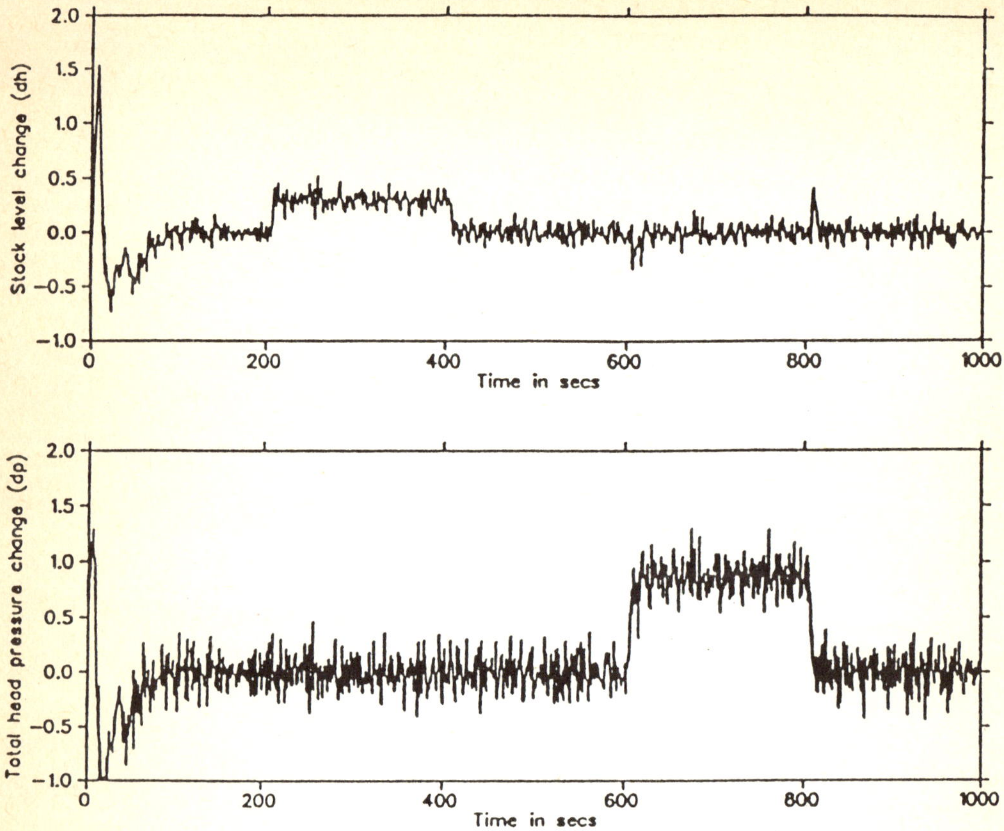

Fig 7– Example 2: Headbox multivariable self-tuning control. Stock level (upper curve), and Total head pressure (lower) changes.

5 Conclusions

Techniques for multivariable deterministic and stochastic adaptive control systems have been described. The adaptive algorithms, characterized by their simplicity, ease of use, and flexible structure, are based on the set of Laguerre orthonormal functions although other orthonormal sets can also be used.

The approach taken is a direct extension of the single-input single-output self-tuning controllers based on orthonormal series representation. A major advantage of the proposed schemes is that of the knowledge of the system delay matrix, *i.e.* of the interactor matrix, is not required. This adds simplicity to the design of the algorithm and provides a better approach to the multivariable self-tuning control problem. Simulation results have been presented showing the applicability of the algorithms.

6 References

1. Åström, K.J., (1970), "Introduction to stochastic control theory", Academic Press, N.Y.

2. Åström, K.J., (1972), "Process control in the pulp and paper industry." Notes from a course given in the Dept. of Auto. Control, Lund University, TFRT–3124.

3. Dumont, G.A. and Zervos C.C., (1986), "Adaptive Controllers based on Orthonormal Series Representation", 2nd IFAC Workshop on Adaptive Systems in Control and Signal Processing, Lund, Sweden.

4. Elliot, H., (1982), "Direct adaptive pole placement with application to non-minimum phase systems", *IEEE Trans. Auto. Control*, vol. AC-27, p. 720.

5. Jury, I.E., (1958), "Sampled-data Control Systems" J. Wiley and Sons, London.

6. Mohtadi, C., Shah, S.L., Clarke, D.W., (1987), "Generalized predictive control of multivariable systems", Proc. of the 5th Yale Workshop on Appl. of Adaptive Systems, Yale Univ.

7. Natarajan, K., (1988), "On the Headbox flow dynamics", personal communication. Pulp and Paper Centre - UBC, Vancouver, BC.

8. Richalet, J., A. Rault, J. Testud and J. Papon, (1978), "Model predictive heuristic control: Applications to industrial processes." *Automatica*, vol. 14, pp. 413-428.

9. Wolovich, W.A. and Falb, P.L., (1976), "Invariants and canonical forms under dynamic compensation", *SIAM Jnl Control and Optimization*, vol. 14, p. 996.

10. Zervos, C.C., P.R. Bélanger, G.A. Dumont, (1985), "On PID controller tuning method using orthonormal series identification". *IFAC Workshop on Adapt. Control Chem. Processes.* Frankfurt, FRG. Also published in *Automatica Journal*, March, 1988.

11. Zervos C.C. and Dumont, G.A., (1988a), "Deterministic Adaptive Control based on Laguerre Series Representation", *International Journal of Control*, vol. 48., no. 6, pp. 2333–2359.

12. Zervos, C.C. and Dumont, G.A., (1988b), "Laguerre Orthonormal Functions in Stochastic Self-Tuning Control" IFAC Workshop on Robust Adaptive Control, Newcastle, NSW, Australia, pp. 102–107.

13. Zervos, C.C. and Dumont, G.A., (1988c), "Laguerre-based adaptive control of pH in an industrial bleach plant extraction stage", IFAC ADCHEM '88 symposium on Adaptive Control of Chem. Proc., August, Lyngby, Denmark.

7 Appendix

By discretizing each block it can be readily shown that a discrete-time state space representation of the Laguerre network can be written in the form:

$$\underline{L}(t+1) = A\underline{L}(t) + \underline{b}u(t) \tag{31}$$

where $\underline{L}(t)$ is the N-dimensional state vector, and $u(t)$ is the system input. A is a lower triangular $N \times N$ matrix where the same elements are found respectively across the diagonal or every subdiagonal. If T is the sampling period and

$$\tau_1 = e^{-pT}$$

$$\tau_2 = T + \frac{2}{p}(e^{-pT} - 1)$$

$$\tau_3 = -Te^{-pT} - \frac{2}{p}(e^{-pT} - 1)$$

$$\tau_4 = \sqrt{2p}\frac{(1-\tau_1)}{p}$$

then,

$$A = \begin{bmatrix} \tau_1 & 0 & \cdots & 0 \\ \frac{-\tau_1\tau_2-\tau_3}{T} & \tau_1 & \cdots & 0 \\ \vdots & \vdots & \vdots & \vdots \\ \frac{(-1)^{N-1}\tau_2^{N-2}(\tau_1\tau_2+\tau_3)}{T^{N-1}} & \cdots & \frac{-\tau_1\tau_2-\tau_3}{T} & \tau_1 \end{bmatrix} \tag{32}$$

and

$$\underline{b}^T = [\ \tau_4 \quad (-\tau_2/T)\tau_4 \quad \cdots \quad (-\tau_2/T)^{N-1}\tau_4\ ] \tag{33}$$

The above state-space system is stable ($p > 0$), observable and controllable.

DIRECT ADAPTIVE CONTROL WITH TIME DELAY MISMATCH

W.R. Cluett
Department of Chemical Engineering, University of Toronto
Toronto, Canada M5S 1A4

S.L. Shah
Department of Chemical Engineering, University of Alberta
Edmonton, Canada T6G 2G6

Abstract

This paper addresses the issue of mismatch between the process delay and the delay selected by the designer for use in the predictive control law. Sufficient conditions are given for global stability of an adaptive predictive control system when delay mismatch is present. It is illustrated that a modification of the normalized parameter estimation system with deadzone proposed for systems with unmodeled dynamics is required. Weighting on the control action is shown to be important when the delay is underestimated in order to avoid high gains in the controller.

1 Introduction

Much of the recent research in adaptive control has focused on the analysis and design of algorithms which are robust in the presence of disturbances and unmodeled dynamics. Recently, a new robust adaptive predictive controller (RAPC) has been developed by Cluett et al (1987,1988) which uses a normalized parameter estimation scheme with a deadzone. The normalization approach permits the modeling errors to be treated as a bounded disturbance.

One condition of the result in Cluett et al (1987,1988) is that the time delay of the process is assumed to be known. However, in many process control applications the dead time is subject to change (e.g. transportation lag associated with fluid flow in a pipe). This paper presents sufficient conditions for the RAPC which ensure global stability when a mismatch exists between the actual process delay and the delay used in the design of the controller.

The contributions of this paper are: (i) the introduction of a new normalization factor designed to handle time-delay mismatch, and (ii) an illustration of how control weighting can be used to prevent high gains in the controller when the delay is underestimated. Related results may be found in Gawthrop and Lim (1982), Kreisselmeier and Anderson (1986) and Ortega (1986).

2 Control System Description

Let the process to be controlled be described by the following discrete-time representation

$$A(q^{-1})y(k) = q^{-d}B(q^{-1})u(k) \tag{1}$$

where

$$A(q^{-1}) = 1 + a_1q^{-1} + ... + a_{n1}q^{-n1}$$
$$B(q^{-1}) = b_0 + b_1q^{-1} + ... + b_{n2}q^{-n2}$$

By successive substitution, (1) may be rewritten as

$$y(k) = \theta^t\phi(k\text{-}d) \tag{2}$$

where

$$\phi(k\text{-}d)^t = [y(k\text{-}d) ... y(k\text{-}d\text{-}n1+1) , u(k\text{-}d) ... u(k\text{-}2d\text{-}n2+1)]$$

Let $\hat{d}$ be an estimate of the true delay d in (1). Consider that the adaptive predictive model estimate $\hat{y}(k|k)$ of the process output at time k is based on a vector of estimated parameters $\hat{\theta}_r(k)$ and an input-output (I/O) vector $\phi_r(k\text{-}\hat{d})$ where

$$\hat{y}(k|k) = \hat{\theta}_r(k)^t\phi_r(k\text{-}\hat{d}) \tag{3}$$
$$\phi_r(k\text{-}\hat{d})^t = [y(k\text{-}\hat{d}) ... y(k\text{-}\hat{d}\text{-}n1+1 , u(k\text{-}\hat{d}) ... u(k\text{-}2\hat{d}\text{-}n2+1]$$

In order to focus on the problem of mismatch in the delay, it has been assumed that the model order is based on the true process order (n1 and n2). If unmodeled dynamics were present due to underestimation of the process order, then n1 and n2 in ϕ_r would be based on the order of the reduced model. The corresponding *a posteriori* estimation error is defined as

$$e(k|k) = y(k) - \hat{y}(k|k) = y(k) - \hat{\theta}_r(k)^t\phi_r(k\text{-}\hat{d}) \tag{4}$$

where the estimated parameter vector is generated by an appropriate adaptive law which verifies the convergence properties that guarantee global stability (Theorem 1). The control input u(k) is generated using adaptive predictive control by making the predicted output at time k+$\hat{d}$ equal to the setpoint, i.e.

$$y_{sp}(k+\hat{d}) = \hat{\theta}_r(k)^t\phi_r(k) \tag{5}$$

3 Global Stability Condition

The following assumptions are used to establish global stability.

A1: An upperbound $\hat{d}$ for d is known.

A2: The sequence $\{\|\Phi(k)\|\}$ is unbounded only if there is a subsequence $\{k_s\}$ such that

$$\text{(a)} \lim_{k_s \to \infty} \|\Phi(k_s-1)\| = \infty, \text{ and}$$

$$\text{(b)} \ |y(k_s)| > \alpha_1 \|\Phi(k_s-1)\| - \alpha_2, \ \forall k_s$$

where

$$\Phi(k-1)^t = [y(k-1) \ldots y(k-\hat{d}-n1+1 \ , u(k-1) \ldots u(k-2\hat{d}-n2+1)]$$

is an I/O vector which contains all of the elements included in $\phi(k-d)$ and $\phi_r(k-\hat{d})$, and $0<\alpha_1<\infty$ and $0\leq\alpha_2<\infty$. The assumption A2 is standard for processes with a stable-inverse. If this problem formulation were to include unmodeled dynamics, n1 and n2 in $\Phi(k-1)$ would be upper bounds on the actual process order.

Theorem 1: Under the assumptions A1 and A2, adaptive predictive control (5) of the process described by (1) guarantees a bounded sequence $\{\|\Phi(k)\|\}$ if the adaptive law of the adaptive predictive model parameters satisfies the following condition

$$\alpha_1 > \|\hat{\theta}_r(k) - \hat{\theta}_r(k-\hat{d})\| + \frac{e(k|k)}{\max(\|\Phi(k-1)\|,c)} \qquad \forall k>k_1>0$$

where $k_1<\infty$ and c is any positive constant and is used to prevent division by zero.

Proof: The proof follows that given in Cluett et al (1988) for Theorem 1 using the fact that $\|\Phi(k-1)\| \geq \|\phi_r(k-\hat{d})\|$.

4 Process Representation for Parameter Estimation

The normalized estimation systems with deadzone proposed by Cluett et al (1987,1988) for systems with unmodeled dynamics may also be used here because the delay mismatch actually leads to unmodeled terms in the predictive model estimation. The variables for the normalized system are defined as follows

$$y^n(k) = y(k)/n(k) \qquad (6a)$$

$$x_r(k-\hat{d}) = \phi_r(k-\hat{d})/n(k) \tag{6b}$$
$$n(k) = \max(\max_{1\le i\le m} |\Phi_i(k-1)|, c) \tag{6c}$$

where Φ_i is the ith element of Φ and m is its dimension. The process equation in (2) may be rewritten as

$$y(k) = \theta_r{}^t\phi_r(k-\hat{d}) + \theta_u{}^t\phi_u(k-d) \tag{7}$$

where $\phi_u(k-d)$ contains the elements in $\phi(k-d)$ not included in $\phi_r(k-\hat{d})$, and θ_r, θ_u contain their respective process parameters. Dividing (7) by n(k) gives

$$y^n(k) = \theta_r{}^t x_r(k-\hat{d}) + \theta_u{}^t\phi_u(k-d)/n(k) \tag{8}$$

The second term on the right-hand side of (8) may be interpreted as a perturbation to the normalized process output $y^n(k)$. This equation is then in a form which is structurally similar to the normalized process representation used in Cluett et al (1987,1988) for the unmodeled dynamics problem, i.e.

$$y^n(k) = \theta_r{}^t x_r(k-\hat{d}) + \Delta^n(k) \tag{9a}$$
$$\Delta^n(k) = \theta_u{}^t\phi_u(k-d)/n(k) \tag{9b}$$

The sequence $\{\Delta^n(k)\}$ is bounded because of the choice of n(k) in (6c). The normalization permits the unmodeled terms created by the delay mismatch to be treated as a bounded disturbance. Therefore, the normalized estimation systems with a deadzone presented in Cluett et al (1987,1988) are valid here.

5 The Importance of Q-Weighting when the Delay is Underestimated

Time delay mismatch leads to I/O elements in $\phi_r(k-\hat{d})$ which do not appear in the true process equation [$\phi(k-d)$]. Therefore the parameters in θ_r associated with these elements are equal to zero. In the case of underestimation of the delay ($\hat{d} < d$), this may lead to high gains in the predictive control law as the estimator attempts to identify a zero parameter for the $u(k-\hat{d})$ term.

Cluett et al (1987) proposed an augmented plant which includes weighting on the output y and input u. The augmented plant will be defined here as in [2] but with d replaced by its estimate $\hat{d}$.

$$z(k) = P(q^{-1})y(k) + q^{-\hat{d}}Q(q^{-1})u(k) \tag{10}$$

where P and Q are weighting polynomials in the delay operator. Substituting (1) into (10) gives

$$Az(k) = q^{-\hat{d}}[q^{\hat{d}-d}PB + QA]u(k) \qquad (11a)$$
$$= q^{-\hat{d}}B'(q^{-1})u(k) \qquad (11b)$$

If P=1, Q=λ (a constant), and $d-\hat{d} \geq 1$, then B' has the following structure

$$B'(q^{-1}) = q^{\hat{d}-d}(b_0+...+b_{n2}q^{-n2}) + \lambda(1+a_1q^{-1}+...+a_{n1}q^{-n1}) \qquad (12a)$$
$$= b_0' + b_1'q^{-1} + ... \qquad (12b)$$

The adaptive controller described in this paper may now be applied to the augmented plant with the stable-inverse assumption applying to B'. The parameter associated with $u(k-\hat{d})$ is b_0' which is nonzero and is in fact selected by the user. The following example will illustrate the effect of Q-weighting on the performance of the RAPC when the delay is underestimated.

Example: Assume that the process to be controlled is given by the following transfer function

$$A(q^{-1})y(k) = q^{-d}B(q^{-1})u(k)$$

with

$$A(q^{-1}) = 1 - .905q^{-1}$$
$$B(q^{-1}) = .67 + .28q^{-1}$$
$$d = 3$$

Assume that the delay is underestimated by one sample interval ($\hat{d}=2$) and that a choice of P=1 and Q=λ is made.

A root locus of (12) shows that the roots of B' lie within the unit circle for λ>.28. Therefore, for this example, a choice of Q=λ>.28 will ensure that the augmented plant satisfies the stable-inverse assumption. For implementation, it is suggested that the user start with a large value for λ, and then decrease its value until a satisfactory speed of response is obtained.

Figure 1 shows the simulation results of applying the adaptive controller described in this paper to the above example with λ=2. The manipulated variable u(k) has been offset by 10 units for clarity. The I/O vector used in the predictive model (3) for this case is

$$\phi_r(k-\hat{d})^t = [z(k-2) \;\; u(k-2) \;\; u(k-3) \;\; u(k-4)]$$

This corresponds to an augmented process model with $\hat{d}=2$ and n1=n2=1. Figure 2 shows the results of applying the same adaptive controller without any Q-weighting. The I/O vector for this case is

$$\phi_r(k-\hat{d})^t = [y(k-2) \;\; u(k-2) \;\; u(k-3) \;\; u(k-4)]$$

which corresponds to a process model with $\hat{d}$=2 and n1=n2=1.

Figure 1 clearly shows that the control system is robust to the underestimation of the delay when Q-weighting is used. Figure 2 indicates that the adaptive controller is not capable of handling the delay mismatch without the Q-weighting due to high gains in the control law. The input and output signals in this example saturated after the second setpoint change at k=100 and the rest of the result has been omitted.

6 Conclusions

This paper has extended the robust adaptive predictive controller of [1,2] to handle mismatch between the process delay and the delay selected for the predictive control law. A new normalization factor is used which requires only an upperbound on the process delay. The importance of control weighting is highlighted for the case when the process delay is underestimated.

References

Cluett, W.R., J.M. Martin-Sanchez, S.L. Shah and D.G. Fisher (1988). Stable discrete-time adaptive control in the presence of unmodeled dynamics. IEEE Trans. Aut. Contr., AC-33, 410-414.

Cluett, W.R., S.L. Shah and D.G. Fisher (1987). A stable robust adaptive controller. Int. J. Contr., 45, 1265-1273.

Gawthrop, P.J. and K.W. Lim (1982). Robustness of self-tuning controllers. IEE Proc.-Pt.D, 129, 21-29.

Kreisselmeier, G. and B.D.O. Anderson (1986). Robust model reference adaptive control. IEEE Trans. Aut. Contr., AC-31, 127-133.

Ortega, R. (1986). Stability analysis of discrete-time adaptive controllers for systems with unknown delays. IEEE Trans. Aut. Contr., AC-31, 980-982.

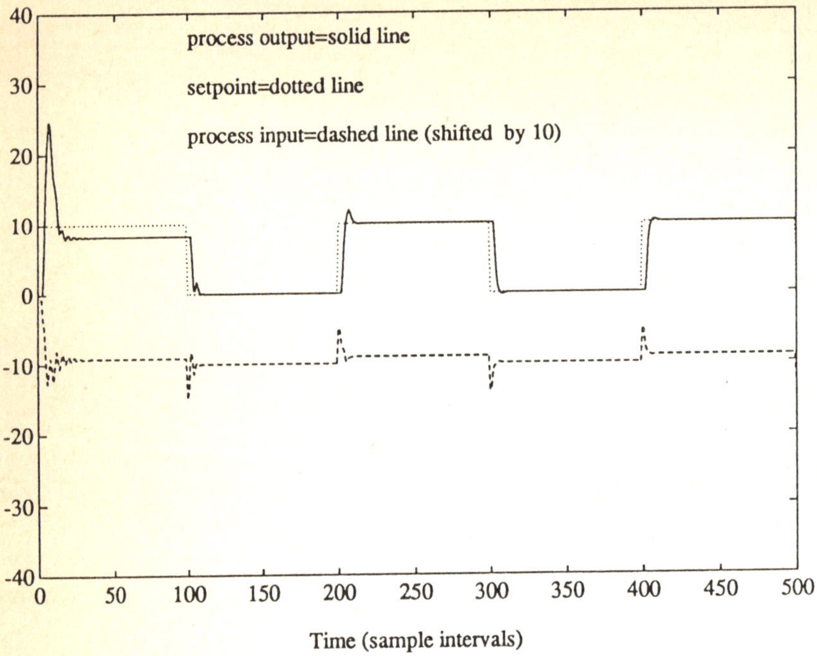

Figure 1 : RAPC with Q-weighting

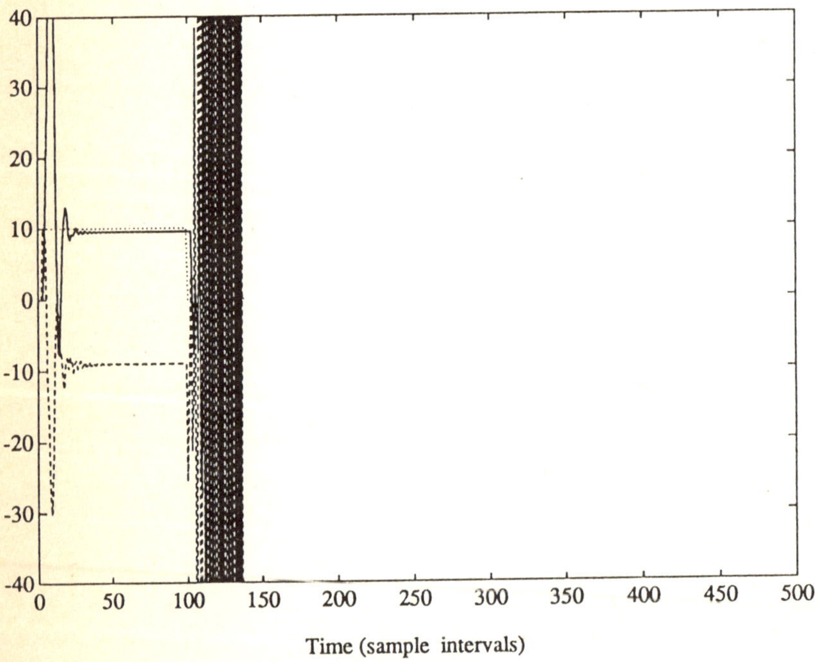

Figure 2 : RAPC without Q-weighting

DIRAC : A FINITE IMPULSE RESPONSE DIRECT ADAPTIVE CONTROLLER

Robain M.C. De Keyser
Automatic Control Lab. University of Ghent
Grotesteenweg Noord 2, B-9710 Ghent, Belgium

Abstract

A process control strategy based on adaptive control by means of a Finite Impulse Response (FIR) controller will be described. The method is called DIRAC (DIRect Adaptive Control). A simulation study indicates that this control algorithm leads to comparable performance characteristics as a well-tuned PID-type regulator. Moreover it has the advantage that it needs no tuning. The underlying adaptation technique is simple. As a special case a self-tuning PID-controller is obtained.

Keywords

Adaptive control, parameter estimation, PID control, process control, self-adjusting systems.

Introduction

During the last decade a vast and growing amount of self-tuning and adaptive control algorithms was developed and published in the control literature. Most of the methods are based on the principle of identification and control, i.e. in a first step the unknown parameters of a postulated process model are estimated and in a second step these estimated parameters are used to design a controller.

Most of the parameter identification algorithms belong to the class of recursive least-squares methods, but as far as the controller part is concerned, a lot of different design principles gave rise to many adaptive controllers. Well-known design strategies are e.g. self-tuning minimum-variance and related algorithms, self-tuning pole-placement, self-tuning LQG, self-tuning predictive control. Nice surveys in adaptive control are given in (Åström, 1983) and (Seborg et al., 1986).

Two rather different classes of adaptive control algorithms were compared to each other by the author. Special attention was paid to the kind of situations that occur in process control systems, such as processes with important, possibly varying dead-time, complex (high-order) dynamics, nonlinear characteristics, non-stationary disturbances consisting of stochastic noise but also random steps and drift.

The two classes of adaptive control strategies were long-range predictive control algorithms on the one hand and a set of simple, almost heuristic control algorithms on the other hand.

A typical member of the model-based predictive controllers is the EPSAC strategy (Extended Prediction Self-Adaptive Control) which has been applied successfully to several real-life processes (De Keyser and Van Cauwenberghe, 1985). Assuming a good process model is available or estimated, it gives excellent control especially on dead-time processes, on processes with difficult dynamic behaviour and for high-performance control loops. Of both types of adaptive control strategies mentioned earlier, EPSAC will be considered as the complex one because the control action is calculated by means of an (estimated) dynamic process model. The on-line identification of this model is not always easy and requires a rather extended supervisory level (as is the case for most current adaptive control algorithms, see e.g. Wittenmark and Åström, 1984).

The results of an extensive simulation study with this class of "complex" adaptive control strategies are described in (De Keyser et al., 1988). Notice that several well-known control strategies belong to this class, such as MAC (Model Algorithmic Control; Richalet, 1980), DMC (Dynamic Matrix Control; Cutler and Ramaker, 1980), EHAC (Extended Horizon Adaptive Control; Ydstie, 1984), GPC (Generalized Predictive Control; Clarke et al., 1987). Most of them were considered in the simulation study, together with EPSAC.

It is remarkable that the only adaptive control strategies that resulted until now in a commercial realisation are rather simple methods (Andreiev, 1981; Hoopes et al., 1982; Kraus and Myron, 1984; Kofahl and Peter 1987). These commercial controllers do generally not lead to the same excellent control performance and for most control loops in industry this is probably not even necessary. Having a stable, robust and easy-to-understand control system is at least of equal importance.

As a result of the success of simple adaptive regulators and parallel to the development of "effective but complex" adaptive controllers, a class of "suboptimal but simplified" self-tuning and adaptive control techniques was developed. Most of them are of the popular PID structure and differ mainly in the way the tuning parameters of the controller are computed from the estimated process parameters. For further details reference is made to the literature (e.g. Åström and Hägglund, 1984; Bànyàsz et al., `1985; Cameron and Seborg, 1983; Clarke, 1986; Gawthrop, 1982; Proudfoot el al., 1983; Radke and Isermann, 1984; Tjokro and Shah, 1985).

COLA (Control in Open Loop by Adaptation) and DIRAC (DIRect Adaptive Control) are two methods belonging to this group of simplified self-tuners. The COLA strategy is described in (De Keyser, 1988) while the DIRAC strategy is the subject of this paper.

This method has an important plus-point for application in industry : it is quite simple compared to some other adaptive strategies because the regulator parameters are estimated directly. Moreover a PI or PID-type controller is just a special case. The adaptation needs practically no supervision and it is not at all sensitive to structural parameters such as dead-time or process order simply because it does not need a dynamic process model. On the other hand the DIRAC method generally does not lead to the fast and nice control response that can be obtained with EPSAC (and similar strategies). The difference in performance depends on the process characteristics itself and no general appreciation of DIRAC vs EPSAC can be stated. The performance of the DIRAC controller is similar to that of a PID controller and is thus sufficiently good for regulation of most processes standard in industry. It has the advantage compared to PID that it needs no tuning.

It is the aim of this paper to evaluate the relative merits of DIRAC and PID controllers on systems with typical dynamical characteristics.

This will be done by computer simulation where both methods have to deal with the same process characteristics and the same nasty disturbances.

The DIRAC Principle

The method is a direct adaptive controller, i.e. with direct estimation of the controller parameters without the need for first estimating a process model. Its structure is thus given in Fig. 1. It is a way of automating the tuning of a regulator. The steps are the following : postulate a suitable model structure for describing the dynamical characteristics of the controller. Estimate the numerical parameters of the model recursively using a suitable parameter estimator.

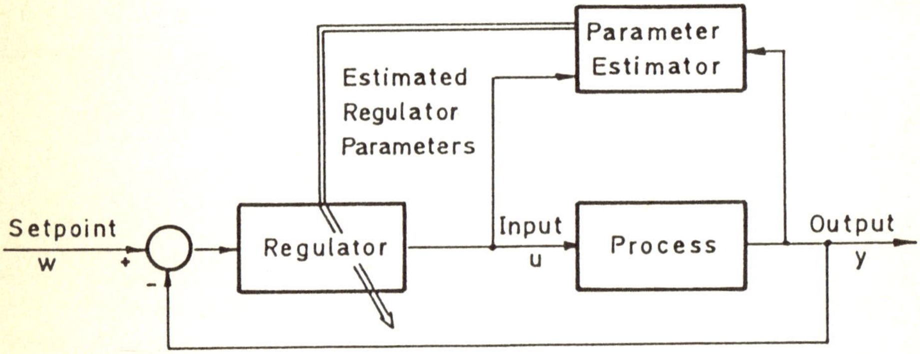

Fig. 1 : General principle of a direct self-adaptive controller.

We introduce the following notation :

- T_s : sampling interval.
- t : discrete-time index.
- q^{-1} : backward shift operator (i.e. $q^{-1}y(t) = y(t-1)$).
- r(t) : reference output.
- u(t) : process input.
- $\Delta u(t)=u(t)-u(t-1)$.
- y(t) : process output.
- w(t) : setpoint.
- e(t) = w(t)-y(t) = control error.
- process model : $P(q^{-1})$.
- controller model : $C(q^{-1})$.
- design model : $R(q^{-1})$.

The relationship between the several models is indicated in Fig. 2.

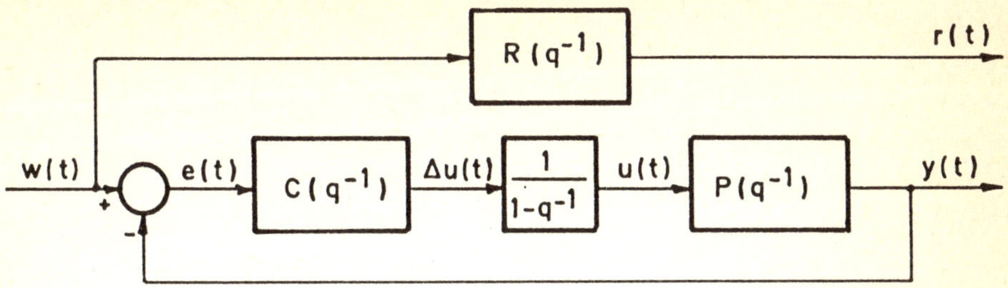

Fig. 2 : Process, controller and design models.

The structure of the process model $P(q^{-1})$ needs no further specification. The model $R(q^{-1})$ is specified by the designer and can generally have the form of a pulse transfer function, i.e.

$$R(q^{-1}) = \frac{R^N(q^{-1})}{R^D(q^{-1})} = \frac{r_0^N+r_1^Nq^{-1}+r_2^Nq^{-2}+\ldots}{1+r_1^Dq^{-1}+r_2^Dq^{-2}+\ldots} \qquad [1]$$

The controller $C(q^{-1})$ is a Finite Impulse Response (FIR) model. It has the structure

$$C(q^{-1}) = c_0+c_1q^{-1}+c_2q^{-2}+\ldots+c_nq^{-n} \qquad [2]$$

The parameters of $C(q^{-1})$ are assembled in the parameter vector

$$\underline{\theta} = [c_0 \quad c_1 \quad c_2 \ \ldots \ c_n]^T \qquad [3]$$

The design task is to calculate these controller parameters. This is done according to the following reasoning.

Suppose we would do a hypothetical experiment were we change the setpoint $w(t)$. Given the design model $R(q^{-1})$ we could compute the reference output $r(t)$. If we want the process output $y(t)$ to vary (approximately) according to the reference output $r(t)$ AND if we would know the process model $P(q^{-1})$ we could compute the required process input $u(t)$ as $u(t) = r(t)/P(q^{-1})$ (we assume here for a moment that $P(q^{-1})$ is invertible although we shall see later that this is not really necessary). We could also compute the signal $e(t) = w(t)-$

$y(t) = w(t)-r(t) = [1-R(q^{-1})]w(t)$. From the time-series $e(t)$ and $\Delta u(t)$ we could then estimate a set of parameters for $C(q^{-1})$ such that, given the restricted structure of $C(q^{-1})$, we approximately have the relationship $\Delta u(t) \cong C(q^{-1})e(t)$.

The problem is that we do not have the model $P(q^{-1})$. We could estimate it with an identification method from real data $u(t)$ and $y(t)$. Rather than going this way we start again with the above hypothetical experiment but now specify the (fictitious) setpoint variations $w(t)$ to vary such that the corresponding reference output $r(t)$ would be equal to the REAL (measured) process output $y(t)$, i.e.

$$w(t) = \frac{r(t)}{R(q^{-1})} = \frac{y(t)}{R(q^{-1})} \qquad [4]$$

Because $y(t)$ is now the real process output, the corresponding $u(t)$ need not be computed as $u(t)=y(t)/P(q^{-1})$, as it is simply the real process input. Thus we can identify the controller parameters from the time-series $e(t)$ and $\Delta u(t)$, where :

- $e(t)=w(t)-y(t)=[(1-R(q^{-1}))/R(q^{-1})]y(t)$ where $y(t)$ is the measured process output.
- $\Delta u(t)=u(t)-u(t-1)$ and $u(t)$ is the measured process input.

The design model $R(q^{-1})$ should be invertible. The process model $P(q^{-1})$ need not be estimated and is not restricted to any form. Of course, due to the restriction on the controller structure, this strategy is not a model following strategy, i.e. the real process output $y(t)$ will only follow approximately the reference output defined by $r(t)=R(q^{-1})w(t)$ where $w(t)$ is now the real setpoint. Notice that the strategy leads to offset-free regulation. The response to setpoint variations and to load disturbances is specified by means of the dynamic characteristics of the design model $R(q^{-1})$.

The DIRAC algorithm is obviously a simple adaptive strategy, easy to use without the need for complex supervision algorithms and quite robust. Its dynamic performance is good (although not excellent), comparable to the performance one expects in an industrial process control system.

Implementation Details

The design model $R(q^{-1})$ is specified as a first order model with time delay :

$$R(q^{-1}) = q^{-d} \frac{1-\alpha}{1-\alpha q^{-1}} \qquad [5]$$

Notice that, in order to fulfill the requirement y=w in static regime, we should specify $R(q^{-1})$ such that R(1)=1.

The design parameter α ($0 < \alpha < 1$) is used to make the control system faster or slower ($\alpha \longrightarrow 0$: fast control; $\alpha \longrightarrow 1$: slow control). A good choice is $\alpha = 0.90$. Most experiments in the next section were done with this value although the processes considered are quite different. This indicates that practically α can be considered as a fixed parameter rather than as a design parameter (assuming as usual a suitably chosen value for the sampling period T_s).

The dead-time index d is another design parameter. From theoretical considerations it is clear that it should be chosen such that the value $d.T_s$ is larger than the actual process time-delay. Simulation studies indicate however that the choice is really not critical and that overestimation of the dead-time generally is safer than under-estimation of this value.

The structure of the controller model has already been postulated as a FIR-model of order n. The control performance is not sensitive to the value of n, which normally can be quite small. In the next section it will be shown that n = 1 corresponds to a PI controller and n = 2 to a PID controller.

As indicated in the previous section we have to estimate the controller $C(q^{-1})$ from the signals $[(1-R(q^{-1}))/R(q^{-1})]y(t)$ as input and $(1-q^{-1})u(t)$ as output. For obvious reasons, in a real-time situation we have to use the delayed signals $[(1-R(q^{-1}))/R(q^{-1})]y(t-d)$ and $\Delta u(t-d)$.

Define :

$$x(t) = \frac{1-R(q^{-1})}{R(q^{-1})} y(t-d) = \frac{q^{-d}[1-R(q^{-1})]}{R(q^{-1})} y(t) = \frac{1-\alpha q^{-1}-(1-\alpha)q^{-d}}{1-\alpha} y(t) \quad [6]$$

$$\underline{\phi}(t) = [x(t) \quad x(t-1) \quad x(t-2) \ \dots \ x(t-n)]^T \quad [7]$$

leading to the estimation model :

$$\Delta u(t-d) = \underline{\phi}^T(t).\underline{\hat{\theta}} + \varepsilon(t) \quad [8]$$

where ε is the estimator residu.

The unknown parameters in $\underline{\theta}$ are then estimated recursively by means of the LMS (Least Mean Squares) method, which is a computationally simple, scalar parameter estimator (Goodwin and Sin, 1984). With this method the relatively large number of parameters in $\underline{\theta}$ (when n is chosen large) is no longer an objection from a computational point of view.

The LMS algorithm applied to our problem gives :

$$\underline{\hat{\theta}}(t) = \underline{\hat{\theta}}(t-1) + \frac{\mu \ \underline{\phi}(t)}{\gamma+\underline{\phi}^T(t)\underline{\phi}(t)} [\Delta u(t-d)-\underline{\phi}^T(t)\underline{\hat{\theta}}(t-1)] \quad [9]$$

with γ a small positive constant (e.g. $\gamma=0.01$) and μ ($0 \leq \mu < 2$) a design parameter which determines essentially the adaptation speed. In the DIRAC strategy the value of μ should not be chosen too high, typically μ = 0.01 à 0.1.

In the case of a constant offset z in the process model (due to a stochastic disturbance with nonzero average value, a step disturbance or a linearisation offset) we have (ref. to Fig. 2) :

$$y(t) = P(q^{-1})u(t) + z \quad [10]$$

Repeating the reasoning we have made in the previous section, we should now do the identification of the controller parameters in $C(q^{-1})$ from the time-series $\Delta u(t)$ and $[(1-R(q^{-1}))/R(q^{-1})][y(t)-z]$. The value of z is unknown. However because $1-R(1) \equiv 0$ (as $R(1) \equiv 1$) the term in z can be omitted.

A special case : PID

Starting from the equation of a continuous-time PID algorithm

$$u(t) = K_c[e(t) + \frac{1}{T_i}\int^t e(s)ds + T_d \frac{de(t)}{dt}] \qquad [11]$$

and replacing time derivatives by backward differences we obtain as discrete-time PID algorithm :

$$\Delta u(t) = K_c[(1-q^{-1}) + \frac{T_s}{T_i} + \frac{T_d}{T_s}(1-2q^{-1}+q^{-2})]e(t) \qquad [12]$$

or

$$\Delta u(t) = K_c[(1 + \frac{T_s}{T_i} + \frac{T_d}{T_s}) - (1 + 2\frac{T_d}{T_s})\ q^{-1} + \frac{T_d}{T_s}q^{-2}]e(t) \qquad [13]$$

This is thus a special case of a FIR-structure of order n=2, with

$$\left.\begin{aligned} c_0 &= K_c(1 + \frac{T_s}{T_i} + \frac{T_d}{T_s}) \\ c_1 &= -K_c(1 + 2\frac{T_d}{T_s}) \\ c_2 &= K_c\frac{T_d}{T_s} \end{aligned}\right\} \qquad [14]$$

A PI-controller corresponds to the case n = 1. The parameters of the continuous-time PID-controller, which normally have some intuitive appeal to process operators, can be computed from the estimated c_0, c_1 and c_2 :

$$\left.\begin{aligned} K_c &= -(c_1 + 2c_2) \\ T_i &= -T_s.(c_1 + 2c_2)/(c_0 + c_1 + c_2) \\ T_d &= -T_s.c_2/(c_1 + 2c_2) \end{aligned}\right\} \qquad [15]$$

Alternatively, these parameters could also be estimated directly based on equation [12]. Define a new parameter vector

$$\underline{\theta}' = [K_c \qquad K_c\frac{T_s}{T_i} \qquad K_c\frac{T_d}{T_s}]^T \qquad [16]$$

and a new data vector

$$\underline{\phi}'(t) = [x(t)-x(t-1) \qquad x(t) \qquad x(t)-2x(t-1)+x(t-2)]^T \qquad [17]$$

where x(t) is given by [6]. The estimator [9] will then estimate the parameters in [16] which are directly related to K_c, T_i and T_d. This can be a more interesting structure from the point of view of supervision, because normally an acceptable region for these parameters can be stated a priori, thus preventing the estimator from identifying meaningless values.

Notice that the impulse response of a PID-controller between e(t) and u(t) (which is also the step response between e(t) and Δu(t), thus $C(q^{-1})/(1-q^{-1})$), has the form given in Fig. 3. For a general DIRAC controller, this has a similar shape, but with more terms.

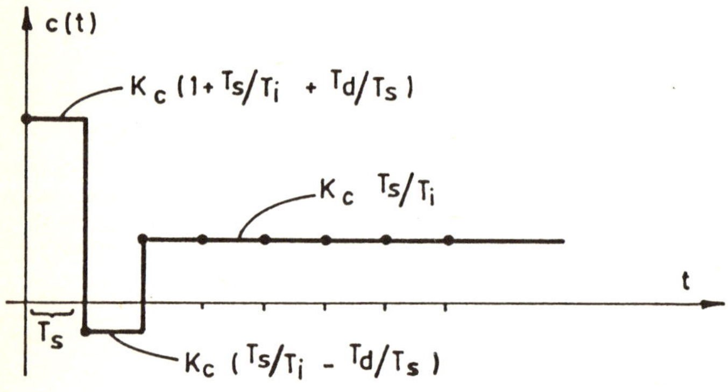

Fig. 3 : Shape of impulse response of discrete-time PID-controller.

Simulation Experiments

A lot of experiments were done with the DIRAC strategy on computer-simulated processes. Some results on several typical processes are summarized in this section.

The processes are characterized by their stepresponse rather than by a transfer function. This visual representation is better to get quickly an idea of the specific process characteristics (e.g. value of time-delay relative to important time constant, possible non-minimumphase behaviour, etc.).

For most experiments the design parameters were set to some default value (unless indicated otherwise) :
- sampling period T_s = 5 time units;
- design model speed parameter α = 0.90;
- adaptation speed parameter μ = 0.01.

Process 1

The process stepresponse is shown in Fig. 4. This S-type response is typical for many units in process control. The design model dead-time index was chosen as d = 6. The FIR model order was set to n = 4.

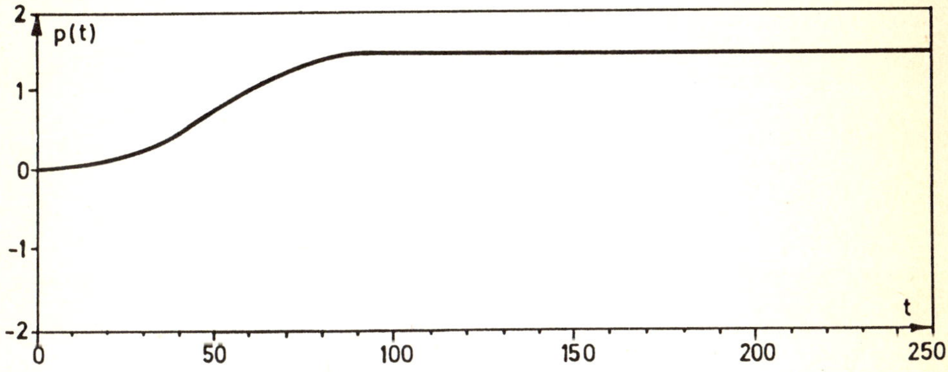

Fig. 4 : Stepresponse of process 1.

In Fig. 5 the response to a setpoint change 30% ⟶ 80% is compared for the DIRAC controller and for a fixed parameter, well tuned PI controller (gain K_c = 0.30; reset time T_i = 25).

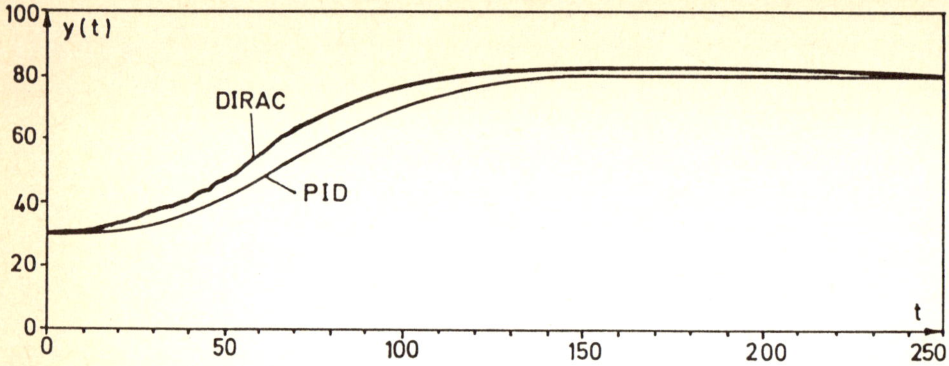

Fig. 5 : Setpoint change for DIRAC and PI controllers.

Figure 6 shows the regulation performance of both controllers. Additive stochastic disturbances where entered at the process input (binary noise +10%/-10%). Notice that for DIRAC the adaptation remains continuously on and with the same adaptation speed during setpoint changes, stochastic disturbances or (later) stepwise disturbances.

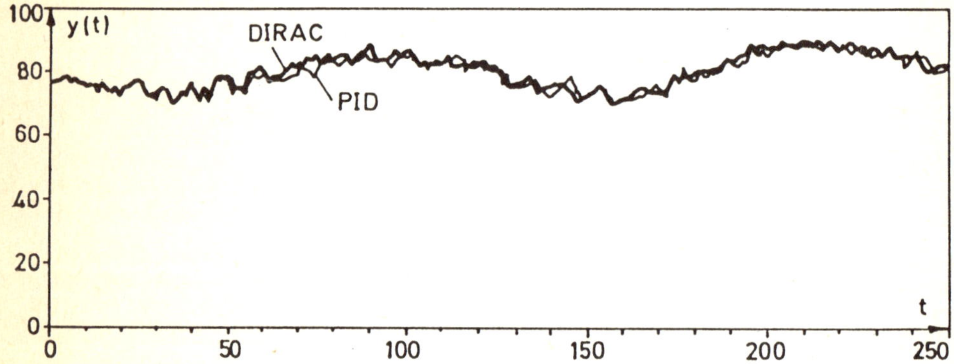

Fig. 6 : Regulation against disturbances.

In Fig. 7 a new setpoint change (80% ⟶ 30%) was realized after a long period of regulation against noise at a constant setpoint of 80%. This experiment indicates that the DIRAC method does not suffer from this frequently occuring situation in process control. Many adaptive control strategies would blow up or would result in a detuned controller (unless the adaptation is switched off by a supervisory level, which was not the case here).

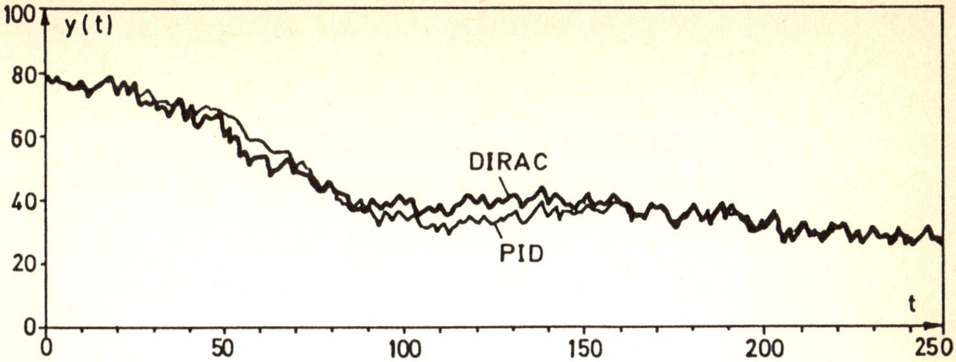

Fig. 7 : Setpoint change in the presence of disturbances.

Fig. 8 shows the results of both controllers for a stepwise load change at the process input (-20%) in the presence of stochastic disturbances. The setpoint was 50%.

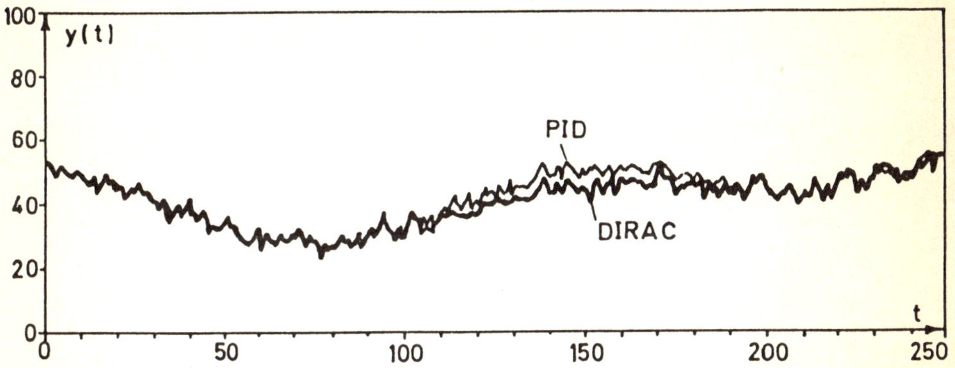

Fig. 8 : Regulation of a load change at the process input.

Process 2

The process stepresponse is shown in Fig. 9. Notice that the simulated process was switched from process 1 to process 2 during 1 sampling period and that both controllers had to deal with this abrupt change!

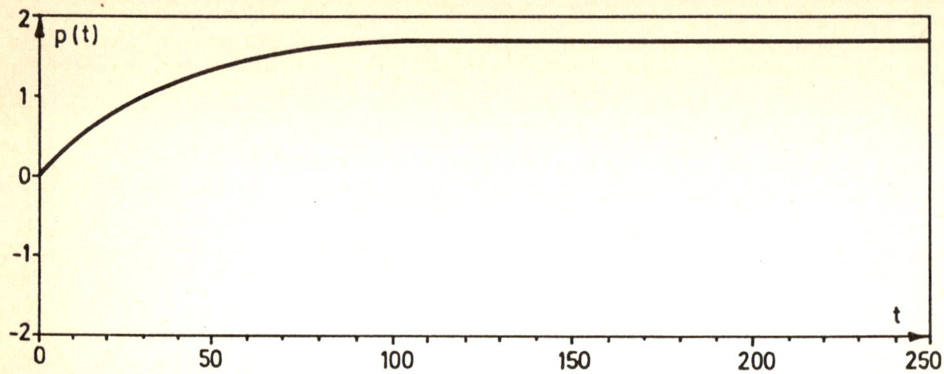

Fig. 9 : Stepresponse of process 2.

Figure 10 shows again a setpoint variation in the presence of stochastic disturbances for both controllers. Notice that the PI controller was not retuned, indicating the robustness of this popular algorithm. Also DIRAC is able to deal with the changed process characteristic. However the retuning of the controller to the new situation of course takes some time.

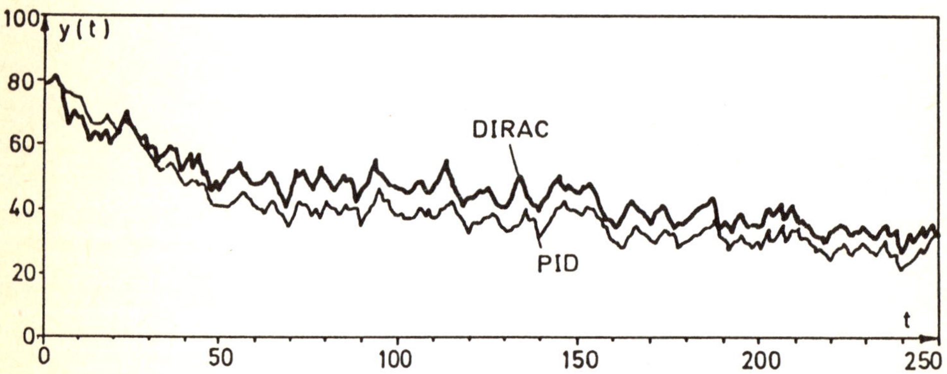

Fig. 10 : Setpoint change on process 2.

Fig. 11 shows a similar setpoint change after retuning (now with d = 1). The response is now faster. Notice that the fixed parameter PI regulator was also retuned manually (K_c = 0.4; T_i = 15).

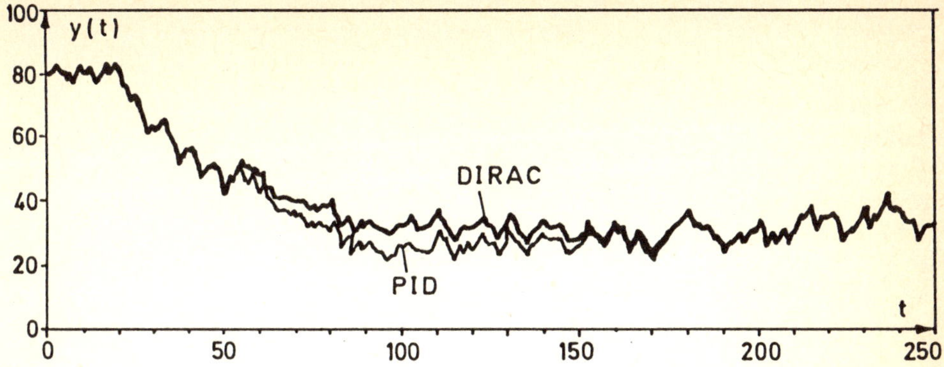

Fig. 11 : Setpoint change on process 2 after retuning.

Process 3

The process stepresponse is shown in Fig. 12. The process has an important transportation time-delay relative to its main time constant. The design model dead-time parameter was set to d = 10. The FIR model order was chosen as n = 1, which corresponds to a PI structure. The fixed parameter PI controller was tuned at K_c = 0.50; T_i = 40.

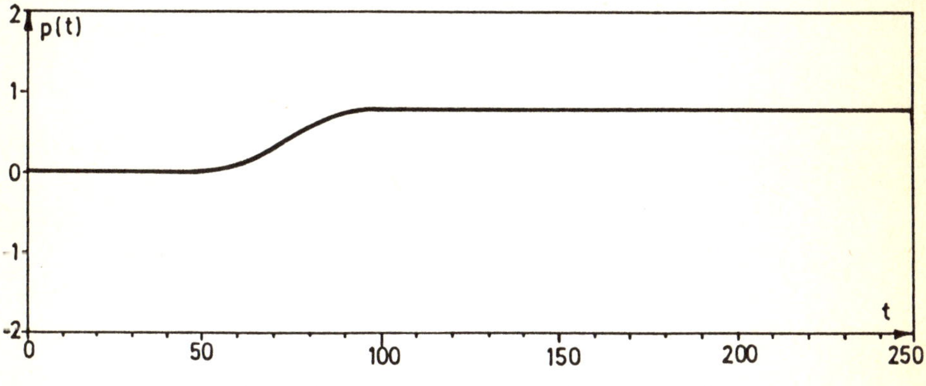

Fig. 12 : Stepresponse of process 3.

Figure 13 shows a setpoint change (60% ⟶ 20%) for both regulators. There were no stochastic disturbances.

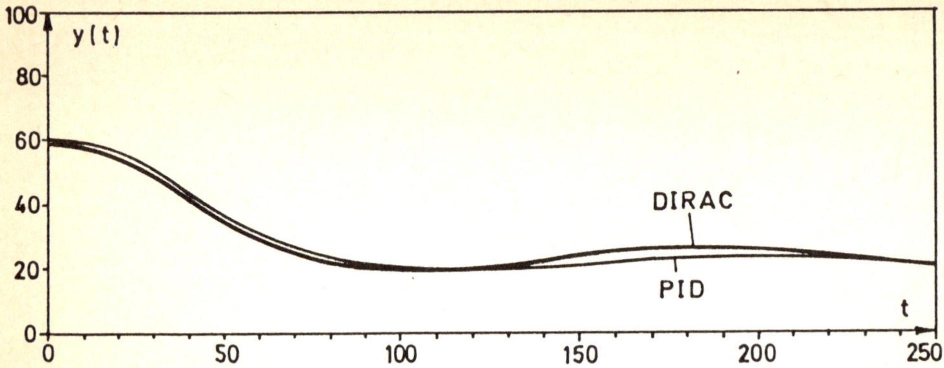

Fig. 13 : Setpoint change on process 3.

The estimated parameters of the DIRAC regulator with PI structure were $K_c = 0.6$; $T_i = 50$.

Process 4

The process stepresponse is shown in Fig. 14. It is a non-minimumphase system. For the fixed parameter PI controller as well as for the DIRAC controller all design parameters were again kept as for process 3, except for the FIR model order which was now increased to n = 2 (PID structure). The result of a setpoint variation (60% $\longrightarrow$ 20%) is shown in Fig. 15 (process inputs) and Fig. 16 (process outputs).

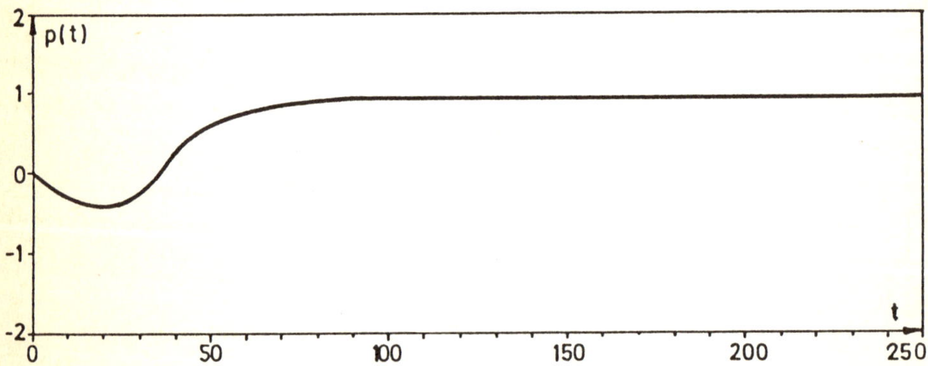

Fig. 14 : Stepresponse of process 4.

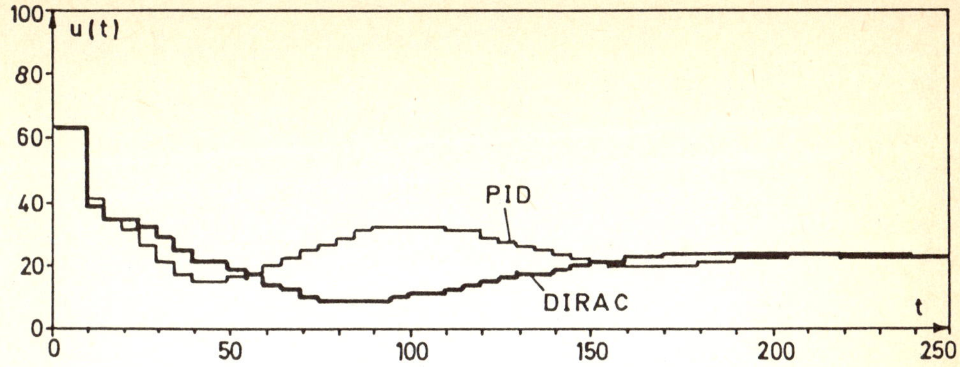

Fig. 15 : Setpoint change on process 4 (process input).

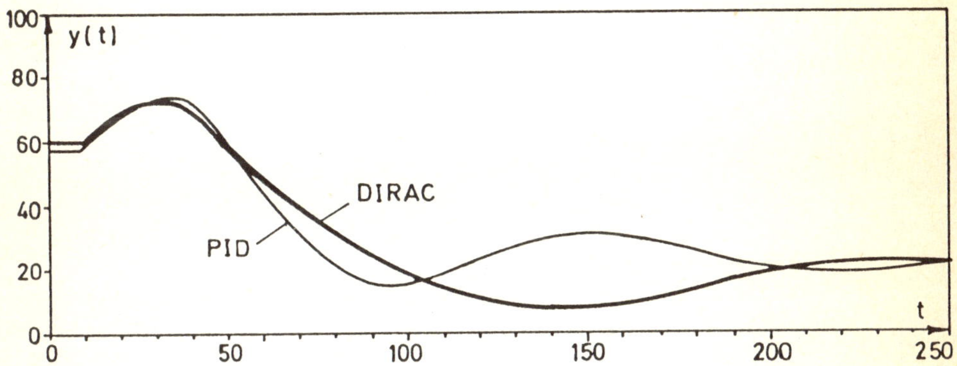

Fig. 16 : Setpoint change on process 4 (process output).

The estimated FIR model is given by its stepresponse in Fig. 17 (ref. to Fig. 3 for the shape). The corresponding PID parameters were $K_c = 0.2$; $T_i = 19.8$; $T_d = 4.5$.

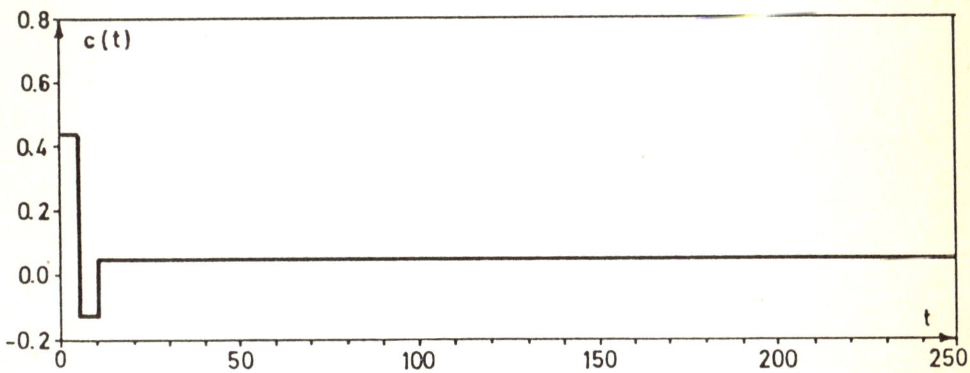

Fig. 17 : Stepresponse of estimated FIR model (n = 2).

When the order of the FIR model was increased to n = 10 (for the same process) the shape of the stepresponse of the FIR model became as illustrated in Fig. 18.

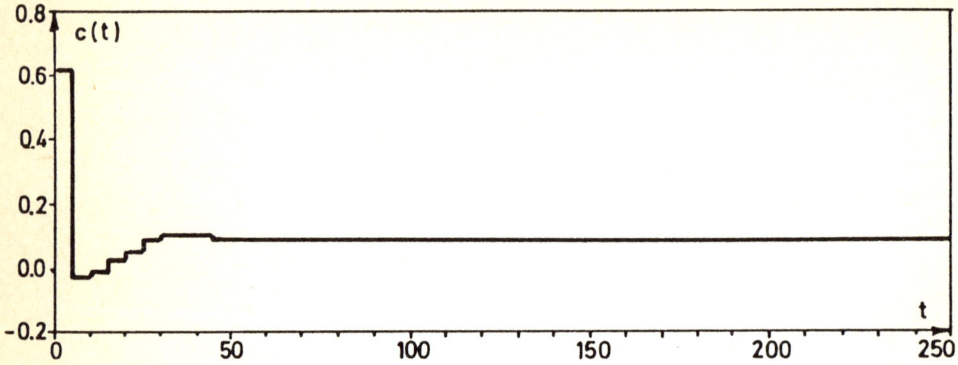

Fig. 18 : Stepresponse of estimated FIR model (n = 10).

Process 5

The process stepresponse is now oscillatory and is shown in Fig. 19. The fixed parameter PI-regulator was initially tuned at $K_c = 0.2$; $T_i = 10$. The DIRAC controller (type PID, i.e. n = 2) estimated the parameters $K_c = 0.13$; $T_i = 4.2$; $T_d = 18.2$, which means a considerable amount of derivative action. The response to a setpoint change 20% ⟶ 80% is shown in Fig. 20 while Fig. 21 shows the response 80% ⟶ 20% after changing the gain of the fixed parameter PI-regulator from $K_c = 0.2$ to $K_c = 0.1$.

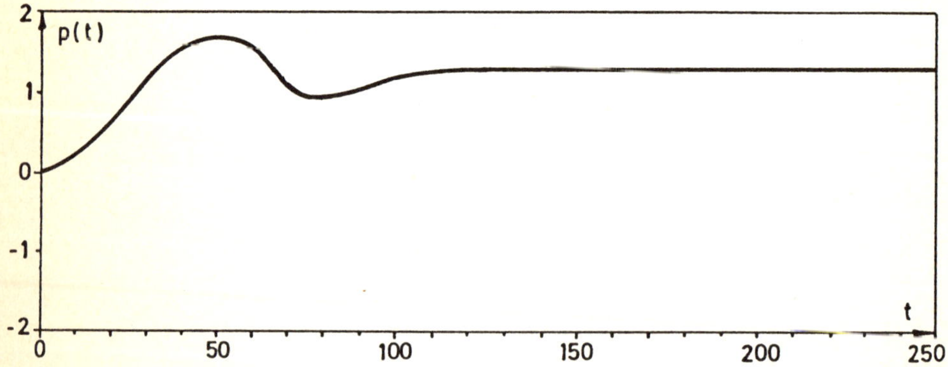

Fig. 19 : Stepresponse of process 5.

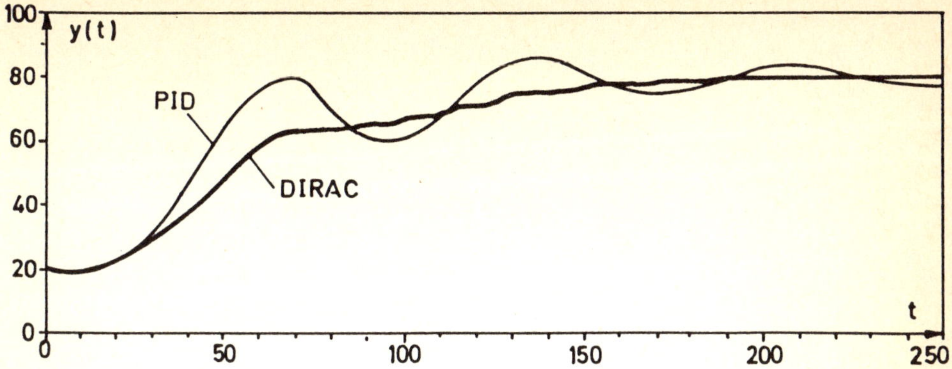

Fig. 20 : Setpoint change on process 5.

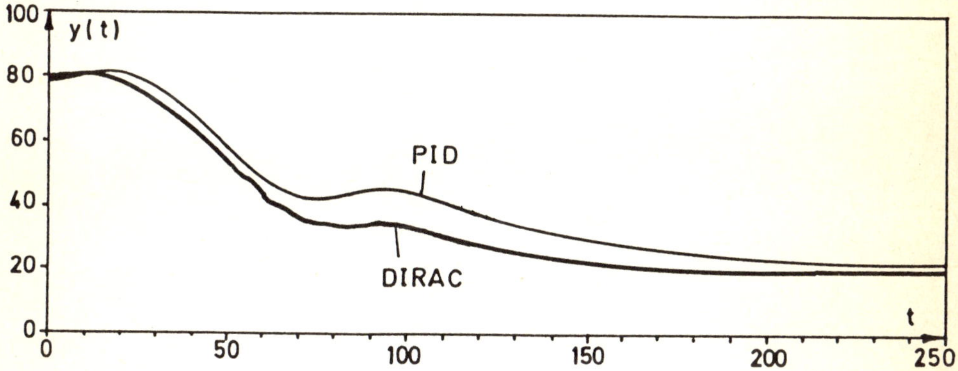

Fig. 21 : Setpoint change on process 5 after retuning the fixed-parameter PI controller.

Finally the speed of the DIRAC control loop was increased by varying the design parameter α from 0.90 to 0.80. After some time needed for retuning, the response to a setpoint variation 20% ⟶ 80% was now as given in Fig. 22 (process inputs) and Fig. 23 (process outputs).

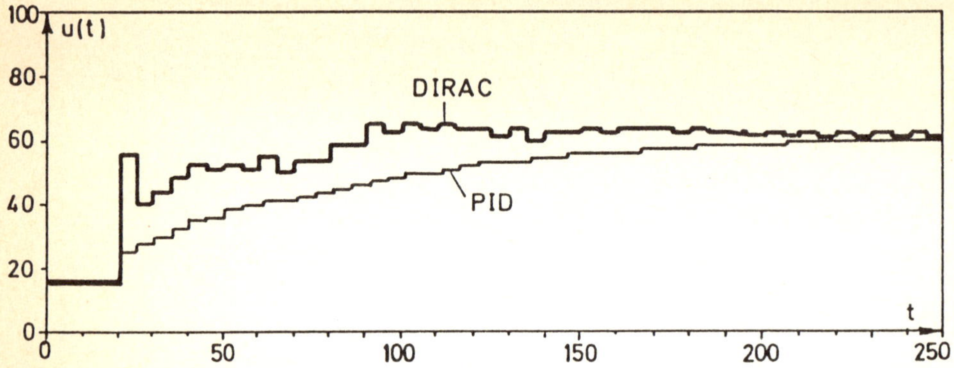

Fig. 22 : Setpoint change on process 5 after speeding up the DIRAC control loop (process inputs).

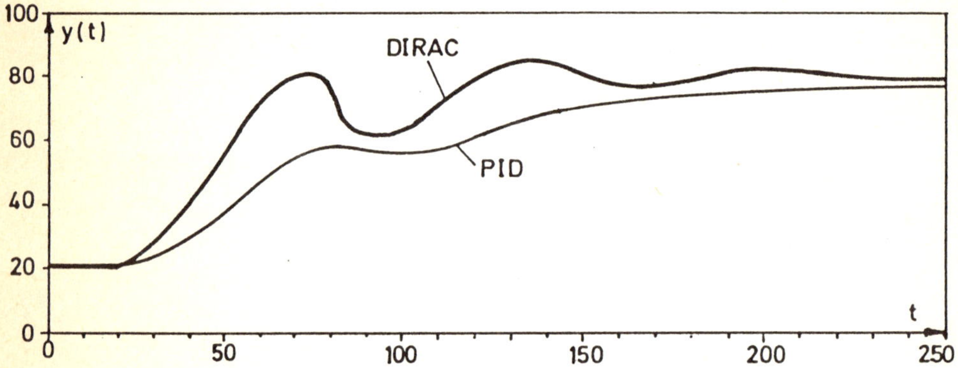

Fig. 23 : Setpoint change on process 5 after speeding up the DIRAC control loop (process outputs).

Comparison with Ziegler-Nichols tuning rules

The Ziegler-Nichols oscillation rule is a popular method for tuning PID controllers. The results of a PID controller tuned according to this rule were compared to those of a PID controller with the DIRAC auto-tuner.

Proces 1

The process step response was given in Fig. 4. The ultimate oscillation period for the ZN tuning rule was $T_u = 100$, and was obtained for a proportional gain of $K_{cu} = 1.75$. This leads to the PID parameters as given in Fig. 24 (indicated by ZNPID). The PID parameters estimated with the DIRAC method are also indicated. The performance of both controllers for a stepwise load change of -20% at the process input is compared. Fig. 24 shows the process output, while Fig. 25 shows the corresponding process input (regulator output). Notice that the oscillations of the ZNPID regulator are due to the high derivative action of this controller (combined with a relative high proportional gain). They could be removed by using low-pass filtered derivative action, as is standard practice in real-life situations.

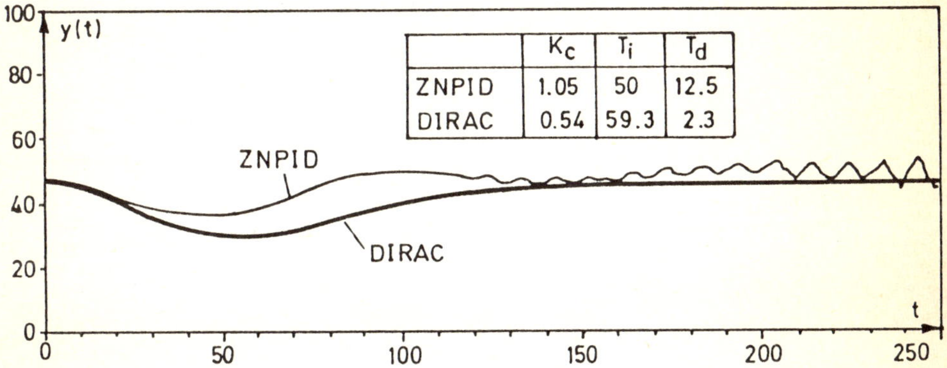

	K_c	T_i	T_d
ZNPID	1.05	50	12.5
DIRAC	0.54	59.3	2.3

Fig. 24 : ZN and DIRAC PID-controllers compared on a load disturbance for process 1 (process output).

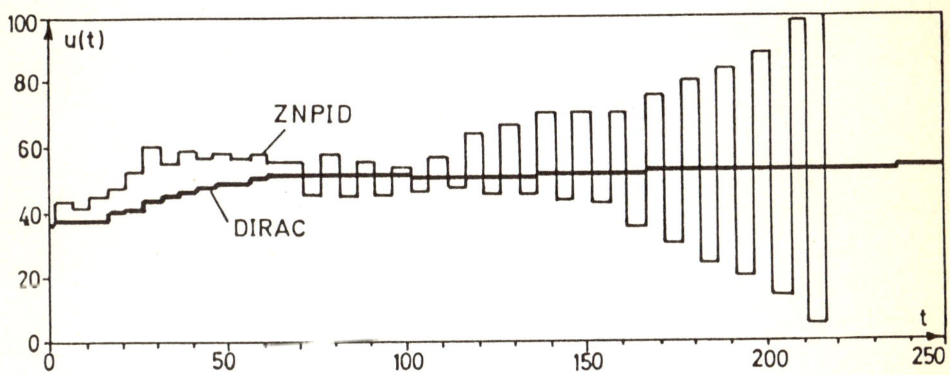

Fig. 25 : ZN and DIRAC PID-controllers compared on a load disturbance for process 1 (process input).

Process 5

The process step response was given in Fig. 19. The ZN oscillation values were T_u = 60 and K_{cu} = 0.75. The ZNPID parameters as well as the estimated DIRAC parameters are given in Fig. 26. The performance of both controllers for a stepwise input load change of -20% is given in Fig. 26 (process outputs) and 27 (process inputs).

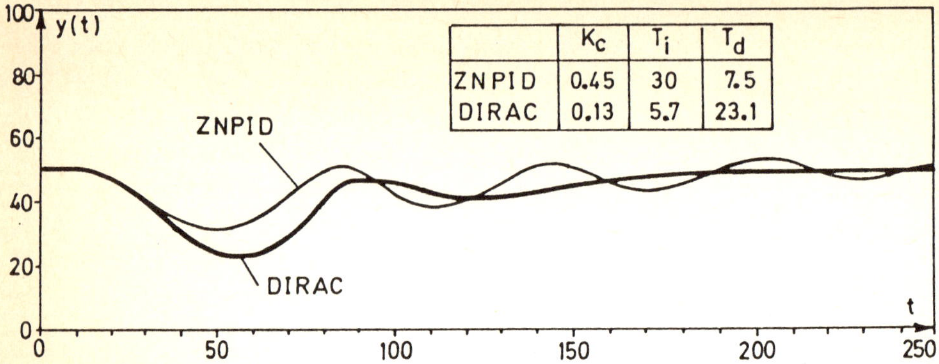

Fig. 26 : ZN and DIRAC PID-controllers compared on a load disturbance for process 5 (process output).

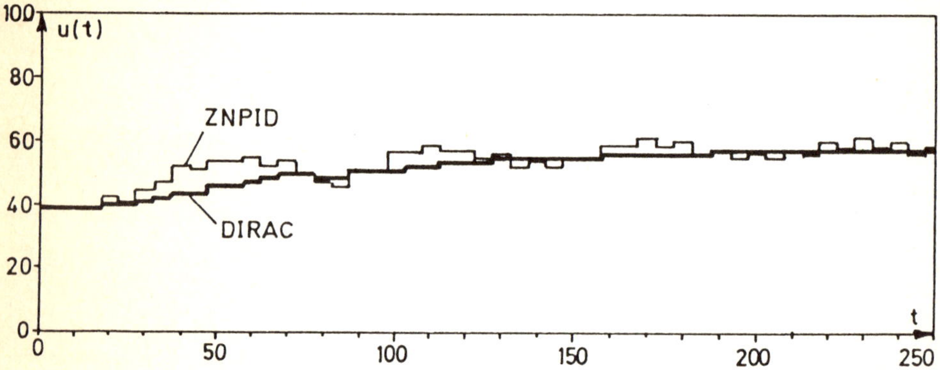

Fig. 27 : ZN and DIRAC PID-controllers compared on a load disturbance for process 5 (process input).

Conclusion

A self-adaptive control strategy has been described. It was called DIRect Adaptive Control (DIRAC). It is characterized by its simplicity and by its robustness which are two important properties for industrial control applications. Although its dynamical response performance could easily be beaten by more complex control strategies (e.g. pole-placement or model-based predictive controllers), this is not always of prime importance in these kind of applications.

The performance of the DIRAC controller is comparable to that of the usual industrial regulator (e.g. PID) and is therefore quite acceptable in this application domain. Moreover it has the advantage of self-tuning and self-adaptation compared to a fixed parameter PID. This is done by means of an algorithm which is both robust (i.e. no need for a huge amount of jacketing software to keep the adaptive core alive and well) and user-friendly (i.e. no real design parameters because they can be fixed at some default value for most types of process characteristics in industry).

Further work on the DIRAC-idea focuses on continuing simulation experiments in order to get more feeling for the behaviour of the strategy. Also a theoretical analysis, which is still lacking now, would give more insight. Finally the method will be further evaluated and refined by means of real-life experiments on an industrial unit.

References

Andreiev, N. (1981). A New Dimension : A Self-Tuning Controller that Continually Optimizes PID Constants. Control Engineering, 28 (8), 84.

Åström, K.J. (1983). Theory and Applications of Adaptive Control - A Survey. Automatica, 19 (5), 471-486.

Åström, K.J. and T. Hägglund (1984). Automatic Tuning of Simple Regulators with Specifications on Phase and Amplitude Margins. Automatica, 20 (5), 645-651.

Bànyàsz, Cs., J. Hetthesy and L. Keviczky (1985). An Adaptive PID Regulator dedicated for Microprocessor-based Compact Controllers. In : H.A. Barker, P.C. Young (Eds.) Identification and System Parameter Estimation, Pergamon Press, Oxford, 1299-1304.

Cameron, F. and D. Seborg (1983). A Self-Tuning Controller with a PID Structure. Int. J. Control, 38 (2), 401-417.

Clarke, D. (1986). Automatic Tuning of PID Regulators. In : A. Mamdani, J. Efstathiou (Eds.) Expert Systems and Optimization in Process Control, Unicom Seminars, The Technical Press.

Clarke, D., C. Mohtadi and P. Tuffs (1987). Generalized Predictive Control. Automatica, 23 (2), 137-160.

Cutler, C. and B. Ramaker (1980). Dynamic Matrix Control - A Computer Control Algorithm. JACC San Francisco, WP5-B.

De Keyser, R. and A. Van Cauwenberghe (1985). Extended Prediction Self-Adaptive Control. In : H. Barker, P. Young (Eds.) Identification and System Parameter Estimation, Pergamon Press, Oxford, 1255-1260.

De Keyser, R., Ph. Van de Velde and F. Dumortier (1988). A Comparative Study of Self-Adaptive Long-Range Predictive Control Methods. Automatica, 24 (2), 149-163.

De Keyser, R. (1988). Simple versus Complex Adaptive Strategies in Process Control. IFAC Symposium ADCHEM '88 (Adaptive Control of Chemical Processes), Lyngby Denmark.

Gawthrop, P. (1982). Self-tuning PI and PID controllers. IEEE Conference on Applications of Adaptive and Multivariable Control, Hull, 158-163.

Goodwin, G. and K. Sin (1984). Adaptive Filtering, Prediction and Control, Prentice-Hall, Englewood Cliffs NJ.

Hoopes, H., W. Hawk Jr. and R. Lewis (1983). A Self-Tuning Controller. ISA Trans., 22 (3), 49-58.

Kraus, T. and T. Myron (1984). Self-Tuning PID Controller uses Pattern Recognition Approach. Control Engineering, 31 (6), 106-111.

Kofahl, R. and K. Peter (1987). Interkama '86 : Adaptive Regler. ATP, 29 (3), 122-131.

Proudfoot, C., P. Gawthrop and O. Jacobs (1983). Self-Tuning PI Control of a PH Neutralisation Process. Proc. IEE-D, 130 (5), 267-272.

Radke, F. and R. Isermann (1984). A Parameter-Adaptive PID Controller with Stepwise Parameter Optimization. IFAC 9th World Congress, Budapest, 284-289.

Richalet, J. (1980). General Principles of Scenario Predictive Control Techniques. JACC, San Fransisco, FA9-A.

Seborg, D., T. Edgar and S. Shah (1986). Adaptive Control Strategies for Process Control : A Survey. AIChE Journal, 32 (6), 881-913.

Tjokro, S. and S. Shah (1985). Adaptive PID Control. Proc. ACC, Boston, 1528-1534.

Ydstie, B. (1984). Extended Horizon Adaptive Control. IFAC 9th World Congress, Budapest, 133-137.

Wittenmark, B. and K.J. Åström (1984). Practical Issues in the Implementation of Adaptive Control. Automatica, 20 (5), 595-605.

ADAPTIVE POLE ASSIGNMENT CONTROL BY MEANS OF ADAPTIVE OBSERVER

M. Ishitobi* and Z. Iwai
Department of Mechanical Engineering, Kumamoto University
Kumamoto 860 Japan

Abstract

This paper considers the problem of constructing the adaptive pole assignment control system for unknown linear time-invariant continuous-time plants using the state estimate from the adaptive observer. A similar problem has been solved by Kreisselmeier and Minamide et al. In this paper, a different type of adaptive algorithm using continuous-time least squares method with a forgetting factor as the parameter adaptation scheme is proposed, so that the adaptive control loop is simplified. The stability of the total closed-loop system is proved under the sufficient richness assumption of the filter signals without assumption of the boundedness of these signals. In particular, it is noted that this algorithm has exponential convergence rate with respect to parameter identification errors and the rate does not depend on the system dynamics. The effectiveness of the scheme proposed in this paper is confirmed by applying it to an experimental thermal system.

1. Introduction

A Luenberger observer allows asymptotic reconstruction of the state variable vector of a linear system from measurements of its input and output, provided that the system parameters are known. If, as usually is the case, the plant parameters are unknown, the state observation is subject to error. This fact leads to the use of the adaptive observer having parameter identification scheme for practical application. In this area of control technique, the most prominent adaptive control method is a model reference adaptive control approach where the controller gains are determined in order that the unknown system asymptotically behaves as a given reference model. However, this approach essentially requires that the plant is a minimum phase system. On the other hand, adaptive control using adaptive observer is a natural extension of the usual state feedback control system synthesis and no assumption such as stable invertibility is made of plant characteristics. Although the control system design concept seems to be simple in case of state feedback control using an adaptive observer, the principal difficulty is the complexity of the proof of the stability of the closed-loop system because it closely depends on the structure of the parameter identification scheme included in the adaptive observer.

---

* Visiting Professor at Department of Chemical Engineering, University of Alberta, Canada, from May 1988 to April 1989.

Up to now, this problem has been considered by several authors. The first important contribution was made by Kreisselmeier (1980a, 1980b) and then Minamide et al. (1983) treated the same problem using Kreisselmeier's parameter identification scheme in the modified form. However, their approaches depend on the adaptive observer using gradient type of parameter identifier with some special auxiliary dynamic system which is necessary for the removal of the uniform signal boundedness assumption of earlier adaptive observer. Thus, the adaptive loop system is made of very complicated control structure and this limits the practical applicability.

The present paper is concerned with an adaptive pole assignment control method using an adaptive observer based on the continuous-time recursive least-squares type parameter adjustment law with a forgetting factor. The results have the following features: structure of the parameter identifier is simple in comparison with previous methods and the exponential convergence of the estimation error with respect to parameters and state variables can be specified easily by the system designers. The adaptive state feedback control scheme proposed in this paper is applied to a laboratory scale thermal system.

2. System Representation and Adaptive Observer

Let us consider an n-th order linear controllable and observable single-input and single-output system:

$$\dot{\underline{x}}(t) = A\underline{x}(t) + \underline{b}u(t), \quad \underline{x}(0) = \underline{x}_0 \tag{2.1a}$$

$$y(t) = \underline{c}^T\underline{x}(t) \tag{2.1b}$$

$$A = \begin{bmatrix} \underline{a}, & \begin{matrix} I_{n-1} \\ \underline{0}^T \end{matrix} \end{bmatrix}, \quad \underline{c}^T = [1,0,\dots,0]$$

where $\underline{a}$ and $\underline{b}$ are n-th unknown but constant parameter vectors. We assume that only input u(t) and output y(t) are available for measurement.

Let K be a stable nxn matrix:

$$K = \begin{bmatrix} \underline{k}, & \begin{matrix} I_{n-1} \\ \underline{0}^T \end{matrix} \end{bmatrix}, \quad \sigma = \min_i[|Re(\lambda_i(K))|]$$

where $\underline{k}$ is a n-th order vector. Then, the solution of (2.1a) and (2.1b) can be written in the form:

$$\underline{x}(t) = P(t)\,\underline{\theta} + \exp(Kt)\,\underline{x}_0 \tag{2.2a}$$

$$y(t) = \underline{z}^T(t)\,\underline{\theta} + \underline{c}^T\exp(Kt)\,\underline{x}_0 \tag{2.2b}$$

where $\underline{\theta}^T = [(\underline{a}-\underline{k})^T, \underline{b}^T]$, and $\underline{z}^T(t) = [\underline{z}_1{}^T(t), \underline{z}_2{}^T(t)]$ is a solution of the following state variable filters:

$$\dot{\underline{z}}_i(t) = K^T\underline{z}_i(t) + \underline{c}u_i(t), \quad \underline{z}_i(0) = \underline{0}, \quad i = 1,2 \tag{2.3}$$

$$u_1(t) = y(t), \quad u_2(t) = u(t)$$

and $P(t) = [P_1(t), P_2(t)]$ is an nx2n matrix defined by

$$P_i(t) = \begin{bmatrix} \underline{c}^T \\ \underline{c}^T K \\ \cdot \\ \cdot \\ \cdot \\ \underline{c}^T K^{n-1} \end{bmatrix}^{-1} \begin{bmatrix} \underline{z}_i^T(t) \\ \underline{z}_i^T(t) K \\ \cdot \\ \cdot \\ \cdot \\ \underline{z}_i^T(t) K^{n-1} \end{bmatrix}, \quad i = 1,2 \tag{2.4}$$

Corresponding to the system representation (2.2a) and (2.2b), let us introduce the following adaptive observer:

$$\hat{\underline{x}}(t) = P(t)\,\hat{\underline{\theta}}(t) \tag{2.5a}$$

$$\hat{y}(t) = \underline{z}^T(t)\,\hat{\underline{\theta}}(t) \tag{2.5b}$$

where $\hat{\underline{\theta}}(t)$ is an estimate of the unknown parameter $\underline{\theta}$, and $\hat{\underline{x}}(t)$ and $\hat{y}(t)$ are estimates of the state $\underline{x}(t)$ and the output $y(t)$, respectively. Then, from (2.2a), (2.2b), (2.5a) and (2.5b), estimation errors become

$$\Delta\underline{x}(t) = \hat{\underline{x}}(t) - \underline{x}(t) = P(t)\,\underline{\zeta}(t) + \underline{f}_x(t) \tag{2.6a}$$

$$e(t) = \hat{y}(t) - y(t) = \underline{z}^T(t)\,\underline{\zeta}(t) + f_e(t) \tag{2.6b}$$

where $\underline{f}_x(t) = -\exp(Kt)\underline{x}_0$, $f_e(t) = \underline{c}^T\,\underline{f}_x(t)$ and $\underline{\zeta}^T(t) = [(\hat{\underline{a}}(t) - \underline{a})^T,$ $(\hat{\underline{b}}(t) - \underline{b})^T]$. The purpose of the adaptive observer is to construct the parameter adjustment law of $\hat{\underline{\theta}}(t)$ which realizes

$$\underline{\zeta}(t) \to \underline{0} \quad \text{and} \quad \Delta\underline{x}(t) \to \underline{0} \quad (t \to \infty) \tag{2.7}$$

3. Identification of Parameters

The basic adaptive adjustment law to obtain the identified parameter vector $\hat{\underline{\theta}}(t)$ is determined from minimizing the weighted square error (Iwai et al., 1985);

$$J(t) = \int_0^t f(\tau)[y(t,\tau) - y(\tau)]^2 \exp(-\phi(t-\tau))d\tau \tag{3.1}$$

$$y(t,\tau) = \hat{\underline{\theta}}^T(t)\,\underline{z}(\tau) \tag{3.2}$$

where $f(\tau)$ is a uniformly bounded positive scalar function and $\phi>0$ is a forgetting factor. At the minimizing value $\hat{\underline{\theta}}(t)$, $\partial J/\partial\,\hat{\underline{\theta}} = \underline{0}$ holds and this implies the normal equation:

$$R(t)\,\hat{\underline{\theta}}(t) = \underline{r}(t) \tag{3.3}$$

$$\dot{R}(t) = -\phi R(t) + f(t)\,\underline{z}(t)\,\underline{z}^T(t) \tag{3.4}$$

$$\dot{\underline{r}}(t) = -\phi \underline{r}(t) + f(t)\,y(t)\,\underline{z}(t) \tag{3.5}$$

where $R(t)$ is a 2nx2n symmetric matrix and $\underline{r}(t)$ is a 2n-dimensional vector defined by

$$R(t) = \int_0^t f(\tau)\,\underline{z}(\tau)\,\underline{z}^T(\tau)\exp(-\phi(t-\tau))d\tau \tag{3.6}$$

$$\underline{r}(t) = \int_0^t f(\tau)\,y(\tau)\,\underline{z}(\tau)\exp(-\phi(t-\tau))d\tau \tag{3.7}$$

If the inverse matrix of $R(t)$ exists, the adaptive law is derived from (3.5)-(3.7) as follows:

$$\dot{\underline{\theta}}(t) = -f(t)\, R^{-1}(t)\, \underline{z}(t)(\underline{z}^T(t)\, \hat{\underline{\theta}}(t) - y(t)) \tag{3.8}$$

Here, consider the following more general form;

$$\dot{\hat{\underline{\theta}}}(t) = \dot{\underline{\zeta}}(t) = -\phi_0\, R^{-1}(t)\, \underline{z}(t)\, e(t)/g_0(t) \tag{3.9}$$

$$\dot{R}(t) = -\phi_1\, R(t) + \phi_2\, \underline{z}(t)\, \underline{z}^T(t)/g_0(t), \quad R(0)>0 \tag{3.10}$$

$$g_0(t) = 1 + \kappa||\underline{z}(t)||^2, \; \kappa>0,\; 2\phi_0 \geq \phi_2>0,\; \phi_1>0$$

Eqns. (3.9) and (3.10) are regarded as generalized recursive least squares identification algorithm of (3.4) and (3.8). In the following discussion, the generalized algorithm (3.9) and (3.10) is considered. Now we formulate the assumption on the system regression vector $\underline{z}(t)$ of the filter (2.3).

Assumption 1

There exist some positive constants t_0 and ρ_0 such that

$$\int_{t-t_o}^{t} \frac{\underline{z}(\tau)\, \underline{z}^T(\tau)}{g(\tau)}\, d\tau > \rho_o\, I_{2n} \tag{3.11}$$

Then we have the following lemma.

Lemma 1

(1) $\exists\rho_1, \exists\rho_2 > 0,\; \rho_2\, I_{2n} \geq R(t) \geq \rho_1\, I_{2n}$ (3.12)

(2) $\exists\rho_3 > 0,\; ||R^{-1}(t)||<\rho_3$ (3.13)

for $t>t_o$.

Proof. See Iwai et al. (1988).

Based on Lemma 1, the following theorem is established.

Theorem 1

Under the Assumption 1, the parameter error $\underline{\zeta}(t)$ is globally exponentially stable and the rate of exponential convergence is indicated by $\bar{\sigma} = \min(\sigma, \phi_1/2)$.

Proof. See Iwai et al. (1988).

It is stressed that the system input and output are not required to be uniformly bounded for the convergence of parameter error $\underline{\zeta}(t)$. However, this result does not guarantee the convergence of the state estimation error $\Delta\underline{x}(t)$ because it is apparent from (2.6a) that the uniform boundedness of $\underline{z}(t)$ is necessary for that purpose.

4. Adaptive Pole Assignment Control System and Its Stability

Adaptive Pole Assignment Control System

Suppose that system parameters $\underline{a}$ and $\underline{b}$ are known and the state $\underline{x}(t)$ is available for measurement in (2.1). Let it be desired to control (2.1) by means of state feedback control law

$$u(t) = \underline{k}^T\, \underline{x}(t) + v(t) \tag{4.1}$$

where $v(t)$ is a uniformly bounded external input which satisfies Assumption 1. If desirable pole configuration is indicated by an n-th stable polynomial $D(s)$, then the optimal feedback controller gain is given by

$$\underline{k} = -D(A^T)\underline{d}, \quad \underline{d} = (Q^T)^{-1}\underline{e}_n$$
$$Q = [\underline{b}, A\underline{b}, \ldots, A^{n-1}\underline{b}], \quad \underline{e}_n^T = [0, \ldots, 0, 1] \tag{4.2}$$

(Ackermann, 1977). Applying (4.1) to (2.1) would result in the ideal stable closed-loop system;

$$\dot{\underline{x}}^*(t) = (A + \underline{b}\underline{k}^T)\,\underline{x}^*(t) + \underline{b}v(t), \quad \underline{x}^*(0) = \underline{x}_0 \tag{4.3}$$

Actually, the system parameter values are unknown and the system states are not available for measurement. Hence, estimates of these parameters and states have to be used instead of the true values; that is

$$\hat{\underline{k}}(t) = -D(\hat{A}^T(t))\,\hat{\underline{d}}(t), \quad \hat{\underline{d}}(t) = (\hat{Q}^T(t))^{-1}\underline{e}_n \tag{4.4}$$
$$\hat{Q}(t) = [\hat{\underline{b}}(t), \hat{A}(t)\hat{\underline{b}}(t), \ldots, \hat{A}^{n-1}(t)\hat{\underline{b}}(t)]$$

and the realizable feedback control law becomes

$$u(t) = \hat{\underline{k}}^T(t)\,\hat{\underline{x}}(t) + v(t) \tag{4.5}$$

For the brevity of the following discussion, we assume that rank $(\hat{Q}(t)) = n$ for all $t \geq 0$ and $\hat{Q}(t)$ is continuously differentiable with respect to parameters $\hat{\underline{a}}(t)$ and $\hat{\underline{b}}(t)$. Note that the above assumption is not restrictive and can be removed by introducing an auxiliary system which generates $\hat{\underline{d}}(t)$ asymptotically (Kreisselmeier, 1980a). It is emphasized that the parameters A and $\underline{b}$ can be estimated with exponential decay regardless of the uniform boundedness of the system signals by using (3.9) and (3.10). That is, the relation; $\hat{\underline{k}}(t) \to \underline{k}(t \to \infty)$ is always realized as far as the external input $v(t)$ satisfies the so-called sufficiently exciting condition; Assumption 1.

<u>Stability</u>

The overall closed-loop system involves the differential equation (2.1), (2.3), (3.9) and (3.10) with state feedback (4.5). To prove the stability of this system, it is necessary to show that the nonlinear system of the differential equations has a unique solution for all $0 \leq t < \infty$. Suppose that there exists a unique solution on the interval $[0, t_0]$. Then, we have the following lemma.

<u>Lemma 2</u>

The closed-loop system derived by the system of differential equations: (2.1), (2.3) (3.9) and (3.10), has a unique solution on the interval $[t_0, \infty)$.

<u>Proof</u>. See Iwai et al. (1988).

Now we will prepare some lemmas which are required for the proof of the stability of the closed-loop system. Proofs are shown in Iwai et al. (1988).

<u>Lemma 3</u>

There exist positive constants M_p and M_x such that

$$\frac{||\Delta \underline{x}(t)||}{(1+M_p||\underline{z}(t)||^2)^{1/2}} < M_x \exp(-\bar{\sigma}(t - t_0)), \ t \geq t_0 \tag{4.6}$$

Lemma 4

Suppose that the initial parameter estimation error $\underline{\zeta}(t_0) = \underline{\zeta}_0$ and initial state variable $\underline{x}_0$ are small. Then, there exist constants M_{z1} and M_{z2} such that

$$||\underline{z}(t)|| < M_{z1} + M_{z2} \sup_{t \geq \tau > t_0} ||\Delta \underline{x}(\tau)|| \tag{4.7}$$

Lemma 5

$||\underline{z}(t)||$ is uniformly bounded in the interval $[t_0, \infty)$.

Lemma 6

There exists a positive constant $\bar{M}_x$ such that

$$||\Delta \underline{x}(t)|| \leq \bar{M}_x \exp(-\bar{\sigma}(t-t_0)), \ t \geq t_0 \tag{4.8}$$

These lemmas correspond to the Key Technical Lemmas of Goodwin et al. (1984). That is, Lemma 3 and 4 give the conditions of the KTL and Lemma 5 and 6 show the result of KTL. In these lemmas, the exponential stability of the system is assured. However, the KTL in Goodwin et al. only guarantees the convergence of $\Delta \underline{x}(t)$. The results obtained in this section are summarized in the form of the theorem.

Theorem 2

Construct the adaptive feedback control law (4.5) using a parameter adjustment law (3.9) and (3.10). Then, if (1) v(t) is uniformly bounded and satisfies the Assumption 1, (2) $\hat{Q}(t)$ in (4.4) is nonsingular and (3) $||\underline{\zeta}(t_0)||$ and $||\underline{x}_0||$ are sufficiently small, then the overall closed-loop system is stable and estimation errors concerning parameters and states converge exponentially and, as a result, actual state $\underline{x}(t)$ approaches ideal state $\underline{x}^*(t)$.

Throughout the discussion, we have assumed that the initial parameter deviation $\underline{\zeta}(t_0)$ and the initial state variable $\underline{x}_0$ are small. However, according to Theorem 1, the exponential convergence of $\underline{\zeta}(t)$ is assured irrespective of the boundedness of $\underline{z}(t)$. Therefore, $\hat{\underline{k}}(t) \to \underline{k}$ as $t \to \infty$. This in turn implies the existence of $t_e < \infty$, such that the conditions stated in the Lemma 6 are satisfied for all $t \geq t_e > t_0$. This guarantees the global stability of the closed-loop system.

5. Application to a Thermal System

The algorithm is evaluated by experimentally implementing it on a minicomputer for a heated tank with a water jacket.

Experimental Plant

The plant is shown schematically in Fig. 1. It is the identical to the one used in Inoue and Iwai (1984), and Ishitobi and Iwai (1988). It has two tanks. The outer tank water flows in through valve A and goes out through valve B with flow rate Q_0. The water is heated by an electric heater placed at the bottom of the outer tank.

The heat from the heater (maximum 118.8 kJ/min) is the manipulating input to the plant. A software program in the minicomputer MELCOM 70/30C implements the adaptive observer (2.3) and (2.5), and the parameter adjustment law (3.9) and (3.10) with numerical integration and gives input value u(t) of (4.5). Effectively, a continuous time observer and controller are realized by numerical integration. Then the computer controls the heater through a triac controller to output heat proportional to value u(t). The output y(t) is the deviation between the measured temperature of the outer tank and the initial temperature. The measured temperature is fed into the computer to be used in the calculation of the adaptive observer and the control law. The experiment is started after the water settles into a steady state; the state in which the outflow rate equals the inflow rate.

Second-order State-space Equation of the Plant

Assume that the heat capacity of the tank wall is negligible and that both tanks are well stirred so that the temperature in each of the tanks is kept uniform. Then the state equation for temperature is of the second-order and is expressed by

$$\begin{bmatrix} \dot{\theta}_1 \\ \dot{\theta}_2 \end{bmatrix} = \begin{bmatrix} a_{11} & a_{12} \\ a_{21} & a_{22} \end{bmatrix} \begin{bmatrix} \theta_1 \\ \theta_2 \end{bmatrix} + \begin{bmatrix} b \\ 0 \end{bmatrix} u \tag{5.1a}$$

$$y = [1 \quad 0] \begin{bmatrix} \theta_1 \\ \theta_2 \end{bmatrix} \tag{5.1b}$$

$$a_{11} = -\frac{1+c_0RQ_0}{c_0V_1R}, \qquad a_{12} = \frac{1}{c_0V_1R}$$

$$a_{21} = \frac{1}{c_0V_2R}, \qquad a_{22} = -a_{21}, \qquad b = \frac{1}{c_0V_1}$$

θ_1 temperature change of water in the outer tank from initial temperature (°C)

θ_2 temperature change of water in the inner tank from initial temperature; output (°C)

u heat from heater; manipulating input (kJ/min)

Q_0 flow rate at valve A and at valve B (kg/min)

V_1 capacity of water in the outer tank (kg)

V_2 capacity of water in the inner tank (kg)

R resistance to heat transfer through inner tank wall (min·°C/kJ)

c_0 specific heat of water (4.1868 kJ/(kg·°C))

The state equation (5.1) is transformed into the observable canonical form

$$\begin{bmatrix} \dot{x}_1 \\ \dot{x}_2 \end{bmatrix} = \begin{bmatrix} a_1 & 1 \\ a_2 & 0 \end{bmatrix} \begin{bmatrix} x_1 \\ x_2 \end{bmatrix} + \begin{bmatrix} b_1 \\ b_2 \end{bmatrix} u \tag{5.2a}$$

$$y = [1 \quad 0] \begin{bmatrix} x_1 \\ x_2 \end{bmatrix} \tag{5.2b}$$

$$a_1 = a_{11} + a_{12},\ a_2 = a_{12}\,a_{21} - a_{11}\,a_{22}$$

$$b_1 = b,\ b_2 = -a_{22}b$$

where the transformation is given by

$$\begin{bmatrix} x_1 \\ x_2 \end{bmatrix} = \begin{bmatrix} 1 & 0 \\ -a_{22} & a_{12} \end{bmatrix} \begin{bmatrix} \theta_1 \\ \theta_2 \end{bmatrix} \tag{5.3}$$

Experiment

Table 1 gives experimental conditions, and design parameters of adaptive observer and controller.

A variable forgetting factor $\phi_1(t)$ is used here in the parameter adjustment law (Ishitobi et al., 1987) though it is assumed to be constant in the theory mentioned above. The factor is automatically determined by the output estimation error; A larger forgetting factor is produced when the output estimation error is large. A large forgetting factor gives the fast convergence speed. On the other hand, a small forgetting factor causes the identification process to be robust with respect to the effect of noise when the output estimation error decreases.

Figure 2 shows the plant output, and the corresponding input is depicted in Fig. 3. Figure 4 indicates the identified parameters normalized by their initial values. All of the identified parameters converge to some constants close to one. As for the plant output, simulation runs suggest that the output is in a steady state mode at 6.4°C when the input does not include the external input v(t). The plant output in experiment drifts around that temperature 6.4°C.

6. Conclusions

An adaptive state feedback control scheme which realizes a desired pole configuration is presented. The method consists of the use of adaptive observer which involves a parameter adjustment law based on continuous-time recursive least squares algorithm. The features of the proposed method are that the exponential convergence of parameter and state estimation errors are guaranteed and the rate can be specified by the designer intuitively. The effectiveness of the scheme was confirmed by its successful evaluation on an experimental thermal system.

References

Ackermann, J.E. (1977). On the synthesis of linear control systems with specified characteristics. Automatica, 13, 89-94.

Goodwin, G.C. and K.S. Sin (1984). Adaptive filtering prediction and control. Prentice-Hall, Inc.

Inoue, A. and Z. Iwai (1984). Design and application of reduced-order adaptive observer with exponential rate of convergence for single-input single-output systems. Int. J. Control, 39-2, 375-393.

Ishitobi, M., Z. Iwai, A. Inoue and H. Tokumaru (1987). Proposal of continuous-time adaptive law with adjusting gain using identification error and its application to an adaptive observer of a thermal experimental system. Trans. of

Society of Instrument and Control Engineers, 23-2, 200-202 (in Japanese).
Ishitobi, M. and Z. Iwai (1988). Adaptive control of a non-minimum phase system tracking a periodic reference output. Preprints of 8th IFAC/IFORS Symposium on Identification and System Parameter Estimation, Beijing, 85-90.
Iwai, Z., A. Inoue, M. Ishitobi and H. Tokumaru (1985). A scheme for digital control and application of a continuous-time adaptive control with exponential rate of convergence. Proc. of the IMACS Meeting, Patras, Greece, 397-402.
Iwai, Z., M. Ishitobi and M. Hino (1988). Design of an adaptive pole placement control system using adaptive observer. Preprints of 8th IFAC/IFORS Symposium on Identification and System Parameter Estimation, Beijing, 161-166.
Kreisselmeier, G. (1980a). Algebraic separation in realizing a linear state feedback control law by means of an adaptive observer. IEEE Trans. on Auto. Cont., AC-25-2, 238-243.
Kreisselmeier, G. (1980b). Adaptive observer and asymptotic feedback matrix synthesis. IEEE Trans. on Auto. Cont., AC-25-4, 717-722.
Minamide, N., P.N. Nikiforuk and M.M. Gupta (1983). Design of an adaptive observer and its application to an adaptive pole placement controller. Int. J. Control, 37-2, 349-366.

TABLE 1 Experimental Conditions and Design Parameters

experimental condition	V_1 : 17.449 kg V_2 : 2.376 kg Q_0 : 1.967 kg/min initial temperature : 17.5°C sampling time : 5 sec
adaptive observer	$\lambda_i(K)$: -0.6, -0.7 ϕ_0 : 0.5, $\phi_1(t)$: $0.05+\lvert e(t)\rvert/(1+\lvert e(t)\rvert)$ ϕ_2 : 1.0 κ : 0.05, $R(0)$: $10^{-1}I_4$ $\hat{a}_1(0)=-0.90980$, $\hat{a}_2(0)=-0.07909$ $\hat{b}_1(0)=0.013688$, $\hat{b}_2(0)=0.009602$
controller	$D(s)$: $(s+0.1)(s+0.9)$ $v(t)$: $60.0+v_0(t)$ $v_0(t): \begin{cases} 10.0 & \text{if } \sin(2\pi t/T)\geq 0 \\ -10.0 & \text{if } \sin(2\pi t/T)<0 \end{cases}$ T: random variable with mean 10.0 and variance 6.0

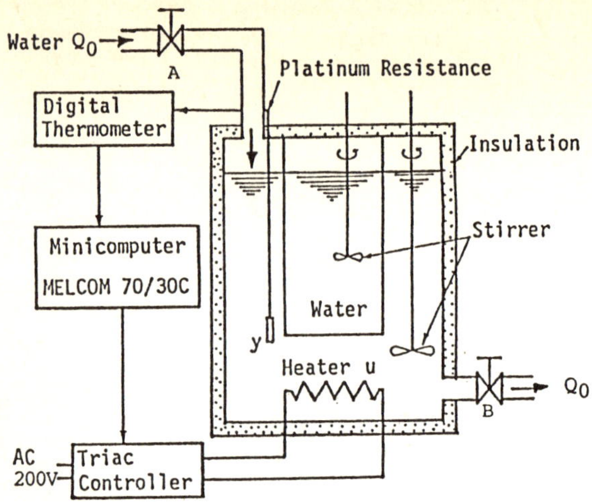

Fig.1. Schematic diagram of experimental heated tank.

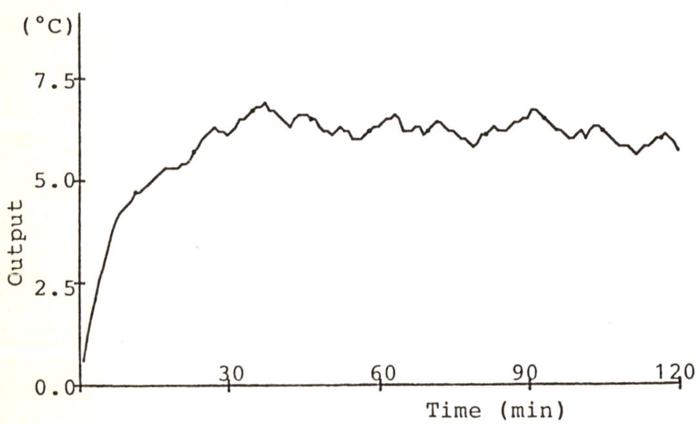

Fig.2. Plant output.

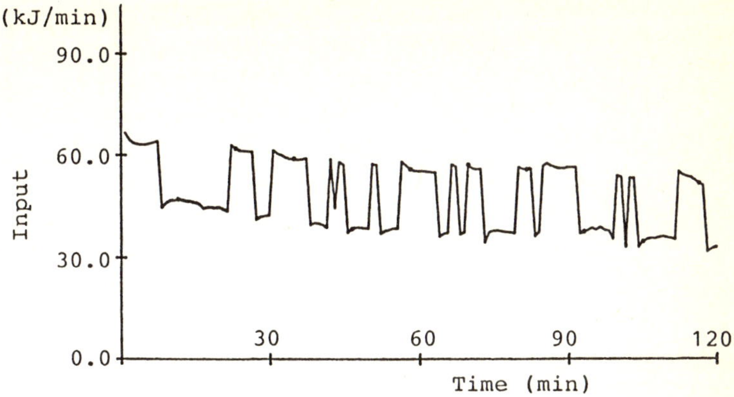

Fig.3. Input.

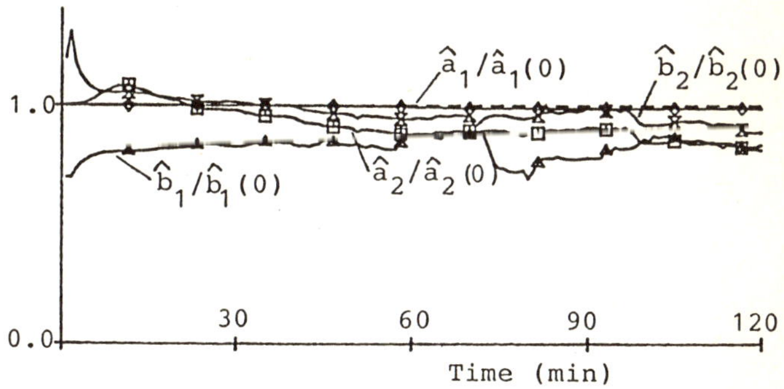

Fig.4. Identified parameters.

Implementation Issues in Adaptive Control

Adaptive Control: Implementation and Application Issues

Björn Wittenmark

Lund Institute of Technology
Box 118, S-221 00 Lund, Sweden

Abstract. There are many things to consider when implementing an adaptive controller. The paper contains a discussion of some of these issues. When an adaptive controller is going to to work in an industrial environment it is of outmost importance that the designer has thought of all situations and combination of conditions that may occur. Special attention is given to the interaction between the controller and the estimator. Some types of applications are discussed to point out when adaptive control can be a good alternative to a fixed gain controller.

1. Introduction

Adaptive controllers are now more and more used in industrial applications, see the surveys Åström (1983b) and Seborg *et al* (1986). An overview of adaptive control methods is given in the book Åström and Wittenmark (1988). Two main design approaches for adaptive controllers are Model Reference Adaptive System (MRAS) and Self-tuning Regulators (STR). These methods have, however, many common features. In the paper we will discuss the main issues of digital implementation of these systems.

The robustness properties of adaptive controllers have been debated intensively during the last years. The consensus of these discussions seems to be that adaptive control is useful, but that care must be taken when implementing the controllers. Different fixes have been suggested based partly on experience and partly on analysis. The implementation and robustness issues are therefore of great importance. This paper discusses some of these aspects. The signal processing, i.e. the choice of different filters, in adaptive controllers is treated. An attempt is made to quantify some of the "folklore" around the implementation issues.

Some of the signal processing issues are the same as for implementation of fixed parameter sampled data controllers. The problems are for instance the choices of the sampling period, the antialiasing filters and the specifications. These choices are, however, also influenced by the process, which now is unknown. The estimator part of the adaptive controller is also of great importance since it must produce accurate process models in appropriate frequency bands. Finally the interaction between the estimator and the controller may cause problems. Practical issues have been discussed for instance in Åström (1983a,b), Rohrs *et al* (1984) and Wittenmark and Åström (1984). The following quotations give some general advices:

Quote 1 (Ljung and Söderström (1983), p 139-140)
"...even if the true system is more complex than our models, the identification procedure will pick the best approximation of the system. ...The recursive algorithm converges to that approximation that is best under the input signal used during the experiment."

Quote 2 (Ljung and Söderström (1983), p 267)
"It is thus good practice to choose an input for the identification experiment that as far as possible is similar to inputs to be used for the system at later occasions."

Quote 3 (Åström (1983a), p 986)
"Do not estimate unless the data is good."

Quote 4 (Åström (1983a), p 986)
"...beneficial to use a design method which gives a high gain at low frequencies and use adaptation only to find the characteristics around the cross-over frequency."

Quote 5 (Rohrs et al (1984), p 653)
"Keep the adaptation gain of the system small and let the adaptation proceed slowly."

Quote 6 (Rohrs et al (1984), p 653)
"Design the nominal control loop so that it is robust and that approximate model matching can be achieved even in the presence of unmodelled dynamics."

In this paper we will try to quantify these advices and arrive at an overall view of the implementation problem. Finally some types of applications will be discussed which show when and when not adaptive control is useful.

2. The Adaptive Control Problem

An adaptive controller can often be structured as in Figure 1. The system consists of two loops: an inner loop with a conventional regulator and an outer loop which changes the parameters in the regulator. Figure 1 also contains the filters that are discussed later in the paper. The regulator parameters are calculated based on the specifications on the closed loop system and the estimated parameters. Figure 1 is typical for the self-tuning regulator approach. Figure 2 shows a block diagram for a typical model reference adaptive system. The system consists also in this case of two loops: a conventional controller and an outer loop for the updating of the controller parameters. In the MRAS the controller parameters are updated based on the error between a desired response y_m and the true response of the system. The desired response is specified using a reference model. In the MRAS in Figure 2 the regulator parameters are updated directly based on the error signal. In the STR in Figure 1 there are two possibilities. The estimated parameters can either be the parameters in a process model or be the parameters in the controller directly. The first case is called an *indirect* adaptive controller, while in

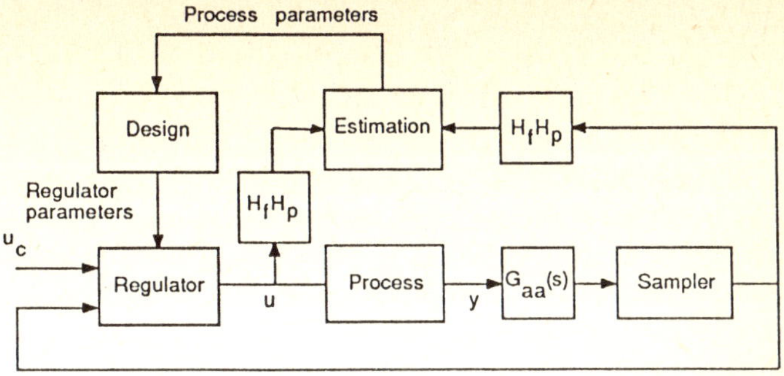

Figure 1. Structure of an adaptive control system based on the idea of Self-tuning Regulators (STR).

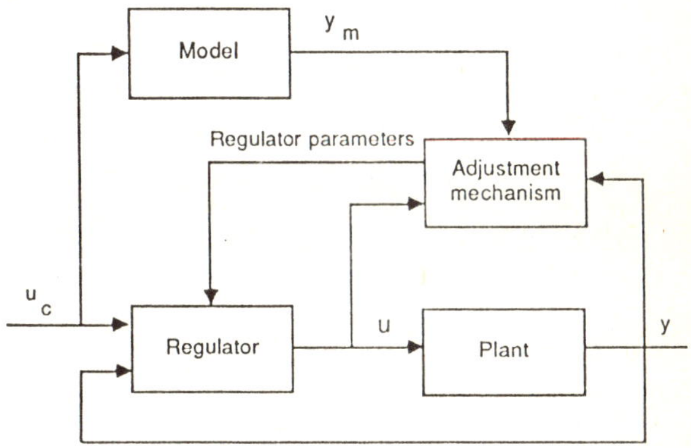

Figure 2. A Model Reference Adaptive System (MRAS).

the second case we talk about a *direct* adaptive controller. In the direct methods the controller parameters are estimated directly. The specifications are in this case used to make a reparameterization of the model, see for instance Åström and Wittenmark (1980, 1988). This implies that the design block in Figure 1 is very simple. In the indirect methods we first estimate the parameters in a process model. The estimated parameters are then used in the design calculations as if they are the true ones.

Even if the system is adaptive the user has to specify for instance what the specifications should be and which design method that should be used. These choices depend on the physical insight into the process and the disturbances acting on the process. An adaptive controller can thus not be implemented without any knowledge about the process. This knowledge is used (in the same way as for the case with a known system) to determine suitable specifications and the structure of the controller. The processes are in practice more complex than the estimated model. This implies that we have to decide over which frequencies we want to have a good approximation, compare Quotes 4 and 6. To be more specific we will assume that the process is described by the single input single output discrete time model

$$y(t) = G_0(q)u(t) + H_0(q)e(t) + d(t) \tag{2.1}$$

where

$$G_0(q) = \frac{B_0(q)}{A_0(q)} \qquad H_0(q) = \frac{C_0(q)}{D_0(q)}$$

and y and u are the output and input of the process. Notice that we allow possible antialiasing filters to be a part of the process $G_0(q)$. The pulse transfer operator $G_0(q)$ is also a function of the sampling period h. This dependence is, however, supressed in the formulas. The disturbance $e(t)$ is assumed to be a sequence of independent random variables with zero mean values and variances σ^2. The second disturbance term $d(t)$ is a deterministic signal of known form, but unknown magnitude. It may be a level, a ramp or a sinusoidal signal. It is assumed that the desired specifications of the closed loop system is given as the bandwidth or in terms of a desired closed loop model

$$G_m(q) = \frac{B_m(q)}{A_m(q)} \tag{2.2}$$

In the following we will discuss how to estimate the process model G_0. The estimated model is used to determine a regulator, which should give desired closed loop properties.

3. Adjustment to the Prejudice Model

When determining the structure and complexity of a controller we always have some more or less well defined "feeling" of how the process can be described. For instance its complexity, amount of time delay, types of disturbances. The purpose of the estimator part of the adaptive controller is to find the parameters of this prejudice model. In this section we will discuss how to adjust the measured data (Sic!) to the prejudice model.

Elimination of Known Types of Disturbances

The known types of disturbances are represented by $d(t)$ in (2.1). $d(t)$ can be thought of as generated by pulses into known dynamic systems. It is assumed that it is not known a priori when the pulses occur and the amplitude of the pulses. This type of disturbance can be generated by

$$d(t) = H_d(q)\delta_s(t) = \frac{D_n(q)}{D_d(q)}\delta_s(t)$$

where $\delta_s(t)$ is a sequence of pulses and $H_d(q)$ a filter. The signal $d(t)$ can be eliminated (except for occasional pulses or finite length disturbances) by filtering the input and the output by $D_d(q)$.

The disturbance annihilation filter $D_d(q)$ is of high pass or notch type. It can then be of the form

$$H_f(q) = \frac{D_d(q)}{D_d'(q)} \tag{3.1}$$

where $D_d'(q)$ is a stable polynomial and H_f has a high attenuation at high frequencies. Filtering by H_f will then remove low frequencies in the signal and eventually also some

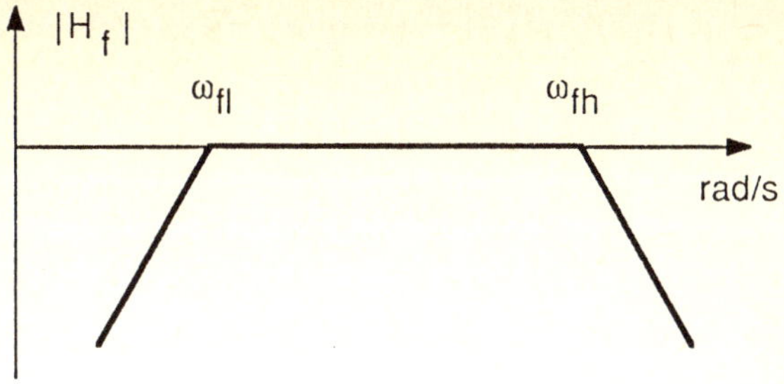

Figure 3. Amplitude curve for the disturbance annihilation filter $H_f(q)$.

other frequencies within a narrow band. The filter $H_f(q)$ should thus have an amplitude curve as in Figure 3. The lower break frequency ω_{fl} depends on the desired cross over frequency of the closed loop system. As a rule of thumb ω_{fl} should be at least one decade below the cross over frequency. The high frequency break frequency will be discussed later in this section. With the filter H_f we will remove $d(t)$ from the data. In several references it is pointed out that it is advantageous for the estimation to eliminate levels in the signals. See for instance Peterka (1984). These recommendations are thus followed when the filter H_f is used. The estimator will thus not be confused by low frequency drift or disturbances at specific frequencies. These disturbances are instead taken into consideration when the controller is designed by using the internal model principle. This implies that a level is removed by having an integrator in the controller and so on. Notice that the regulator is still using the unfiltered sampled signal.

The Bias Distribution Problem

The influence of the filters on the estimation will be discussed heuristicly in this section. By making some coarse assumptions it is possible to obtain an expression that gives an understanding for the influence of different filters. This will make it possible to analyze how the asymptotic estimates of the parameters in the process are influenced by design variables such as

- input signal and reference signal spectrum
- model orders and noise structure
- prefilters
- desired closed loop performance

We will follow the lines of thought for open loop estimation given in Wahlberg and Ljung (1986). Their results are modified to suit the adaptive control problem. One result in this section is a limiting expression for the loss function that is minimized by the estimator. This result can then be used for a more quantitative discussion *in the frequency domain* of the influence of the design variables. Similar aspects are also discussed in Gevers and Ljung (1986). Only the final result is given here. A more detailed discussion is given in Wittenmark (1986). The purpose of the estimator is to provide an estimate of the process and eventually also the noise characteristics. Let the prejudice model be described by

$$y(t) = G(q, \theta)u(t) + H(q, \theta)e(t) \tag{3.2}$$

where the pulse transfer operators G and H are parameterized by the unknown parameters θ. It is assumed that $d(t)$ in (2.1) has been eliminated by the filter H_f in (3.1). The structure such as model orders and time-delays are parts of the parameterization. The model (3.2) includes for instance ARMAX models of different orders. It is assumed that the estimation is done using a prediction error method. If the parameterization is such that the true process model $\{ G_0(q), H_0(q) \}$ is among the model set $\{ G(q,\theta), H(q,\theta) \}$ then under weak conditions the estimate will converge to the true parameters independent of the filtering and the frequency content of the input signal as long as it is persistently exciting. However, if the model set does not include the true system then the estimate will be biased. The bias distribution will be a function of the design variables. We refer to Wahlberg and Ljung (1986) for a thorough discussion of this important property for the case of open loop estimation.

Discussion of the Choice of Prefilter

To get further insight into the bias distribution problem and guidance to choose the prefilter we simplify the problem further. Let the true process be described by

$$A_0(q)y(t) = B_0(q)u(t) \tag{3.3}$$

i.e. the noise free case is considered. The A and B polynomials are estimated using the least squares method. The input and output are filtered through a filter H_p before the data is used in the estimation. See Figure 1. The input signal to the process is generated by the fixed control law

$$R(q)u(t) = -S(q)y(t) + T(q)u_c(t) \tag{3.4}$$

where we have assumed that the controller is chosen such that $G_m(q) = B_m(q)/A_m(q)$ is the desired pulse transfer operator from u_c to y. We are thus assuming that the control signal has the optimal (= desired) properties, compare Quote 2. Some calculations now give that the asymptotic loss function to be minimized is

$$\bar{V}(\theta) = \int_{-\pi/h}^{\pi/h} |A(\theta)G_0 - B(\theta)|^2 |G_0^{-1} G_m|^2 |H_p|^2 \phi_{uc}(\omega)\, d\omega \tag{3.5}$$

where ϕ_{uc} is the spectrum of the reference signal u_c. The minimization is done with respect to the unknown parameters in the A and B polynomials. Equation (3.5) gives a frequency interpretation of the estimation problem. The part $|G_0^{-1} G_m H_p|^2 \phi_{uc}$ is a frequency dependent weighting, which determines the frequency band where the estimate should be good. The desired closed loop transfer operator G_m is in general constant for low frequencies and the gain decreases for frequencies above the bandwidth. Further G_0 is in general of low pass character. This implies that G_0^{-1} will be of high pass character. For the design of controllers it is important to have a good model around the cross over frequency. This implies that the prefilter H_p should be chosen such that the weighting is largest around the desired cross over frequency and low at frequencies above the bandwidth. H_p should thus in general be of low pass type or band pass type with the pass band around the desired cross over frequency. If $H_p = 1$ then the estimator will concentrate on too high frequencies, which is a well known property of least squares estimation, see Ljung and Söderström (1983). It can be shown that inclusion of the noise in the process (2.1) will introduce a bias and more weight on higher frequencies. This makes it even more important that a proper weighting is introduced through prefiltering of the data before it is used in the estimator. Also the desired bandwidth of the closed loop system will influence the bias through G_m and A_m. A high desired bandwidth implies that the model must be accurate over a large frequency band. It may then be necessary to increase the order of the model. Finally increased input power or excitation will decrease the relative influence of the noise terms. This again in accordance with rules of thumb in identification.

The Automated Design Procedure

Since the design procedure is automatic, it is necessary to cover all possibilities and pitfalls in the logic of the design. For instance, it can be necessary to test if the estimated process model is minimum phase or if there are common factors in the estimated polynomials. The direct adaptive controllers have the advantage that the design step is eliminated, since the parameters of the regulator are estimated directly. This can be made possible by a reparameterization of the process model. A typical example of a direct algorithm is the moving average self-tuner, see Åström and Wittenmark (1985). The direct methods are, however, limited to special design procedures.

The direct algorithms usually have more parameters to estimate than the indirect algorithms, particularly if there are long time delays in the process. Simulations and practical experiments indicate, however, that the direct algorithms in many situations are more robust. One reason why this may be correct is given in Quote 1. Direct methods are essentially based on prediction of the output and direct estimation of the parameters in the predictor. Prediction error methods are robust even for low order models. Also numerical difficulties in the design calculations are avoided in direct methods. The direct algorithms usually have the disadvantage that all process zeros are cancelled. This implies that the direct methods can only be used for processes with stable inverse or minimum phase systems. Sampling of a continuous time system often gives a discrete time system with zeros on the negative real axis, inside or outside the unit circle. It is not good practice to cancel these zeros even if they are inside the unit circle. Cancellation of these zeros will give rise to 'ringing' in the control signal. The direct methods with cancellation of process zeros cannot be used when the process is nonminimum phase. It is thus advisable not to cancel any process zeros when using direct methods. There is no way to test if the process is minimum phase or not, if only the controller polynomials are known. Indirect methods have the advantage that the B polynomial in the process model is estimated. The stability of this polynomial can then easily be tested and the design procedure can be changed if the estimated model is nonminimum phase.

EXAMPLE 1—Influence of desired closed loop bandwidth
The example is taken from Wittenmark and Åström (1980). The process is a fourth order system.

$$G(s) = \frac{1}{(s+1)^4}$$

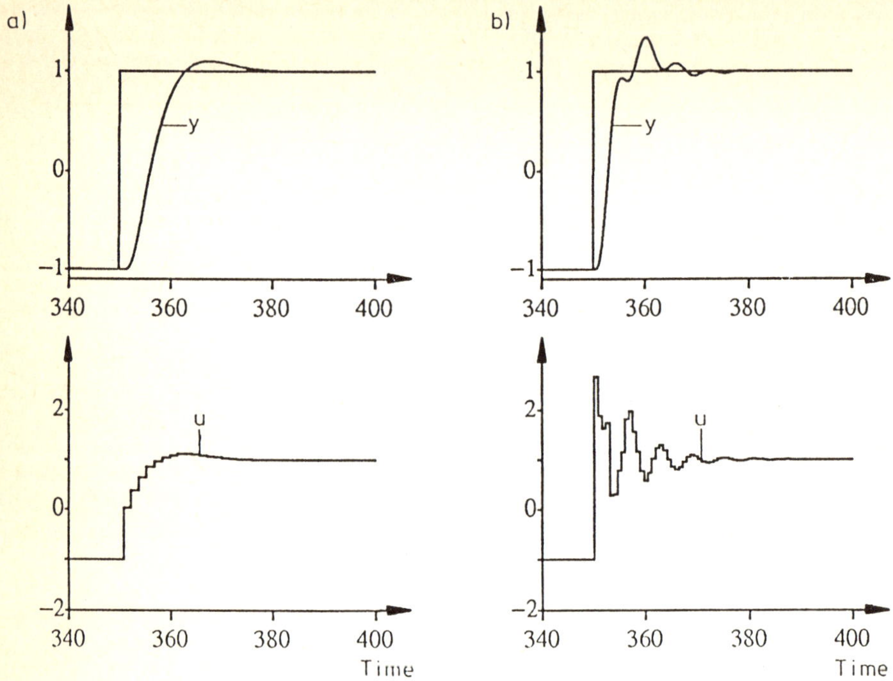

Figure 4. The output and control signals for the system in Example 1 when (a) $\omega_m = 0.3$; (b) $\omega_m = 0.6$.

An indirect adaptive pole-placement controller is used. The process is identified as a second order sampled data system using least squares estimation without any prefiltering of the signals. The controller is of the form (3.4) and has the same structure as a digital PID-controller. The desired closed loop response is given as the model

$$G_m(s) = \frac{\omega_m^2}{s^2 + 2\zeta_m\omega_m s + \omega_m^2} \tag{3.6}$$

with $\zeta_m = 0.7$. The system and the controller was simulated for different values of ω_m. The sampling interval is chosen such that $\omega_m h = 0.3$. Figure 4 shows the output and the control signal when the parameters have converged. For $\omega_m = 0.3$ the control is good. The behavior starts to deteriorate when ω_m is increased to about 0.4. The reason is the difficulty for the estimator to find a good second order model of the process in the desired frequency band. The example shows that too simple models can restrict the performance of the system, due to modelling errors. □

4. The Estimator Algorithm

In this section we will comment on implementation of recursive estimation algorithms. A straightforward implementation of an estimator with exponential forgetting can result in *estimator windup*, see Åström and Wittenmark (1988), if the input is not sufficiently exciting. Different ways to avoid this problem have been discussed in the literature:

- Covariance resetting, see Goodwin and Sin (1984)
- Time variable forgetting factors, see Fortescue *et al* (1981), Wellstead and Sanoff (1981)
- Constant trace algorithm, see Irving (1979)
- Directional forgetting, see Hägglund (1985)

The main idea with these modifications is to ensure that the covariance matrix stays bounded. Following the advice in Quote 3 it has in practice and theory been shown that it is advantageous to introduce a dead-zone in the estimator and turn it off when the residuals are sufficiently small. See Egardt (1979) Kreisselmeier and Narendra (1982) and Middleton *et al* (1985).

5. Sampled Data Control Aspects

Design and implementation of digital controllers are treated in books on digital control. See for instance Åström and Wittenmark (1984). Some important aspects are briefly discussed in this section.

Antialiasing Filters

Due to the aliasing problem connected with the sampling procedure it is necessary to eliminate all frequencies above the Nyquist frequency before sampling the signals. High frequencies may otherwise be interpreted as low frequencies. This implies that the signal conditioning is related to the choice of the sampling interval. Suitable choices of antialiasing filters are second or fourth order Butterworth, ITAE (Integral Time Absolute Error) or Bessel filters. They consist of one or several cascaded filters of the form

$$G_{aa}(s) = \frac{\omega^2}{s^2 + 2\zeta\omega s + \omega^2}$$

Let ω_B be the desired bandwidth of the filter. Tables for how to choose the damping ζ and the frequency ω can be found for instance in Åström and Wittenmark (1984). The antialiasing filter will naturally low pass filter the outputs of the process. The filter will also make it possible for the estimator to get good models in correct frequency range. Compare the discussion of H_f in Section 2. How will the antialiasing filter influence the sampled data model? The Bessel filter is a good approximation of a time delay for frequencies up to around the bandwidth ω_B. For instance, the fourth order filter can be well approximated by a time delay of

$$T_d = \frac{2\pi}{3\omega_B}$$

This implies that the sampled data model including the antialiasing filter can be assumed to contain an additional time delay compared to the process. Assume that the bandwidth of the filter is chosen as

$$|G_{aa}(i\omega_N)| = \beta$$

Table 1. The approximate time delay T_d due to the antialiasing filter as a function of the desired attenuation β at the Nyquist frequency for a fourth order Bessel filter. h is the sampling period.

β	ω_N/ω_B	T_d/h
0.05	3.1	2.1
0.1	2.5	1.7
0.2	2.0	1.3
0.5	1.4	0.9
0.7	1.0	0.7

where $\omega_N = \pi/h$ is the Nyquist frequency and $G_{aa}(s)$ is the transfer function of the filter. The attenuation β may be in the range 0.1 – 0.5. Table 1 gives some values of T_d as a function of β. The table shows that, for a fixed sampling period or Nyquist frequency, the delay will increase if a higher attenuation is desired. This also implies that the antialiasing filter must be taken into account when estimating the process model and in the design. For reasonable values of β the filter can be approximated with a delay of one or two sampling periods. The price for having the antialiasing filter is that one or two extra parameters have to be estimated in the model. However, if the desired closed loop bandwidth is less than 0.05 – 0.1 times the bandwidth of the antialiasing filter then the filter will have very little influence on the process model around the desired crossover frequency.

EXAMPLE 2—The effect of the antialiasing filter
Let the process to be controlled be given by

$$G(s) = \frac{1}{s(s+1)}$$

which can be a model for a normalized motor. The system is sampled with $h = 0.5$ and the desired closed loop system is assumed to be described by a second order system (3.6) with $\omega_m = 1$ rad/s and $\zeta_m = 0.7$. A *fixed parameter* pole placement controller is determined. The output of the system is disturbed by a sinusoidal signal i.e. the measured signal is

$$y_m(t) = y(t) + a_d \sin(\omega_d t)$$

This signal is filtered through a fourth order Bessel filter with the bandwidth ω_B. Figure 5 shows the influence of the disturbance and the antialiasing filter. Figure 5(a) shows the influence of the disturbance when $\omega_B = 25$ rad/s, which is far too high for the disturbance frequency. The Nyquist frequency is 6.28 rad/s. The disturbance frequency, which is $\omega_d = 11.3$ rad/s, is folded and gives rise to a low frequency disturbance in the sampled measurement. The controller then tries to compensate this "imaginary" frequency. Figure 5(b) shows the effect of the antialiasing filter when $\omega_B = \omega_N$. The disturbance is smaller but still noticeable in the input and the output. The filter will, however, also influence the closed loop performance. The overshoot is increased. Figure 5(c) shows the result when $\omega_B = \omega_N/2.5$ and when the controller is modified. The controller is now determined for the true process and when the antialiasing filter is

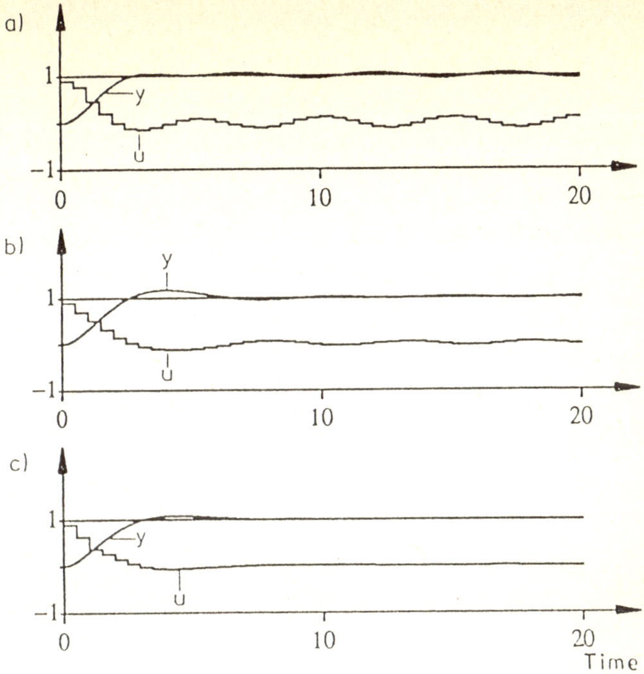

Figure 5. Output, reference value and control signal for the system in Example 2. The measurement disturbance has the frequency $\omega_d = 11.3$ rad/s and $a_d = 0.1.(a)\omega_B = 25$ rad/s; (b) $\omega_B = 6.28$ rad/s; (c) $\omega_B = 2.51$ rad/s and the regulator compensated for a delay of $1.7h$.

approximated by a delay of $1.7h$, compare Table 1. The closed loop response is improved compared to Figure 5(b). □

The example shows that it is important to use an antialiasing filter and that the filter sometimes has to be considered in the design. It is, however, sufficient to approximate the filter by a time delay. This simplifies both estimation and design in the adaptive case.

Choice of Sampling Period

The choice of the sampling period depends on the design algorithm as well as of the desired closed loop performance. One rule of thumb that is useful for deterministic design methods is to let the sampling interval h be chosen such that

$$\omega h \approx 0.25 - 1 \tag{5.1}$$

where ω is the natural frequency of the dominating poles of the desired closed loop system. If the dominating pole is real one can use

$$h/T_0 \approx 0.25 - 1 \tag{5.2}$$

where T_0 is the time constant of the dominating pole. These rules imply that there will be 5–20 samples on a step response of the closed loop system. The choice of the sampling interval must also be influenced by the disturbances acting on the system. If the process is disturbed by occasionally large disturbances it may be necessary to shorten the sampling interval to ensure that the disturbance is detected and compensated for as soon as possible. For stochastic design methods it is of importance how the performance of the closed loop system is measured. The sampling period can be chosen as in (5.1) and (5.2) with a preference for the higher limit of the sampling frequency, see Lennartson (1986).

The choice of the sampling interval also influences if the antialiasing filter should be taken into account in the design. First assume that the sampling frequency is high (20–100 times) compared with the desired closed loop bandwidth. It is then not necessary to take the filter into account when making the design. The amplitude of the filter is about unity and the phase lag is neglectable at the cross over frequency. The influence of the filter can, however, not be neglected in the design if the Nyquist frequency is only a little bit larger than the desired cross over frequency. This in general implies that the estimated model must be of higher order than if a higher sampling rate is chosen. This is therefore one motivation why the sampling period should be quite short.

Antireset Windup

In the discussion in Section 3 it was pointed out that unknown levels, ramps etc are taken care of by introducing integrators in the controller. An integrator is an unstable system and it may happen that the integral term in the controller can assume very large values if the control signal is limited when there still is an error. This is called *reset windup* or *integrator windup*. Special precautions must be taken in order to avoid this. Different ways are discussed in Åström and Wittenmark (1984) for different controller structures.

Consider a regulator described by

$$R^*(q^{-1})u(t) = -S^*(q^{-1})y(t) + T^*(q^{-1})u_c(t)$$

The regulator may contain unstable modes. One way to solve the reset windup problem is to rewrite the regulator by adding $A_o^*(q^{-1})u(t)$ to both sides. This gives

$$A_o^*(q^{-1})u(t) = (A_o^*(q^{-1}) - R^*(q^{-1}))u(t) - S^*(q^{-1})y(t) + T^*(q^{-1})u_c(t) \tag{5.3}$$

Antireset compensation is then obtained by using

$$\begin{aligned} A_o^*(q^{-1})v(t) &= (A_o^*(q^{-1}) - R^*(q^{-1}))u(t) - S^*(q^{-1})y(t) + T^*(q^{-1})u_c(t) \\ u(t) &= \mathrm{sat}(v(t)) \end{aligned} \tag{5.4}$$

where sat$(\cdot)$ is the saturation function. This regulator is equivalent to (5.3) when the control signal does not saturate. A_o^* is chosen as a stable polynomial and can be interpreted as the observer dynamics of the controller. A block diagram of (5.4) is shown in Figure 6.

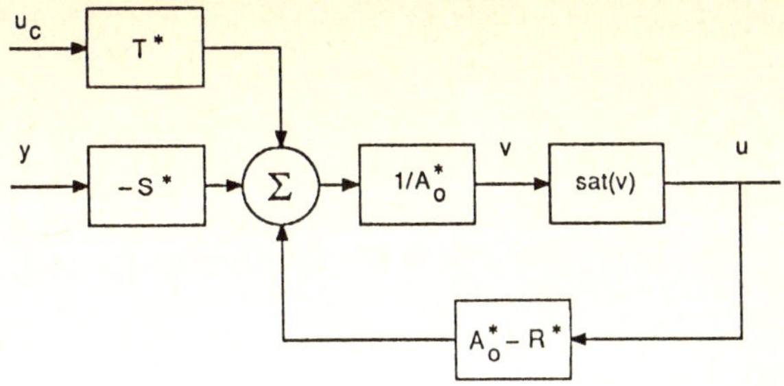

Figure 6. Block diagram of (5.4), which avoids integrator windup.

6. Estimation and Control Interaction

The estimation, design and controller implementation have been discussed as separate subjects in the previous sections. We have, however, also seen that there is a strong interaction between these parts. For instance the desired closed loop performance will influence the frequency content of the input and output signals. This will then influence the properties of the estimator. This interaction is complex as discussed above. One way to decrease the interaction is to follow the advice in Quote 5 and use a low gain in the adaptation loop. This will exaggerate the two time scale property indicated in Figure 1. A low gain in the adaptation loop makes it, however, more difficult to follow fast variations in the parameters of the process. We will now illustrate the interaction through a simulated example.

EXAMPLE 3
Consider again the process and output disturbance in Example 2. The antialiasing filter is a fourth order Bessel filter. The parameters in the process are estimated from the model

$$y(t) + a_1 y(t-1) + a_2 y(t-2) = b_1 u(t-1) + b_2 u(t-2) + b_3 u(t-3) + e(t) \qquad (6.1)$$

i.e. it is assumed that the antialiasing filter can be approximated by a delay that is less than one sampling period. The controller is designed based on the estimated parameters such that the corresponding continuous closed loop system is described by the natural frequency ω_m and the damping ζ_m. Figure 7(a) shows the output and reference value when $\omega_B = 25$ rad/s, $\omega_m = 1$, $\zeta_m = 0.7$, and $a_d = 0.1$. Compare Figure 5(a). Figure 7(b) is the same as Figure 7(a) but when $\omega_B = \pi/h = \omega_m$. Compare Figure 5(b). These two figures show that the antialiasing filter is necessary and that it is automatically compensated for by using the model structure (6.1). Figure 7(c) shows a simulation of the adaptive controller when $\omega_B = 2.512$. The closed loop response is not satisfactory since (6.1) will not give an adequate model. To be able to use a lower ω_B it would be necessary to change the structure of the estimated model and introduce an extra time delay. Compare Table 1. Figure 7(d) shows the same as Figure 7(c) but when $\omega_m = 0.5$. By decreasing the desired closed frequency it is possible to again get a good model using

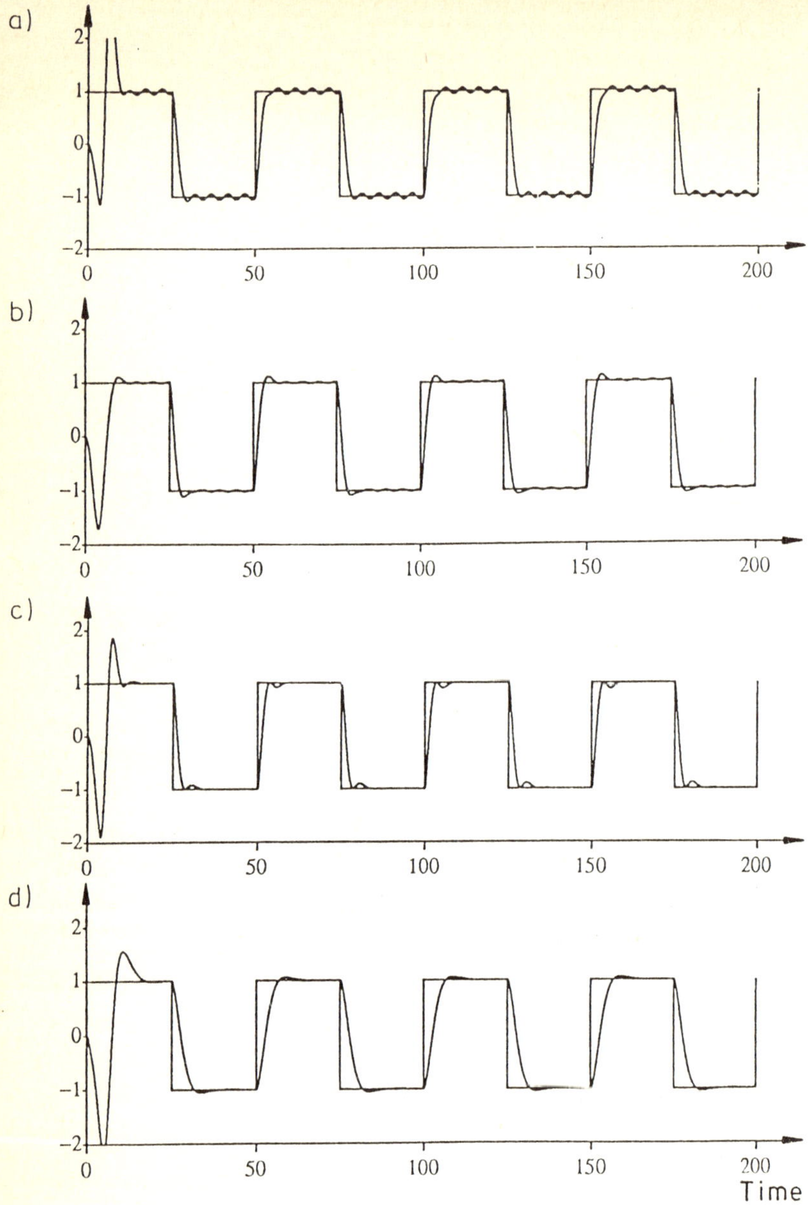

Figure 7. Output and reference value when using an adaptive controller on the process in Example 3, when $h = 0.5$ and $a_d = 0.1$. (a) $\omega_m = 1$, $\omega_B = 25$; (b) Same as (a) but with $\omega_B = 6.28$; (c) Same as (a) but with $\omega_B = 2.512$; (d) $\omega_m = 0.5$ and $\omega_B = 2.512$.

(6.1). This implies that there is an intricate relationship between the desired response, the model structure, and the frequency content of the input original. □

7. Application of Adaptive Control

Adaptive control now has a great impact on the practice of process control. With the event of commercial adaptive controllers adaptive control has been used on many different processes. It has been estimated that more than 50,000 loops are under adaptive control in 1988. Many of these applications show that adaptive control can be very useful, but they also show that the adaptive controllers often have to be "tuned" to the specific application. See, for instance, Åström and Wittenmark (1988). It is very difficult to judge the need for an adaptive controller based on the behavior of the open loop system. The open loop system may in some processes vary very much, but a fixed feedback controller can give a very robust closed loop behavior. It is therefore important to have a feel for the interaction between specifications and closed loop performance, as discussed in the previous sections.

Many of the applications show that it can be advantageous to use an adaptive controller even if the process is time-invariant, since the adaptive controller will result in better tuned regulators and improved performance. One manufacturer has even gone so far that they only have one control block, an indirect adaptive controller, in their DDC-package. By proper choice of the structure of the estimated model it is for instance possible to obtain a controller with PID-structure.

Commercial adaptive controllers can be divided into:

- Tuners for PID-controllers
- Adaptive tool-boxes
- Adaptive controllers for specific applications

To the first category belongs Foxboro's Exact controller and SattControl's auto-tuner. These are controllers that tune the parameters in conventional PID-controllers. This type of tuning aid is very useful at commissioning, but also as a tool to maintain good tuning of the plant. The tuning is usually done at the request of the operator, who then can supervise the tuning. This implies that the safety net around the algorithm can be simplified.

Foxboro's Exact controller is watching the process all the time and detects natural disturbances, which can be used to decide if the controller settings should be modified. The design procedure is based on a large set of empirical rules. About 20,000 Exact controllers are in operation in December 1987.

The SattControl auto-tuner is based on relay-tuning and the controller is tuned by pressing a button. See Åström and Hägglund (1988). A relay controller is connected to the system, which starts to oscillate. The period and amplitude of the limit-cycle oscillation are measured. Logic is used to limit the amplitude to a predetermined level. The controller parameters are determined by a modified Ziegler-Nichols rule. There is also some logic to determine if derivative action is needed or not. The advantage with the controller is that there are no parameters that have to be set a priori. About 2,000 auto-tuners are in operation in December 1987. Simplicity is the major advantage of the auto-tuner, since it can be used by inexperienced personnel.

There are several general purpose adaptive controllers on the market, for instance Asea's Novatune and First Control's Firstloop. These are DDC-packages where an adaptive control unit is included. The adaptive controller can be used as a conventional regulator block in the system. This gives a great flexibility, but it also requires a better knowledge about adaptive control from the operator. These types of adaptive controllers can be regarded as toolboxes for adaptive control. The block oriented structure makes it possible for the process engineer to make a general and flexible system with adaptation. The Novatune controller is a direct self-tuner based on least squares estimation and a modified minimum variance controller. One of the closed loop poles is a design parameter. The Firstloop system contains an indirect pole placement algorithm.

The Novatune and Firstloop systems differ from the auto-tuners because they are not based on PID-controllers. The algorithm is instead based on a general transfer function model. This implies that the systems can be used to make efficient dead-time compensation and adaptive feedforward.

Finally, special purpose adaptive controllers are available for instance for paper machines, autopilots for ships, and ultrafiltration. In those controllers much of the process knowledge is built into the system and there are only few parameters that have to be selected by the operator. For instance regulator structure, time scales, and desired performance can be more or less prespecified in the design.

The many applications of adaptive control indicate some occasions where it can be very advantageous to use adaptation. Some of these cases are:

- Systems with long time delays.
- Systems where feedforward can be used.
- Systems with time varying character of the disturbances.

In the first two cases it is necessary to have good process models to design good controllers. It can then be beneficial to use an adaptive controller, which automatically can find the relevant process model. In many industrial processes the characteristics of the disturbances are changing with time. The change may come from changes in raw material or may be disturbances from earlier process units. To maintain good control it is necessary to retune the controller in those cases. Using adaptation makes it possible to increase the quality of the control.

Analysis of products and applications clearly shows that although the key principles are straightforward, there are many details and fixes that have to be done in order to make an adaptive system work well under all possible conditions. Experience has, however, shown that only modest efforts are required to make a standard adaptive system to work well.

8. Conclusions

In this paper we have discussed different filters that are used in the implementation of adaptive controllers. The filters play an important role in order to obtain a robust implementation of an adaptive controller. The adaptive control algorithm will contain the following steps:

- Filter the continuous time process output using an antialiasing filter with bandwidth below or around the Nyquist frequency π/h, where h is the sampling period.
- High pass filtering of the sampled signal to remove low frequency disturbances such as levels and ramps. Known sinusoidals are removed using notch filters.

- Lowpass filter with H_p to get a weighting in the estimator in an appropriate frequency band.
- Estimate a low order model using an algorithm with time variable exponential forgetting, regularized constant trace or directional forgetting. The estimator should also contain a dead zone. Finally the estimator may contain a "switch", which detects if the system is sufficiently excited.
- The design method for the controller should be robust against unmodelled dynamics. Level and ramp disturbances are eliminated by introducing integrators in the controller. The control signal should be limited and the controller should include antireset windup.

The block diagram in Figure 1 includes the added filters. All these points contain choices of parameters. The discussions in the previous sections contain guidelines that can be used for the selection of the parameters.

Acknowledgements

The first version of the paper was written during a visit at University of Newcastle, Australia. The author's visit in Australia has been possible through support by Australian Research Grant Scheme, Swedish Board of Technical Development (STU), Troedssons Forskningsfond and Rydins Stiftelse.

9. References

Åström, K. J. (1983a): "Analysis of Rohrs counter examples to adaptive control," *Reprints 22nd IEEE Conference on Decision and Control, San Antonia TX*, pp. 982–987.

Åström, K. J. (1983b): "Theory and applications of adaptive control—A survey," *Automatica*, **19**, 471–486.

Åström, K. J. and T. Hägglund (1988): *Automatic Tuning of PID Regulators*, ISA, Triangle Research Park, NC.

Åström, K. J. and B. Wittenmark (1980): "Self-tuning controllers based on pole-zero placement," *Proc. IEE, Part D*, **128**, 120–130.

Åström, K. J. and B. Wittenmark (1984): *Computer Controlled Systems*, Prentice Hall Inc., Englewood Cliffs, NJ.

Åström, K. J. and B. Wittenmark (1985): "The self-tuning regulator revisited," *Preprints 7th IFAC Symposium on Identification and System Parameter Estimation*, pp. *xxv–xxxiii*.

Åström, K. J. and B. Wittenmark (1988): *Adaptive Control*, Addison-Wesley, Reading, MA.

Egardt, B. (1979): *Stability of Adaptive Controllers*, Springer-Verlag, New York.

Fortescue, T. R., L. S. Kershenbaum and B. F. Ydstie (1981): "Implementation of self-tuning regulators with variable forgetting factors," *Automatica*, **17**, 831–835.

Gevers, M. and L. Ljung (1986): "Optimal experiment design with respect to the intended model application," *Automatica*, **22**, 543–555.

Goodwin, G. C. and K. S. Sin (1984): *Adaptive Filtering Prediction and Control*, Prentice-Hall, Englewood Cliffs, N.Y.

Hägglund, T. (1985): "Recursive estimation of slowly time varying parameters," *Preprints 7th IFAC Symposium on Identification and System Parameter Estimation*, York, pp. 1137-1142.

Irving, E. (1979): "New developments in improving power network stability with adaptive control," *Yale Workshop on Applications of Adaptive Systems*.

Kreisselmeier, G., and K. S. Narendra (1982): "Stable model reference adaptive control in the presence of bounded disturbances," *IEEE Trans. Automatic Control*, **AC-27**, 1169–1175.

Lennartson, B. (1986): "On the design of stochastic control systems with multirate sampling," Ph.D. Thesis, Control Engineering Laboratory, Chalmers University of Technology, Gothenburg, Sweden.

Ljung, L. and T. Söderström (1983): *Theory and Practice of Recursive Identification*, MIT Press, Cambridge, MA.

Middleton, R. H., G. C. Goodwin, D. J. Hill and D. Q. Mayne (1988): "Robust adaptive control. Convergence, stability and performance," *IEEE Trans. Automatic Control*, **AC-33**.

Peterka, V. (1984): "Predictor-based self-tuning control," *Automatica*, **20**, 39–50.

Rohrs, C. E., M. Athans, L. Valavani and G. Stein (1984): "Some design guide lines for discrete-time adaptive controllers," *Automatica*, **20**, 653–660.

Seborg, D. E., T. F. Edgar and S. L. Shah (1986): "Adaptive control strategies for process control: A survey," *AIChE Journal*, **32**, 881–913.

Wahlberg, B., and L. Ljung (1986): "Design variables for bias distribution in transfer function estimation," *IEEE Trans. Automatic Control*, **AC-31**, 134–144.

Wellstead, P. E., and S. P. Sanoff (1981): "Extended self-tuning algorithm," *Int. J. Control*, **345**, 433–455.

Wittenmark, B (1986): "On the role of filters in adaptive control," Technical report EE8662, University of Newcastle, N.S.W., Australia.

Wittenmark, B., and K. J. Åström (1980): "Simple self-tuning controllers," *International Symposium on Adaptive Control*, Bochum.

Wittenmark, B., and K. J. Åström (1984): "Practical issues in the implementation of adaptive control," *Automatica*, **20**, 595–605.

On the Role of Prefiltering in Parameter Estimation and Control

Coorous Mohtadi
Department of Chemical Engineering
University of Alberta
on leave of absence from
Department of Engineering Science
University of Oxford

Abstract

Much of the current theory of predictive control deals with 'optimality' of predictions and controls based on assumptions which are rarely verifiable in practice. Practitioners on the other hand know the value of simple signal conditioning when dealing with real data. This paper examines the effects of the simplest form of signal conditioning; namely that of a prefilter in the parameter estimator and a so-called 'observer' polynomial in the associated control calculations. It is demonstrated that these filters are absolutely vital in the design and implementation of control algorithms. The theoretical results are supported by some illustrative simulations. Two simple counter examples are used to demonstrate the necessity of the observer polynomial even when the unmodelled dynamics is very small.

1 Introduction

Results of current applications of the Generalized Predictive Control (GPC) algorithm to industrial processes (Lambert, M. 1987, Lambert E. 1987, Al-Assaf, Y. 1988) indicate very strongly the need for having

1. a prefilter for paremeter estimation and,
2. an observer polynomial in the control law.

It is somewhat surprising that relatively few applications reported so far emphasise the need for the prefilter and the observer polynomial (Wittenmark and Åström 1984). The provisions for the prefiltering and the observer polynomial have been discussed in the original derivation of the GPC strategy (Mohtadi 1986, Clarke, Mohtadi and Tuffs 1987a,b) but their use has been primarily in providing asymptotic 'optimality' for the appropriate prediction equation.

The role of prefiltering in parameter estimation has been examined by many authors (eg. Wahlberg and Ljung 1986, Ljung 1987 in the context of frequency response interpretation of the least squares estimator and the bias distribution of the estimates and Anderson *et al* 1986 in the context of satisfying the positive real condition necessary for stability of adaptive algorithms in cases where there is plant–model mismatch). The role of the observer polynomial on the other hand has been explored by others (Åström and Wittenmark 1984, Clarke 1982, in the context of disturbance rejection properties; Doyle and Stein 1979 and Lim 1982 with a view to robustness of the model-based controllers). It is with the GPC law and the more realistic issue of providing robsut stability as opposed to the traditional aspect of optimality that we examine the role of prefiltering in parameter estimation and control design.

The organization of the paper is as follows. Section 2 examines the role of $T_{est}(q^{-1})$ filter in parameter estimation. Section 3 examines the role of the observer polynomial in GPC control. Section 4 provides a number of simulated examples illustrating the importance of these design filters.

2 The role of the prefilter in parameter estimation

Real industrial systems rarely allow an accurate mathematical model let alone a linear Black–Box type CARMA (Controlled Auto Regressive and Moving Average) or CARIMA (Controlled Auto Regressive and Integrated Moving Average) model of possibly low order (usually ≤ 4) as is usually assumed in self-tuning control. Real processes are fraught with unstructured disturbances and non-linearities (internal saturations, slew-rating, or e.g. pH-type nonlinearities), yet there are numerous reports of successful applications of self-tuning controllers (see e.g. Åström 1983, Seborg *et al* 1986) which point to the usefulness of the approaches.

Successful employment of parameter estimators is frequently attributable to appropriate signal conditioning used to 'iron out' the effects of unmodelled dynamics, disturbances and nonlinearities. These techniques typically include:

1. Low-pass and band-pass digital filtering
2. Averaging and FIR intersample filtering
3. Analogue Nyquist filtering
4. Logical spike filtering
5. Inverting nonlinearities (e.g. squaring or taking square roots)
6. Disabling estimation when either the residual is below a lower dead-band or above and upper one.

It is to the first category of digital filtering that we next divert our attention.

In order to gain a better understanding, albeit in a qualitative sense but not as a substitute for realism, we make some idealized assumptions about the underlying process that:

$$y(t) = G_0(q^{-1})u(t-1) + H_0(q^{-1})\xi(t) \tag{1}$$

where $y(t)$, $u(t)$ and $\xi(t)$ are output (measured variable), input (control signal or manipulated variable) and the unmeasurable disturbances of a single-input single-output plant respectively. G_0 and H_0 are linear transfer–functions (not necessarily of finite order) in q^{-1} the backward shift operator. A linear model of the form:

$$\hat{A}(q^{-1})y(t) = \hat{B}(q^{-1})u(t-1) + e(t) \tag{2}$$

is usually adopted for the self-tuning algorithm where $\hat{A}$ and $\hat{B}$ are of some finite predetermined order (usually $\deg \hat{A} \leq 4$ and $\deg \hat{B} \leq 6$ for most applications). Typically a recursive estimator which is some variant of the RLS (Recursive Least Squares) is employed. A simple block diagram of fig.1 depicts such a set-up.

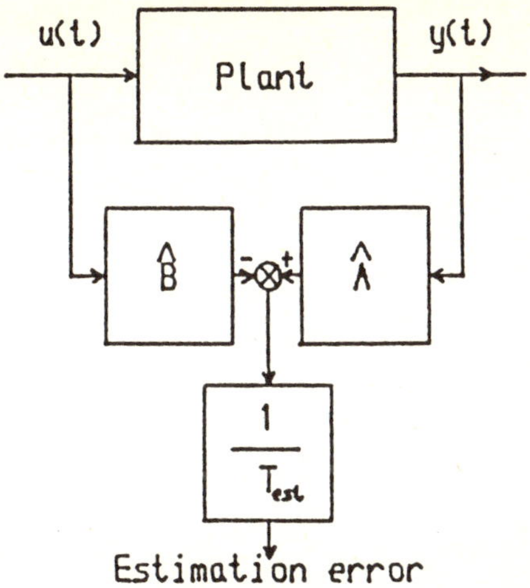

Figure 1: The Recursive Estimator

Many of the theoretical derivations of convergence of these estimators do not include the prefilter T_{est}. Although recent theoretical (Anderson *et al* 1986) and experimental (Lambert, M. 1987, Lambert, E. 1987) results point to the need for such filtering from the robustness point of view. In such a set-up the parameters (coefficients of polynomials $\hat{A}$ and $\hat{B}$) are chosen such that:

$$J_{est} = \frac{1}{N} \sum_{i=0}^{N} \left(\frac{e(t-i)}{T_{est}} \right)^2 \tag{3}$$

is minimized. As $N \rightarrow \infty$, with a stationary sequence of $u(t)$ and $\zeta(t)$ we can use Parseval's theorem which gives (Ljung, 1987):

$$J_{est} = \frac{1}{2\pi} \int_0^{2\pi} |\frac{1}{T_{est}}|^2 \Phi_e(\omega) d\omega \tag{4}$$

where Φ_e is the power spectral density and $\omega = 2\pi f/f_s$ where f is the frequency in Hz and f_s is the sampling frequency. A typical scenario frequently used by practitioners, for commissioning a self-tuner consists of either

- some open-loop identification perhaps with PRBS or step inputs to obtain some idea about the parameters of the system followed by closing the loop with the controller *in situ* or;
- identifying the process in a so-called 'piggy-back' mode where a P or PI control is used to stabilize the system and set-points are varied to help identify the plant.

This leads to the two important cases considered next.

2.1 Open-loop estimation

With open-loop estimation $u(t)$ and $\xi(t)$ are uncorrelated. Eqn.4 then becomes:

$$J_{est} = \frac{1}{2\pi}\int_0^{2\pi} |\frac{\hat{A}}{T_{est}}|^2 \left\{ |G_0 - q^{-1}\frac{\hat{B}}{\hat{A}}|^2 \Phi_u + |H_0|^2\sigma^2 \right\} d\omega \tag{5}$$

where Φ_u is the power spectral density of the input signal and σ^2 is the variance of $\{\xi(t)\}$. Clearly, in open-loop estimation two design 'knobs' can be used: Φ_u the input spectrum and T_{est} the prefilter. Eqn.5 indicates that they perform very similar tasks (in the noise free case they are almost interchangeable). But clearly T_{est} is a more flexible tool than Φ_u: every time a new Φ_u is required a whole new set of experiments have to be performed whereas with T_{est} the same set of data can be used over and over again!

In order to obtain some guidelines for the choice of T_{est} and Φ_u we attempt to examine eqn.5 more closely. The optimization of eqn.5 constitutes a compromise between two matching exercises:

1. $|\frac{\hat{A}}{T_{est}}|^2|G_0 - q^{-1}\frac{\hat{B}}{\hat{A}}|^2\Phi_u$ the weighted modelling error;

2. $|\frac{\hat{A}}{T_{est}}|^2|H_0|^2\sigma^2$ the weighted noise spectrum.

Whenever (1) is larger than (2) the modelling error is reduced and wherever (2) is greater than (1) $\hat{A}$ is adjusted to match the characteristics of the disturbances (i.e. $|\hat{A}| \propto |\frac{T_{est}}{H_0}|$ at those frequencies). This is only acceptable under very stringent circumstances (i.e. exact parameterization, exact coefficients for T_{est} to match the noise model and whiteness of $\xi(t)$) and should be avoided at all costs. Clearly, if T_{est} is chosen in such a way as to attenuate the weight of the frequencies in the integral of eqn.5, wherever the disturbances dominate, $\hat{A}$ will not be adjusted to match the noise dynamics (which is usually unrelated to the I/O dynamics) and the cost will largely minimize (1). This leads us to the first guideline regarding the choice of T_{est}.

Guideline 1 *T_{est} must be chosen so as to attenuate the effect of disturbances wherever they are dominant. This usually implies attenuating very low and very high frequencies.*

Some low frequency disturbances stem from:

- Units of measurement (e.g. temperatures may be measured in degrees centigrade and for a zero control we usually get the ambient temperature!; also signals in control loops may be currents between 4 – 20 mA)
- Offsets in sensor measurement
- Gradual (low-frequency) degradation of the sensor measurement (or for example fouling) before a recalibration is necessary
- Low frequency load and feed changes in chemical processes

and some high frequency disturbances may be attributable to:

- Electrical noise due to poor connections
- Quantization noise (usually low intensity and dominant at higher frequencies)
- Spikes in measurements

This will lead to the choice of

$$T_{est} = \frac{(1-\alpha q^{-1})^2}{1-q^{-1}}$$

say as a simple example. The choice α has to be made by performing the appropriate continious-time design given some idea about the disturbances. This may be performed by collecting open-loop data and performing some form of spectral estimation (e.g. FFT) to obtain an approximate bandwidth for the noise.

It is imperative to have some idea about the end-use of the model to choose the weighting. But before examining the effect of T_{est} on the appropriate frequency matching a special case must be considered where the error $|G_0 - q^{-1}\frac{\hat{B}}{\hat{A}}|$ can be made very small almost everywhere (i.e. there is no structural error). In this case:

$$\begin{aligned} |T_{est}| &= |A_0||H_0| \\ G_0 &= q^{-1}\frac{B_0}{A_0} \end{aligned} \tag{6}$$

This allows no choice in weighting appropriate frequencies other than that of optimality as pointed out by Wahlberg and Ljung (1986). Recall that in the more usual case where an exact match is not possible everywhere we minimize:

$$J \approx \frac{1}{2\pi}\int_0^{2\pi} |\frac{\hat{A}}{T_{est}}|^2 |G_0 - q^{-1}\frac{\hat{B}}{\hat{A}}|^2 \Phi_u d\omega \tag{7}$$

The 'best' choice for T_{est} or Φ_u is functionally dependant on the end-use of the resulting model (Ljung, 1987). This constitutes the main idea behind the so-called optimal input design extensively considered in the '70s (Goodwin and Payne 1977 and Ljung 1987 for a recent appraisal). In these designs the criterion of fitting is arbitrarily chosen as:

$$J_{end-use} = \int_0^{2\pi} C(\omega)|G_0 - q^{-1}\frac{\hat{B}}{\hat{A}}|^2 d\omega \tag{8}$$

where $C(\omega)$ is determined by the end-use (see Ljung 1987 chapters 12–14). The value of $C(\omega)$ in conjunction with other constraints on the input (finite available power for example) will determine the appropriate so-called 'optimal' input. Suffice it to say that wherever $C(\omega)$ is large then Φ_u and T_{est} should also emphasise those frequencies. Here therefore we concentrate on a few examples of $C(\omega)$ to study the choice of different end-uses.

2.1.1 Open-loop plant step-response

For GPC (Clarke *et al* 1987a,b) or DMC (Cutler and Ramaker 1980) methods the first N parameters of the step-response of the process are required. This will lead to the special case of the simulation considered by Ljung (1987), where the open-loop plant's step-response is truncated after N samples. This can be achieved through the use of a unit step input for N samples followed by a set of input signals as if the plant was controlled using a minimum-variance controller with a zero set-point. As here we only require an *approximate* relationship assume that:

$$u(q) = \frac{1-q^{-N}}{1-q^{-1}}$$

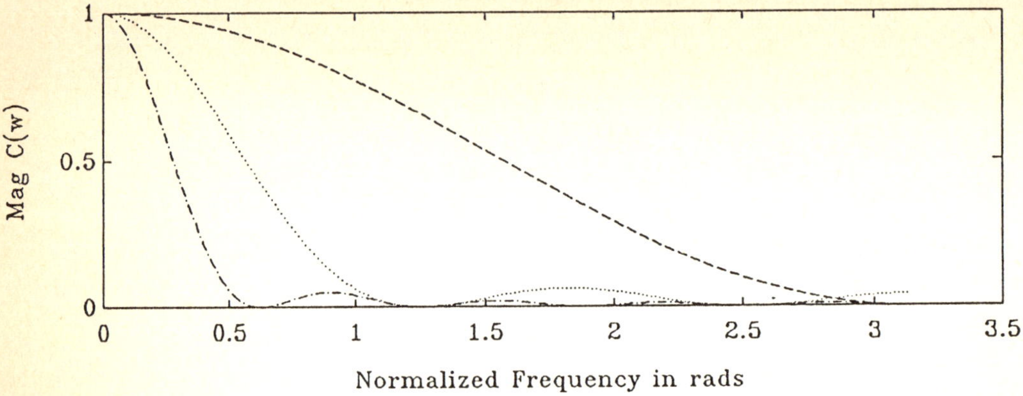

Figure 2: Frequency response of $C(\omega)$ for varying prediction horizon. $N = 1$ for the solid line ; $N = 2$ for the dashed line; $N = 5$ for the dotted line; $N = 10$ for the dot-dashed line.

$$C(\omega) = \left| \frac{1 - e^{-j\omega(N)}}{1 - e^{-j\omega}} \right|^2$$

This points to a stronger emphasis on the lower frequency components when using GPC or DMC control, bearing in mind that in reality there will also be higher frequencies present due to the truncation. This leads us to the view that:

Guideline 2 *Emphasis should be given to lower frequencies as N increases in order to obtain better fits for the ordinates of the step-response of the process.*

2.1.2 d-step-ahead prediction (*known future inputs*)

Consider the case of the $d - step - ahead$ prediction where $d \geq k$ the plant time-delay. Using a simple Taylor series expansion of the error upto the first term gives:

$$C(\omega) = |\frac{H_0^d}{H_0}|^2 \Phi_u$$
$$H_0 = H_0^d + q^{-d}\overline{H}$$

This also indicates that as $d \to \infty$ then the emphasis is shifted to the lower frequencies. It provides yet another reason for deemphasis of high frequency components of the errors.

2.1.3 Feedback Control

Consider the case where a feedback law of the form:

$$u(t) = -F_1 y(t) + F_2 w(t)$$

is to be applied where F_1 and F_2 are transfer-functions in q^{-1} and $w(t)$ is the set-point. The closed-loop transfer-function and the weighting are given by:

$$y(t) = \frac{F_2 G_0 w(t)}{1+F_1G_0} + \frac{H_0\xi(t)}{1+F_1G_0}$$
$$C(\omega) = \frac{|F_2|^2\Phi_w}{|1+F_1G_0|^4} + \frac{|F_1H_0|^2\sigma^2}{|1+F_1G_0|^4}$$

If the desired closed-loop response is given by R then:

$$C(\omega) = \frac{|R|^2\Phi_w}{|G_0|^2|1+F_1G_0|^2} + \frac{|F_1H_0|^2\sigma^2}{|1+F_1G_0|^4}$$

In cases of high signal to noise ratio, this points to three main areas where a good match is necessary:

- where Φ_w is large;
- where $|R/G|$ is large (i.e. closed-loop system bandwidth is larger than the open-loop bandwidth and we require to inject gain into the loop to maintain perfromance);
- where the controller sensitivity is poor.

In cases of poor signal to noise ratio the two areas are where there are large transmission gains between $\xi(t)$ and $u(t)$ and also where the sensitivity is poor.

Finally a note has to be added regarding the order of T_{est}. The overall weighting on the error is given by:

$$|\frac{\hat{A}}{T_{est}}|^2$$

This indicates that if the weighting is to have bandpass properties then T_{est} in its simplest form is given by:

$$T_{est} = \frac{T_N}{\Delta}$$

where the T_N is at least of order $\delta A + 1$. It must however be pointed out that the current experimental results (Lambert, 1987) do not indicate the need for a T_N filter of degrees higher than 2.

2.2 Closed-loop identification

Next we examine some of the properties of feedback in parameter estimation. This is because again as mentioned earlier, a typical commissioning routine for setting-up self-tuners consists of some open-loop probing followed by closing the loop with the self-tuning controller in place. In some cases, however, the plant is identified when under conventional PI control. In any event, the asymptotic expression of the cost here is only valid for fixed parameter controls and where $w(t)$ the set-point and $\xi(t)$ are not correlated:

$$J_{est} = \frac{1}{2\pi}\int_0^{2\pi} |\frac{\hat{A}}{T_{est}}|^2 \frac{1}{|1+G_0F_1|^2}\left\{|G_0 - q^{-1}\frac{\hat{B}}{\hat{A}}|^2\Phi_w + |1+q^{-1}F_1\frac{\hat{B}}{\hat{A}}|^2|H_0|^2\sigma^2\right\} d\omega \quad (9)$$

Note the major differences between eqn.5 and the expression of eqn.9.

- The term $1/|1+G_0F_1|$, the closed-loop sensitivity weights Φ_e implies that the estimator will attempt to minimize the errors where the control is poor (i.e. in the case of self-tuning control this means that we estimate where it really matters!).
- Φ_u has been replaced by Φ_w. It is therefore imperative for Φ_w to have frequency components upto the desired bandwidth of the closed-loop.
- An extra term $|1+q^{-1}\frac{\hat{B}}{\hat{A}}F_1|$ has been added to the contribution to the error due to the noise term. This has the unfortunate consequence that in the frequency range where the second term dominates the model closely resembles the inverse of the controller! This is the well known *identifiability* problem and points more emphatically to the important choice of T_{est} in guideline 1 which is to attenuate the effect of low and high frequency disturbances. In a more positive way, perhaps a reasonable approach is to disable estimation altogether when there are no set-point changes.

3 The role of the observer polynomial in control

GPC was first equipped with $T_{con}(q^{-1})$ polynomial as a natural extension of the associated GMV controller. The full derivation of the GPC control with $T_{con}(q^{-1})$ is given elsewhere (Mohtadi, 1986, Clarke *et al* 1987b). Suffice it to say that in GPC a set of output predictions $\{\hat{y}(t+N_1|t),\cdots,\hat{y}(t+N_2|t)\}$ is obtained assuming that there are no changes in the future control (i.e. open-loop prediction) and based on these predictions a set of control increments $\Delta\hat{u}(t+j)$ are postulated such that the cost:

$$J_{con}=\sum_{i=N_1}^{N_2}[\hat{y}(t+i)-w(t+i)]^2+\sum_{i=1}^{NU}\lambda\Delta\hat{u}^2(t+i-1) \tag{10}$$

is minimized. This gives:

$$\Delta\hat{\mathrm{u}}=[\mathrm{G}^T\mathrm{G}+\lambda\mathrm{I}]^{-1}\mathrm{G}^T(\mathrm{w}-\mathrm{f}) \tag{11}$$

as a set of postulated signals where G is the $N_2-N_1+1\times NU$ matrix of the step response of the system based on the model $q^{-1}\hat{B}/\hat{A}$, w is the vector of future set-point sequence and

$$\mathrm{f}=[\hat{y}(t+N_1|t)\cdots\hat{y}(t+N_2|t)]^T$$

Recall that the $j-step-ahead$ prediction of the output of the plant is given by:

$$\begin{aligned}\hat{y}(t+j|t)&=\tilde{G}_j\frac{\Delta u(t-1)}{T_{con}(q^{-1})}+F_j\frac{y(t)}{T_{con}(q^{-1})}\\ \frac{G_j}{T_{con}(q^{-1})}&=\overline{G}_j+q^{-j}\frac{\tilde{G}_j}{T_{con}(q^{-1})}\\ G_j&=E_jB\\ T_{con}(q^{-1})&=E_j\hat{A}\Delta+q^{-j}F_j\end{aligned} \tag{12}$$

There are many ways of computing the control signal with the GPC law (Mohtadi 1986). As the purpose here is to examine the properties of the controller, we obtain the controller coefficients of the type:

$$w(t)-S\frac{y(t)}{T_{con}(q^{-1})}=R\frac{\Delta u(t)}{T_{con}(q^{-1})} \tag{13}$$

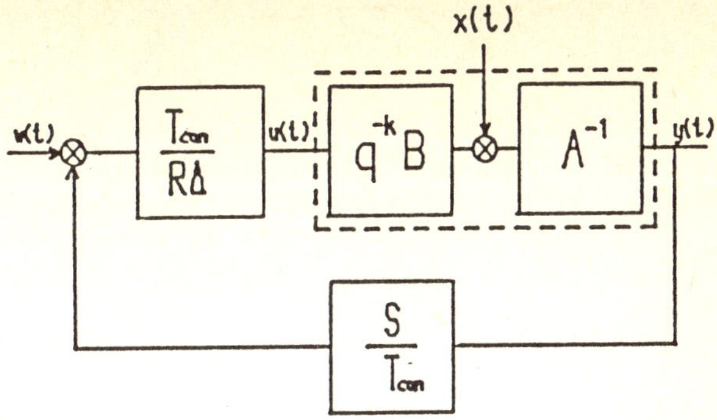

Figure 3: The Standard Control Loop

via a direct calculation based on an extension of the standard nonminimal state-space representaion of the I/O system. A **PC-MATLAB** version of the algorithm is given in the appendix. See also fig.3 for the roles of R and S polynomials. It is straight forward to show that:

$$R\hat{A}\Delta + q^{-1}S\hat{B} = P_cT_{con}(q^{-1}) \tag{14}$$

where P_c is indirectly determined by the choices of (N_1, N_2, NU, λ) and the model parameters $(\hat{A}, \hat{B})$. In pole-placement control (Wellstead *et al* 1979) the designer chooses P_c.

3.1 $T_{con}(q^{-1})$ and disturbances

The closed-loop equation in the nominal case is given by:

$$y(t) = q^{-1}\frac{\hat{B}}{P_c}w(t) + \frac{R\hat{A}\Delta d(t)}{P_cT_{con}(q^{-1})} \tag{15}$$

$$u(t) = \frac{\hat{A}}{P_c}w(t) - \frac{S\hat{A}d(t)}{P_cT_{con}(q^{-1})} \tag{16}$$

$$d(t) = x(t)/\hat{A} \tag{17}$$

Note that the $T_{con}(q^{-1})$ polynomial only affects the disturbance rejection properties of the system (this is the main reason why it is refered to as the observer polynomial). Clearly if $d(t)$ (the combined disturbances reflected at the output) has high frequency components they will be attenuated through the $S/T_{con}(q^{-1})$ filter, if $S/T_{con}(q^{-1})$ has low-pass characteristics. In cases where

$$d(t) = \frac{C}{\hat{A}\Delta}\xi(t)$$

the 'optimal' value of $T_{con} = C$ which is the usual choice in the literature. More often however the choice is simply dictated by the requirements on the bandwidth of the closed loop.

The following example demonstrates the need for $T_{con}(q^{-1})$ even in simple control problems. Consider the plant:

$$G(q^{-1}) = \frac{0.1q^{-1} + 0.2q^{-2}}{1 - 0.9q^{-1}}$$

This gives the controller parameters (for a GPC design with $N_1 = 1, N_2 = 10, NU = 1$) as:

$$\begin{aligned} R(q^{-1}) &= 1.3996 - 0.3929q^{-1} - 1.0066q^{-2} \\ S(q^{-1}) &= 5.5299 - 4.5299q^{-1} \end{aligned}$$

for the case where no T_{con} is used. Note that $S(1) = 1$ whereas $S(-1) = 10.0598$: a ten fold increase in gain! A similar controller with $T_{con} = 1 - 0.8q^{-1}$ then gives:

$$\begin{aligned} R(q^{-1}) &= 1.3996 - 1.11q^{-1} - 0.2896q^{-2} \\ S(q^{-1}) &= 1.5033 - 1.3033q^{-1} \end{aligned}$$

$S(-1)/T_{con}(-1) = 1.5592$ in this case which is considerably smaller than the case without the T_{con} polynomial!

3.2 $T_{con}(q^{-1})$ and plant model mismatch

The GPC calculations are based on $(q^{-1}\hat{B}/\hat{A})$ which is in general different from G_0 the *true* plant dynamics. This leads to the question of how does $T_{con}(q^{-1})$ affect the response of the system when there is plant model mismatch?

A cursory analysis of the situation gives:

$$y(t) = q^{-1}\frac{\hat{B}}{P_c}\frac{1 + \frac{\hat{A}}{P_c}\tilde{G}}{1 + \frac{\hat{A}\tilde{G}S}{P_cT_{con}}}w(t) + \frac{R\hat{A}\Delta}{P_cT_{con}}\frac{1}{1 + \frac{\hat{A}\tilde{G}S}{P_cT_{con}}}d(t) \tag{18}$$

The loop remains stable if the transfer-function $\hat{A}\tilde{G}S/P_cT_{con}$ satisfies the Nyquist criterion. In cases where there are nonlinearities present the appropriate version of the small gain theorem has to be applied instead. From eqn.18 the nominal response is preserved until the cut-off point of $|\frac{S}{T_{con}}|$. In order to ensure robust stability $|\frac{S}{T_{con}}|$ *must* have low-pass characteristics. This leads to the choice that at least $\deg T_{con} = \deg \hat{A}$ to ensure reasonable high frequency properties.

Consider the variation of $|S/T_{con}|$ for the model:

$$y(t) = \frac{0.1 + 0.2q^{-1}}{1 - 0.9q^{-1}}u(t-1)$$

with the standard settings for GPC, as in fig.4 for different values of $T_{con}(q^{-1})$. Note that for the case without T_{con} the gain at high frequencies is 10 times that of low frequencies (previous section). Moreover the transfer-function is only low-pass when the T_{con}'s time constant is larger than that of the plant. For a second order $T_{con} = (1 - \beta q^{-1})^2$ the magnitude of S/T_{con} is uniformly decreasing except when $T_{con} = (1 - 0.99q^{-1})^2$. This indicates that in so far as robustness to unstructured perturbations are concerned this choice of T_{con} is *worse* than the case with $T_{con} = (1 - 0.95q^{-1})^2$! (see fig.5.) This counter intuitive result arises from the fact that the zeros of S lie outside the unit circle for $\beta >$

0.95 (i.e. the controller and therefore the open-loop transfer function is nonminimum-phase). Bode's Integral Theorem for discrete-time plant (Mohtadi, 1988) predicts this loss of performance due to the presence of these zeros. For higher order T_{con} similar behaviour is observed but the loss of performance is sometimes due to the appearance of unstable poles (zeros of $R(q^{-1})$) of the controller which have similar effects in so far as the performance bounds are concerned. This has a rather disconcerting influence on the self-tuning design as even in simple cases decreasing the cut-off frequency of T_{con} can have ***deleterious*** effects!

From eqn.12 it is clear that wherever $T_{con} \approx \hat{A}\Delta$ at high frequencies then $E_j \approx 1$ and $F_j \approx 0$ (i.e. the loop gain is reduced as $|S/T_{con}| \approx 0$ at these frequencies). Clearly, this exact match is impossible in incremental control but if a CARMA model (positional) model is used instead, we have:

$$T_{con} = E_j^p \hat{A} + q^{-j} F_j^p \tag{19}$$

which demonstrates that $T_{con} = \hat{A}$ ensures zero loop gain and the controller is 'optimally' robust to all stable unmodelled dynamics! In fact, with a first order T_{con} polynomial the incremental control will at best be equivalent the positional control without the T_{con} polynomial in so far as the robustness of the controller is concerned.

The discussion above leads us to the following guideline.

Guideline 3 *To ensure good robustness properties choose* $T_{con} = (1 - \beta q^{-1})^n$ *such that the cut-off frequency* β *coincides with the dominant time-constant of* $\hat{A}$ *and* n *is the* $\deg \hat{A} + 1$.

3.3 GPC and IMC strategies

The shortcoming of the choice of T_{con} in GPC control lies mainly in the inability to choose the transfer function S/T_{con} freely. In order to allow ourselves this extra degree of freedom we first look at the closed-loop characteristic equation (eqn.14). Clearly, the general solution is given by:

$$\begin{aligned} R' &= R - q^{-1}\hat{B}X \\ S' &= S + \hat{A}\Delta X \end{aligned}$$

for all stable X. Choosing

$$\frac{S'}{T_{con}} = F$$

where F is a unity dc gain low-pass filter gives:

$$X = \frac{F T_{con} - S}{\hat{A}\Delta T_{con}} \tag{20}$$

If $\hat{A}$ is ***stable*** the condition for X is satisfied. This leads us to the well known Internal Model Control (IMC) of Garcia and Morari 1982. See fig.6 for the block diagram. To guarantee stability $\hat{A}$ must be stable but now F can be chosen freely and it can be shown that there always exist a F which will stablize the system. The short comings of the method lie mainly in the fact that the disturbance rejection is poor especially if $\hat{A}$ has very slow poles. It is for this reason the structure although useful for dealing with a special class of problems (open-loop stable and perhaps somewhat coarse sampling) is not in general beneficial for the high performance role and the standard structure is preferred.

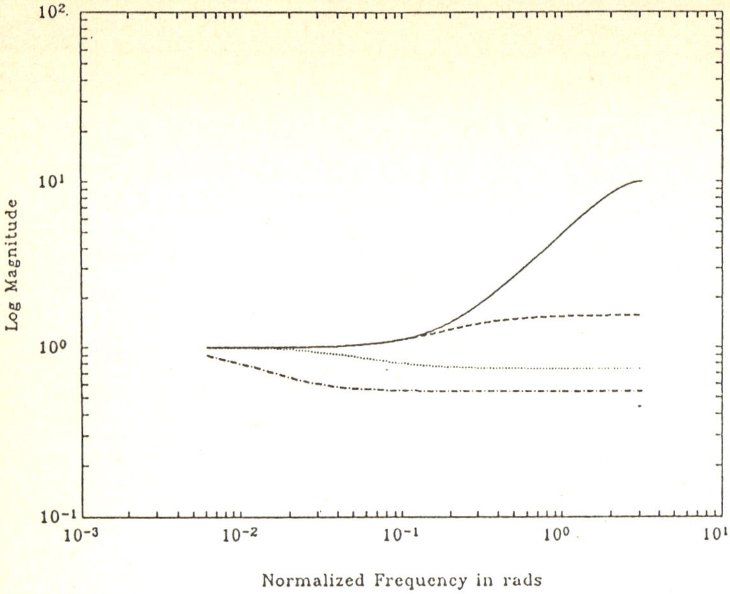

Figure 4: Variation of $|S/T_{con}|$ for a first order T_{con}: solid $T_{con} = 1$, dashed $T_{con} = 1 - 0.8q^{-1}$, dotted $T_{con} = 1 - 0.95q^{-1}$, dot-dash $T_{con} = 1 - 0.99q^{-1}$

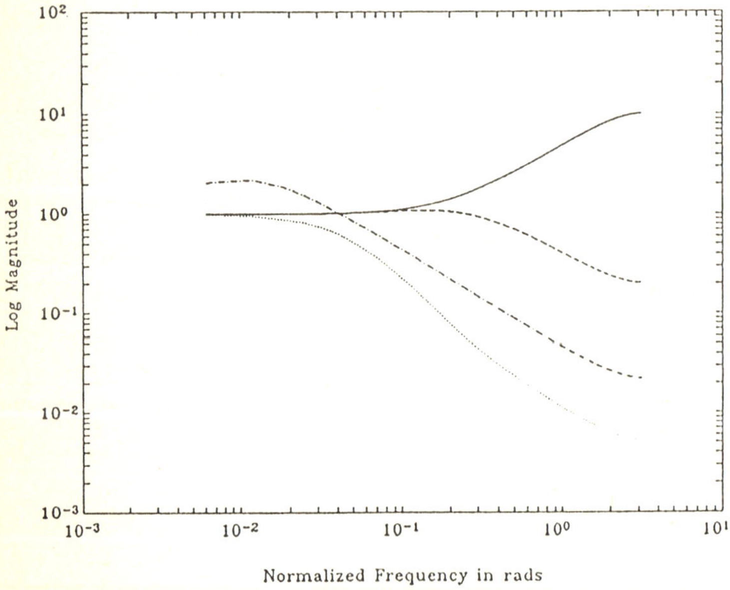

Figure 5: Variation of $|S/T_{con}|$ for a second order T_{con}: solid $T_{con} = 1$, dashed $T_{con} = (1 - 0.8q^{-1})^2$, dotted $T_{con} = (1 - 0.95q^{-1})^2$, dot-dash $T_{con} = (1 - 0.99q^{-1})^2$

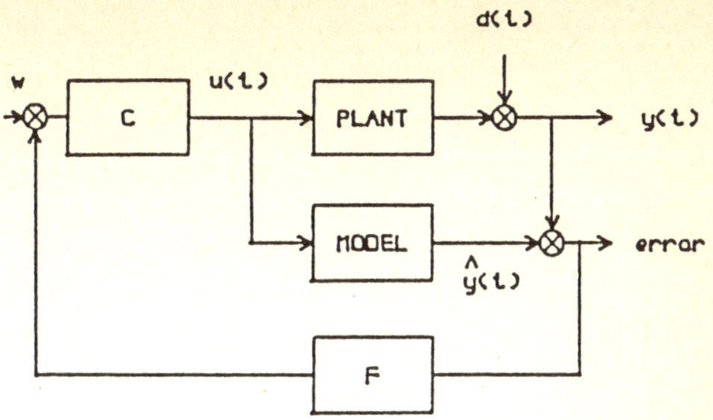

Figure 6: Internal Model Control

4 Some illustrative Simulations

Here we examine some of the ideas already discussed in the previous section as regards the choice of the prefilter and the matching of the plant and model frequency responses.

The plant considered throughout is:

$$G(s) = \frac{5}{(s+1)^2(s+5)}$$

sampled at 0.1 seconds using a second order model. For the first set of experiments the plant is identified in open loop using a white input sequence of 512 samples. Fig.7 shows the frequency and step responses of the plant and the model. Three distingushing features can be seen:

- the low frequency match is quite poor (especially at dc);
- the matching is concenterated at higher frequencies;
- the model exhibits resonance where the actual plant is over-damped!

The first two points have already been discussed. The final point is a frequent occurrence although it is rarely commented upon. A simple explanation of this phenomenon can be as follows. Consider a plant of the form:

$$G_p(s) = \frac{1}{(1+sT)^2(1+\alpha sT)}$$

which is modelled as:

$$G_m(s) = \frac{K_m\omega_n^2}{s^2 + 2\zeta\omega_n s + \omega_n^2}$$

If the gains of the two systems match at a frequency $\omega_0 = r/T$, from Bode's realization theory (Bode, 1945) we know that the rate of change of phase around ω_0 for both

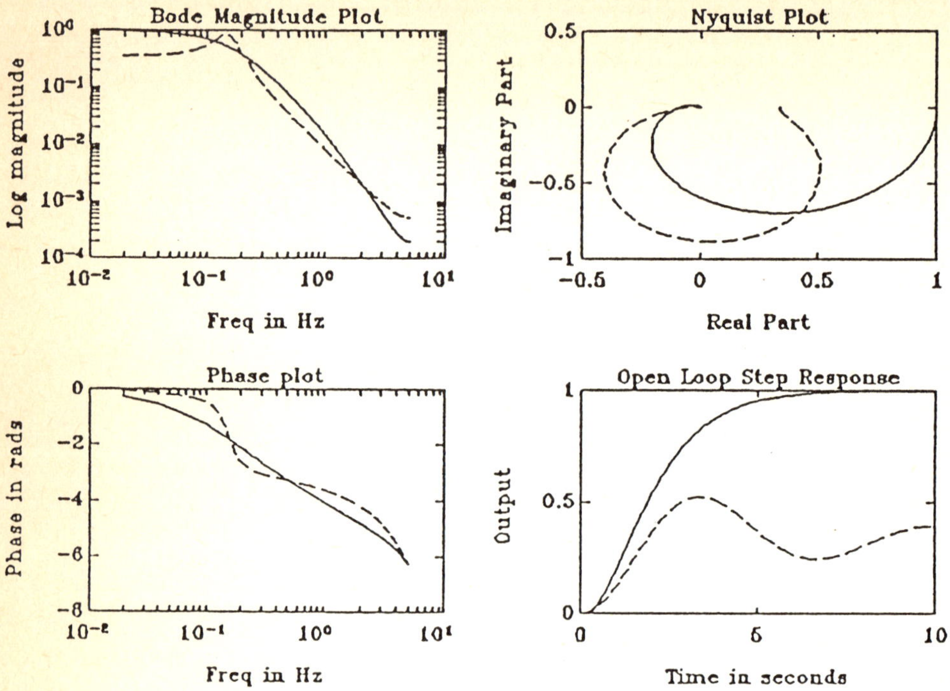

Figure 7: Model and plant frequency and time responses

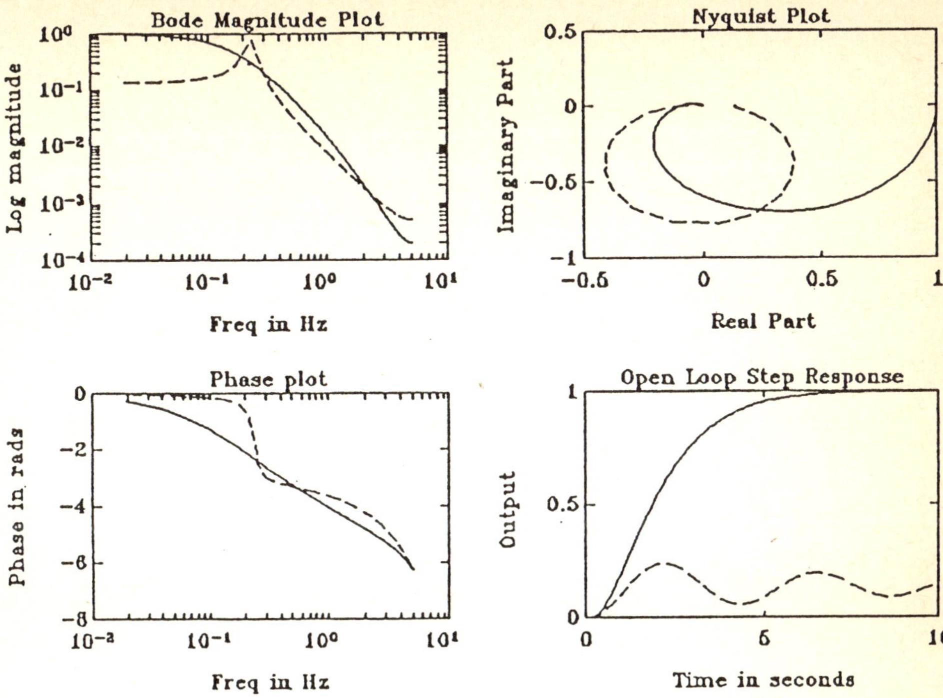

Figure 8: Model and plant frequency and time responses with 1st order T_{est}

systems must be the same, which in turn implies that:

$$\begin{aligned} \omega_n &= \frac{1}{T\sqrt{1+2\alpha}} \\ \zeta &= \frac{1+\frac{\alpha}{2}(1-r^2)}{\sqrt{1+2\alpha}} \end{aligned}$$

Note that $\zeta \leq 1$ for most values of α. Moreover this indicates that as r increases $\zeta \longrightarrow 0$. This is verified in fig.8 with $T_{est} = \frac{(1-0.8q^{-1})}{1-q^{-1}}$. Note the much sharper resonance peak. Other phenomena again attributable to matching plant's frequency responses at high frequencies are:

- obtaining *unstable* models when the real plant is stable (see the second counter example at the end of the section);
- estimating dc gains with *opposite signs* and frequently quite small in magnitude;

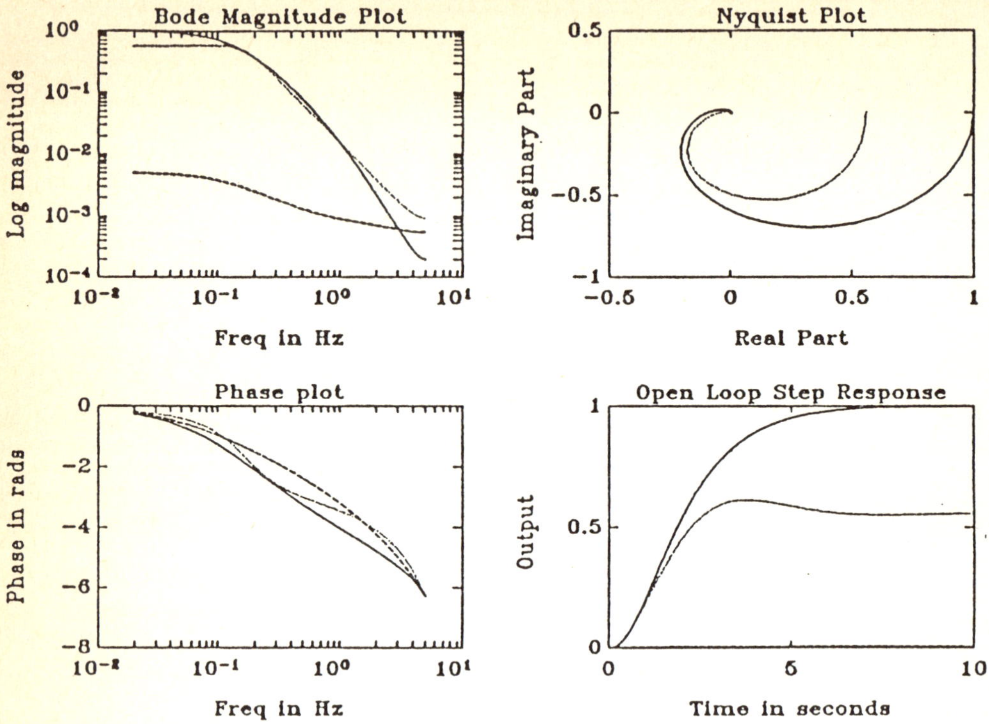

Figure 9: Model and plant frequency and time responses changing T_{est}

with first order models the opposite may occur where dc gain may be somewhat higher than that of the plant;

- finding nonminimum-phase (negative going transient) zeros where there are really either time-delays or lags.

All of the above observations can be explained (albeit qualitatively) via the Bode's realization theorem.

Recall that the predominant reason for using incremental estimation and control is presence of low frequency disturbances. Here we simulate one as $\frac{\xi(t)}{\Delta}$. Fig.9 shows the results using standard LS) with $T_{est} = \frac{1-0.9q^{-1}}{1-q^{-1}}$ and $T_{est} = \frac{(1-0.9q^{-1})^2}{1-q^{-1}}$. Note the marked improvements in each case.

Clearly the aim of estimation is to obtain models for control say using GPC, and as such the quality of the models can not be assessed without this in mind. Fig.10 shows the step response of the 'actual system' when a reduced order model of fig.9 was used.

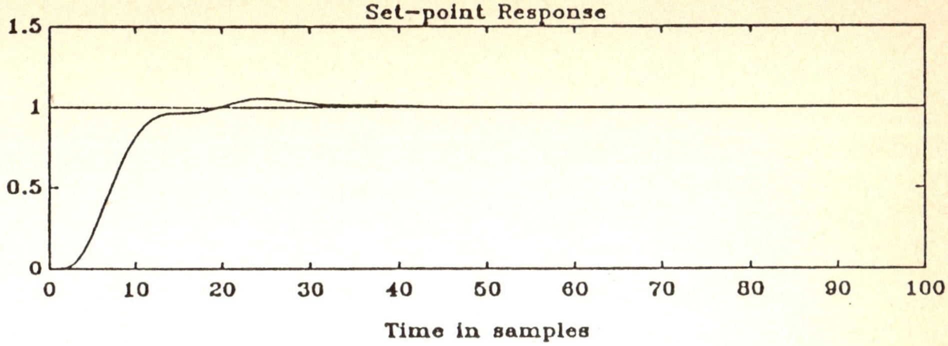

Figure 10: Fixed parameter GPC with T_{con}

$T_{con}(q^{-1})$ was then set to $(1-0.8q^{-1})$. Note the much improved step-response. Although the guidelines indicate the use of a higher order $T_{con}(q^{-1})$ it is seen that the first order observer is quite adequate.

4.1 Two counter-examples

Recall that GPC control uses the step-response of the process for control calculation. The noise free system was identified using $T_{est} = (1-0.9q^{-1})^2$. The results are shown in fig.11. Note the close match of the Nyquist and step-responses. The model was then used for control with $T_{con}(q^{-1}) = 1$ and the standard GPC. This results in an *unstable* closed-loop witha pole at -1.8022 ! The open-loop model has poles at $0.9246 \pm j0.0409$. Using a pole-placement controller with two closed-loop poles at $0.9246 \pm j0.0409$, the actual loop has an unstable pole at -1.3524. Although the model *appears* very reasonable neither of the controllers are stable! Both controllers work well with an observer polynomial of the type $T_{con}(q^{-1}) = (1-0.8q^{-1})$. See fig.12. This is an important issue and implies that

Guideline 4 *An observer polynomial has to be used even when the unmodelled dynamics appear to be negligible.*

Another interesting example is when the same model is identified with $T_{est} = \frac{1}{1-q^{-1}}$ and $T_{con} = 1$. The estimated model has complex conjugate unstable poles although the closed loop response is not drastically different from when the parameters are known exactly (see fig.13 and fig.14: a somewhat counter intuitive result!

5 Concluding remarks

The relevance of using appropriate prefiltering in parameter estimation is discussed. It is shown that in order to have an effective and reliable estimation and control scheme the prefiltering can not be ignored. The results are further verified by a series of simple

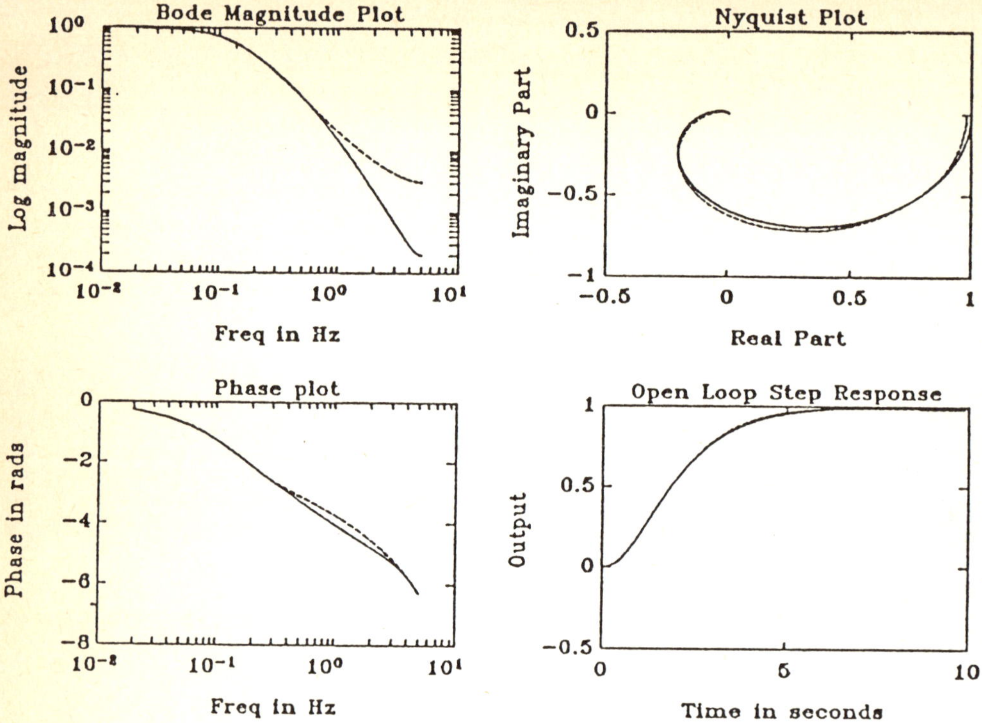

Figure 11: Model and plant frequency and time responses low-pass filtered data : solid line actual, dashed line model

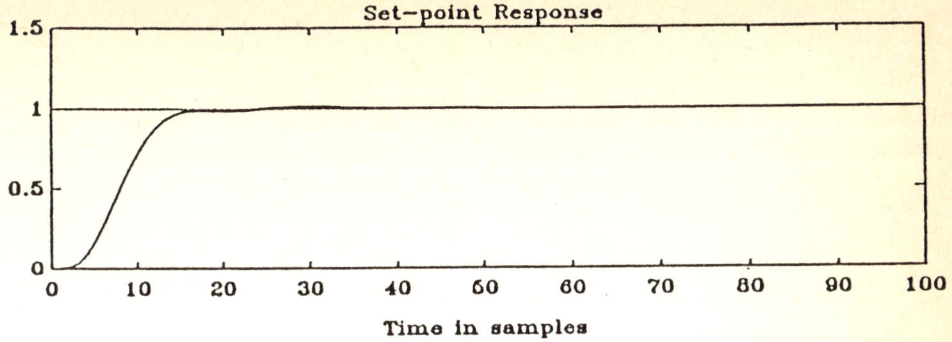

Figure 12: Step response of the GPC control with $T_{con}(q^{-1}) = 1 - 0.8q^{-1}$

simulated examples. It is also demonstrated that the incremental control which in the case of Lelic and Wellstead (1987) was ruled out for the control of a robot arm can be made robust through the use of the observer polynomial. There the authors attribute the poor control using incremental models to high sensor noise which incidentally is not visible in the plots of their experimental results. It is the contention of the present writer that the main reason for the poor performance of the controller in their case and in most applications of incremental estimation and control is the absence of the estimator band-pass *prefilter* and the *observer* polynomial in the control design.

6 Acknowledgements

The author was supprorted by the W.W. Spooner research fellowship at New College, Oxford, during the preparation of this paper, whose funding is gratefully acknowledged. Many thanks also goes to all my colleagues at Oxford University Self-tuning Control Group for many hours of discussion on the perennial T – polynomial!

7 References

Al-Assaf,Y. (1988). Self-tuning control: theory and applications. D.Phil Thesis, Oxford University.

Anderson B.D.O., R.R. Bitmead, C.R. Johnson, Jr., P.V. Kokotovic, R.L. Kosut, I.M.Y. Mareels, L. Praly, B.D. Riedle. (1986). Stability of adaptive systems. MIT Press. Cambridge, Massachusetts.

Åström K.J. (1983). Theory and applications of adaptive control - a survey., *Automatica*, 19, 471-486.

Åström K.J. and B. Wittenmark. (1984). Computer controlled systems. Prentice Hall, Englewood Cliffs, N.J.

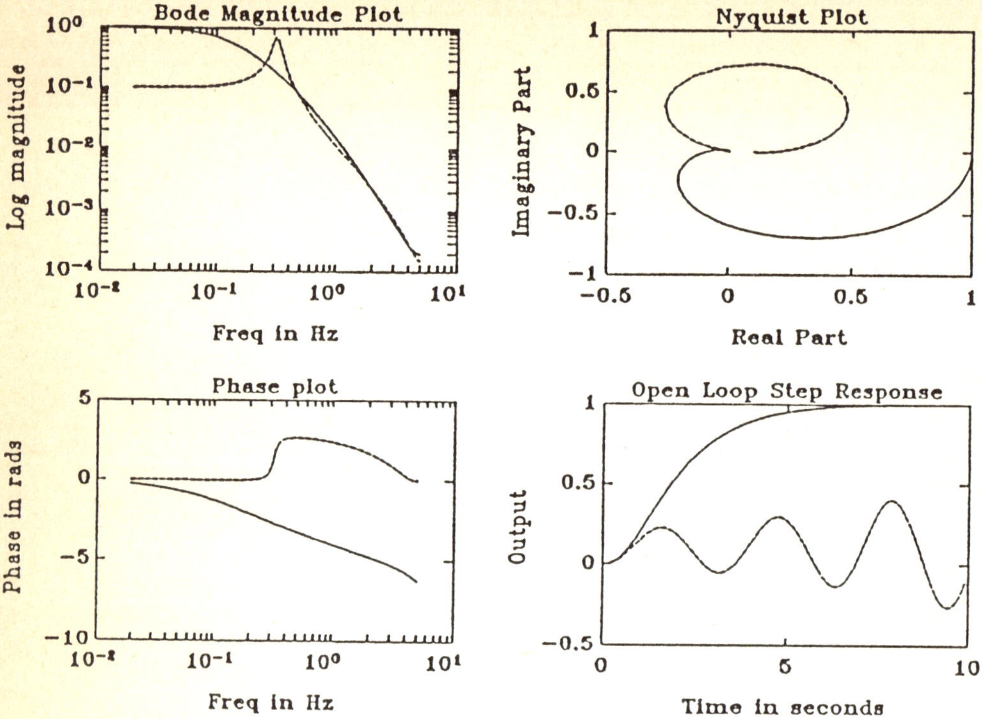

Figure 13: Model and plant frequency and time responses with differenced data

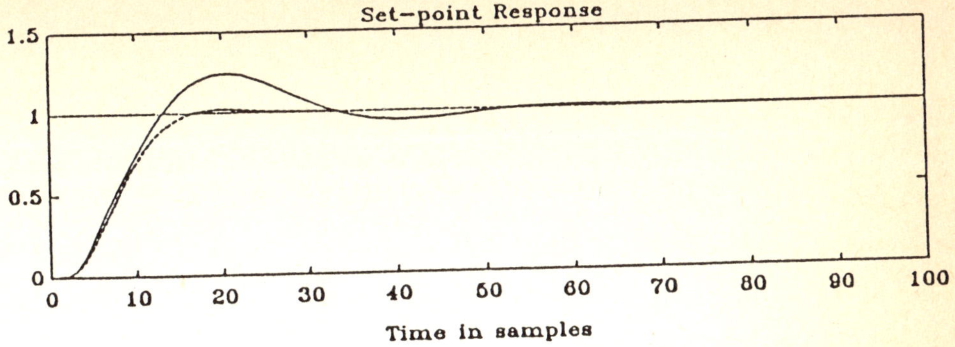

Figure 14: Step response of the GPC control with an unstable model

Bode, H. (1945). Network Analysis and Feedback Amplifier Design. Van Nostrand, Princeton, New Jersey.

Clarke,D.W., (1982). The application of self-tuning control. *Trans Inst. M. C.*, 5, No. 2, 59-69.

Clarke,D.W., Mohtadi,C. and Tuffs,P.S. (1987). Generalized predictive control. Parts 1 and 2. *Automatica*, 23, 137-160.

Cutler C.R., and Ramaker B.L. (1980). Dynamic Matrix Control - as computer control algorithm, *JACC*, San Francisco.

Doyle, J.C. and G. Stein (1979). Robustness with observers. *IEEE Trans AC*, AC-24

Garcia,C.E. and Morari,M. (1982). Internal Model Control. 1.A unifying review and some new results. *I & EC Process Design and Development*, 21, pp.308-323.

Goodwin G.C. and R.L. Payne (1977). Dynamic system identification: experiment design and data analysis, Academic Press, New York.

Lambert,E. (1987). The industrial application of long range prediction: D.Phil Thesis, Oxford University.

Lambert,M. (1987). Adaptive control of flexible systems: D.Phil Thesis, Oxford University.

Lelic,M.A. and Wellstead,P.E. (1987). A generalized pole-placement self-tuning controller. Part II - an application to robot manipulator control. *Int.J.Control*, 46, 569-602.

Lim, K.W. (1982), Robustness of self-tuning controllers: D. Phil thesis, Oxford University.

Ljung, L. (1987). System Identification:Theory for the User. Prentice Hall, Englewood Cliffs, N.J.

Mohtadi,C. (1986). Studies in advanced self-tuning algorithms: D.Phil thesis, Oxford University.

Mohtadi,C. (1988). Bode's Integral Theorem for discrete time systems, submitted to Proc. IEE.

Seborg D.E., S.L. Shah and T.F. Edgar (1986). Adaptive control strategies for process control: a survey, *AIChemE Journal*, 32, No.6, 881-913

Wahlberg, B. (1987). On the identification and approximation of linear systems: PhD thesis, Linkoping University.

Wahlberg, B. and L. Ljung (1986). Design variables for bias distribution in transfer-function estimation. *IEEE Trans. Automatic Control*, AC-31, 134-144

Wellstead,P.E., Prager,D. and Zanker,P. (1979). Pole assignment self-tuning regulator. *Proc.IEE*, 126, 781-787.

Wittenmark, B. and Åström , K.J. (1984). Practical issues in implementation of adaptive control, *Automatica*, 20, 595-605

8 Appendix

```
function [r,s,pc]=gpc(a,b,n1,n2,nu,lambda,t,pnz)
%
% A PC-MATLAB macro to calculate the transfer fuction form
% of the discrete time GPC Controller. The computations
% are done via the state-space as it
% is generally easier to do so. The function call is:
%
%        [r,s,pc]=gpc(a,b,n1,n2,nu,lambda,t,pnz)
%
% Inputs:
%                     -1          -1
% a and b are the A(q   ) and B(q   ) polynomials
% n1 is the initial costing horizon
% n2 is the  final  costing horizon
% nu is the control horizon
% lambda is the control costing parameter on what's left
% t is the perennial observer polynomial !
% pnz is the P of Py auxiliary output
%
% Outputs:
%                                             -1         -1
% r and s are the coefficients of  w(t) - S(q   )y(t) = R(q   )u(t)
% note that r contains the controller integrator as well !
% pc is the closed-loop pole polynomial
%

% set up the matrices for the Riccati equation update
%
if lambda==0, lambda=eps;end
```

```
adelta = conv(a,[1 -1]);
l=length(b);bpol = b(2:l);
np = max(size(pnz)); pnz=pnz/sum(pnz);
pt = conv(pnz,t); npt = max(size(pt));
na = max(size(adelta)); if na < npt+1,..
adelta = [adelta zeros(1,npt-na+1)];end;
na = max(size(adelta));
nb = max(size(bpol)); if nb < na, bpol=[bpol zeros(1,na-nb)];end;
nb = max(size(bpol));
nt = max(size(t)); t0=t(1); t=t/t0; b0=bpol(1);
if nb==nt, bpol = bpol- b0*t;
  elseif nb > nt, bpol = bpol - b0*[t, zeros(1,nb-nt)];
  else
  bpol = [bpol, zeros(1,nt-nb)] - b0*t;
end;
nb = max(size(bpol));
amat   = [-adelta(2:na), bpol(2:nb);
            eye(na-2),zeros(na-2,1),zeros(na-2,nb-1);
          zeros(1,na+nb-2)];
  if nt >=2, amat(na,na:na+nt-2) = -t(2:nt); end;
  if nb > 2,  amat = ..
 [amat; zeros(nb-2,na-1),eye(nb-2),zeros(nb-2,1)];end;
bmat   = b0;
  if na > 2, bmat = [bmat, zeros(1,na-2)];end;
bmat = [bmat, 1];
  if nb > 2, bmat = [bmat, zeros(1,nb-2)];end;
bmat = bmat';
cmat   = [pt, zeros(1,na+nb-npt-2)];
%
% now for the control calculation bit
%
p = cmat'*cmat;
for i = n2:-1:1;
  pstar = p;
    if nu >= i,
    pstar = p - p*bmat*inv(lambda + bmat'*p*bmat)*bmat'*p;
    end;
  k = inv(lambda + bmat'*p*bmat)*bmat'*p*amat;
  p     = amat'*pstar*amat;
    if i > n1, p = p + cmat'*cmat; end;
end;
r0=sum(t)/sum(k(1:na-1));
k = k*r0;
%
% closed-loop poles
%
if nt==nb, r = r0*t + [0, k(na:na+nb-2)];
  elseif nt > nb, r = r0*t + [0, k(na:na+nb-2), zeros(1,nt-nb)];
```

```
  else;
   r = [r0*t , zeros(1,nb-nt)] + [0 k(na:na+nb-2)];
end;
s=k(1:na-1);
pc1 = conv(r,adelta) ;pc2= conv(b,s);
npc1=max(size(pc1)); npc2=max(size(pc2));
if npc1 > npc2, pc2=[pc2 zeros(1,npc1-npc2)];end;
if npc2 > npc1, pc1=[pc1 zeros(1,npc2-npc1)];end;
pc=pc1+pc2; root=abs(roots(pc)); n=length(root);
 for i=1:n,
  if root(i) > 1,
     disp('closed-loop unstable');
     roots(pc),
     return;
  end;
 end;
nr=max(size(r)); nr1=nr;
for i=nr1:-1:1;
 if abs(r(i)) < sqrt(eps), nr=nr-1; end;
end;
r=r(1:nr);
ns=max(size(s)); ns1=ns;
for i=ns1:-1:1;
  if abs(s(i)) < sqrt(eps), ns=ns-1; end;
end;
s=s(1:ns); r=conv(r,[1 -1]); if (ns==0), s=0; end;
return
```

EXPERIMENTAL EVALUATION OF ADAPTIVE CONTROL IN THE PRESENCE OF DISTURBANCES AND MODEL-PLANT MISMATCH

A.R. McIntosh, S.L. Shah and D.G. Fisher
Department of Chemical Engineering, University of Alberta
Edmonton, Alberta, Canada T6G 2G6

Abstract

For successful application of adaptive controllers in a chemical process environment, consideration must be given to the types of disturbances and modelling errors likely to be present. For effective model parameter estimation, a user-specified polynomial may be used to filter the regressor and thereby focus attention on the frequency range of interest. The use of this filter polynomial is demonstrated to allow the identification of suitable models which match the true process at low frequencies, in the presence of nonzero mean disturbances, unmodeled dynamics, and measurement noise. A related controller design polynomial may be introduced into both the Generalized Predictive and Pole Placement controllers in order to "tailor" the rejection of disturbances and improve robustness to modelling errors. In the ideal case where the process model is exact, this polynomial has no influence on the setpoint tracking properties of the closed-loop system. Proper selection of the estimator filter polynomial and controller design polynomial are shown to be necessary for stable control of a pilot plant stirred tank heater.

1 INTRODUCTION

An automatic controller in a typical industrial plant must be able to handle a wide range of unmeasurable disturbances, measurement noise, interactions from other control loops, actuator limitations, etc. If a model-based control strategy is employed, the controller must also be robust in the presence of model-plant mismatch. In applications where there are time-variations in the process (ie catalyst decay, heat exchanger fouling) or frequent changes in operating conditions over nonlinear regions, adaptive controllers which adjust their parameters to counter these changes have the potential for improved performance. Time delays are frequently encountered in chemical process control as a result of transportation of products or the time required to sample and analyze a stream. Two alternative strategies for model-based adaptive control of industrial plants of this type appear promising: long-range prediction and pole placement.

Long-range predictive controllers consider a range of future predictions of the output with the expectation that extending the horizon beyond the longest expected time delay will yield detuned but robust behavior. Early algorithms such as DMC (Cutler and Ramaker, 1980) were based on determinisitic impulse or step response models. Long-range multistep predictive controllers based on autoregressive moving average (ARMA) models (DeKeyser and Van Cauwenberghe, 1982; Peterka, 1984; Mosca et al., 1984) have far fewer parameters in the model representation. These controllers may be made adaptive (self-tuning) if combined with an appropriate recursive parameter estimation routine. Generalized Predictive Control (Clarke et al., 1987a,b) is a recent generalization of the previous multistep predictive algorithms as well as being the natural long-range extension of the well-known Generalized Minimum Variance controller of Clarke and Gawthrop (1979).

The objective of Pole Placement controllers is to position the closed-loop poles at prespecified locations so as to yield the required transient response. No attempt is made to cancel open-loop process zeros. Compensation for time delays is achieved by overparameterizing the model numerator dynamics. A discrete-time self-tuning pole placement regulator was first described by Wellstead et al. (1979). Extensions of the concept to include both setpoint following and regulation were derived independently by Wellstead and Sanoff (1981) and Clarke (1982).

In section 2, an appropriate process model for industrial plants is presented and subsequently, in sections 3 and 4, used to derive Generalized Predictive and Pole Placement controllers with inherent integral action. The simulation example in section 5 and the experimental results presented in section 6 are intended to demonstrate the importance of using design polynomials to handle model-plant mismatch and provide desired rejection of disturbances. This is the main contribution of this paper.

2 MODELLING AND ESTIMATION

In most chemical process control applications, the first requirement of any fixed gain or self-tuning regulator is the ability to eliminate offset between the measured variable and the setpoint due to nonstationary load disturbances. If an appropriate internal model of the disturbances is employed the necessary integral action will be introduced into the controller in a natural way. The CARIMA (Controlled Autoregressive Integrated Moving Average) model has been suggested by several authors (Belanger, 1983; Tuffs and Clarke, 1985) for this purpose.

2.1 CARIMA Process Model

The CARIMA model for a locally linearized sampled-data system is given by

$$A(q^{-1})y(t) = B(q^{-1})u(t-1) + C(q^{-1})\xi(t)/\Delta \qquad (1)$$

where A, B and C are polynomials in the backward shift operator q^{-1}:

$$A(q^{-1}) = 1 + a_1 q^{-1} + \ldots + a_{na} q^{-na}$$
$$B(q^{-1}) = b_0 + b_1 q^{-1} + \ldots + b_{nb} q^{-nb}$$
$$C(q^{-1}) = 1 + c_1 q^{-1} + \ldots + c_{nc} q^{-nc}$$

$u(t)$ is the control input, $y(t)$ is the measured variable or output, $\xi(t)$ is an uncorrelated random sequence and Δ is the differencing operator $1-q^{-1}$. If the process has physical deadtime (other than the unit sampling delay) the leading elements of $B(q^{-1})$ are zero. By various assumptions relative to the form of $C(q^{-1})$ and $\xi(t)$ the disturbance model can be interpreted as representing either stochastic Browian motion or random steps occurring at random times.

2.2 Parameter Estimation

In principle any controller derived from the above model may be made adaptive by estimating the parameters on-line using a suitable recursive identification routine. In practice, however, it is found that identifying a disturbance model is difficult. Since industrial plants are subject to disturbances from different sources acting at different times the polynomial C is expected to vary with time. In addition, algorithms developed for the purpose of identifying C are subject to convergence problems or are computationally burdensome. For example, certain positive-real conditions must be satisfied for the Extended Least Squares (ELS) method to work (Ljung and Söderström, 1983) and it is found that the $\hat{C}$ parameters typically converge at a rate which is an order of magnitude slower than that for the coefficients of the $\hat{A}$ and $\hat{B}$ polynomials (Isermann, 1981).

An alternative approach involves the use of a design polynomial representing prior knowledge of the noise structure, $C_e(q^{-1})$. Multiplying equation (1) by Δ/C_e the incremental model used for parameter estimation becomes:

$$A(q^{-1})\Delta y^e(t) = B(q^{-1})\Delta u^e(t-1) + C(q^{-1})\xi(t)/C_e(q^{-1}) \qquad (2)$$

where the superscript e denotes a signal filtered by $1/C_e(q^{-1})$.

If C_e is equal to the true noise term C, an ordinary Recursive Least Squares (RLS) estimator using the filtered data will give unbiased $\hat{A}$ and $\hat{B}$ parameters. While the estimated parameters will be somewhat biased if this is not the case, selecting $1/C_e$ as a low pass filter will limit the frequency range over which the model is intended to coincide with the process.

A similar user-specified polynomial may be incorporated into the control law for the purpose of modifying the rejection of disturbances. Such a *controller* design polynomial will be denoted by $C_c(q^{-1})$ to distinquish it from the *estimator* filter polynomial $C_e(q^{-1})$ already introduced. In many situations it will be convienient for C_c and C_e to be selected as the same polynomial, although this need not be the case.

3 GENERALIZED PREDICTIVE CONTROL

Generalized Predictive Control (GPC) represents an amalgamation of a wide class of multistep predictive control techniques into a unified algorithm with a computationally simple approach (Clarke et al., 1987a). The use of long-range prediction over a finite horizon beyond the deadtime of the process enables compensation for variable and unknown time delays. The recursion of the Diophantine equation is a computationally efficient approach which allows for the inclusion of a $P(q^{-1})$ weighting polynomial to modify the predicted output trajectory. The choice of a control horizon, NU, (as in the DMC approach of Cutler and Ramaker, 1980) after which projected control increments are set to zero significantly reduces the computational effort and simplifies control of nonminimum phase plants.

3.1 Standard Form of Control Law

To be reasonably general in the following derivation of GPC, an auxiliary function of the output is considered:

$$\psi(t) = P(q^{-1})y(t)$$

where P is a polynomial in q^{-1} with P(1) set to unity to ensure offset-free control. The P weighting polynomial can be used to achieve model-following as described in (Clarke et al., 1987b).

3.1.1 Output Prediction

To derive a j-step ahead predictor of Py(t+j) based on equation (1) consider the Diophantine identity:

$$C_c P = E_j A\Delta + q^{-j}F_j \tag{3}$$

where E_j and F_j are uniquely defined polynomials. (the argument q^{-1} has been dropped for clarity)

Multiplying (1) by $E_j\Delta q^j$ and substituting for $E_jA\Delta$ from (3) yields:

$$C_c Py(t+j) = F_j y(t) + G_j\Delta u(t+j-1) + C_c E_j \xi(t+j)$$

or

$$Py(t+j) = F_j y^c(t) + G_j\Delta u^c(t+j-1) + E_j\xi(t+j) \tag{4}$$

where $G_j = E_jB$ and the superscript c designates a signal filtered by $1/C_c$.

E_j is of degree j-1 so the noise components are all in the future and the optimal predictor, given measured outputs up to time t and u(t+i) for $i \le j-1$, is clearly,

$$P\hat{y}(t+j|t) = F_j y^c(t) + G_j \Delta u^c(t+j-1) \qquad (5)$$

The GPC algorithm is based on the minimization of a multistep quadratic cost function so a range of values for j from the minimum output horizon, N_1, to the maximum output horizon, N_2, must be considered. For j>d, the physical deadtime of the process expressed as an integral multiple of the sample time (d≥0), the predicted outputs depend upon present and future control actions (which have yet to be determined) as well as past known control increments and past and present measured outputs.

Let f(t+j) be that component of $P\hat{y}(t+j|t)$ composed of signals known at time t (ie f is the open-loop prediction of Py assuming no control action is taken now or in the future). In order to determine f(t+j) and express $P\hat{y}(t+j|t)$ in terms of unfiltered present and future control actions consider the identity:

$$G_j = \tilde{G}_j C_c + q^{-j}\overline{G}_j \qquad (6)$$

where the polynomials $\tilde{G}_j$ and $\overline{G}_j$ are of degree j-1 and max(nb-1,nc-1), respectively.
Substituting into equation (5) gives:

$$P\hat{y}(t+j|t) = f(t+j) + \tilde{G}_j \Delta u(t+j-1) \qquad (7)$$

with,

$$f(t+j) = F_j y^c(t) + \overline{G}_j \Delta u^c(t-1) \qquad (8)$$

3.1.2 Multistep Cost Function Minimization

The objective of the predictive control law is to drive future plant outputs close to a setpoint sequence $[w(t+j);\ j=N_1,N_1+1,\dots,N_2]$. In the majority of applications w(t+j) will be equal to w(t), the current setpoint, but in batch process control or robotics applications, future 'programmed' setpoints may be known. GPC is based on minimization of the multistep, quadratic cost function:

$$J = E\left\{ \sum_{j=N_1}^{N_2} [Py(t+j) - w(t+j)]^2 + \sum_{j=1}^{NU} \lambda[\Delta u(t+j-1)]^2 \right\} \qquad (9)$$

where λ is the control weighting. The expectation in (9) is conditioned on data up to time t assuming no future measurements are available. Note that after the control horizon, NU, projected control increments are constrained to be zero:

$$\Delta u(t+j-1) = 0 \quad j>NU \tag{10}$$

Substituting equation (7) into the cost function and carrying out the minimization as in (Clarke et al., 1987a) yields the control law:

$$\begin{bmatrix} \Delta u(t) \\ \vdots \\ \Delta u(t+NU-1) \end{bmatrix} = \left[G_r^T G_r + \lambda I \right]^{-1} G_r^T \begin{bmatrix} w(t+N_1) - f(t+N_1) \\ \vdots \\ w(t+N_2) - f(t+N_2) \end{bmatrix} \tag{11}$$

where,

$$G_r = \begin{bmatrix} \tilde{g}_{N_1-1} & \cdots & \tilde{g}_0 & 0 \\ & & & \ddots \\ \vdots & & & \tilde{g}_0 \\ & & & \vdots \\ \tilde{g}_{N_2-1} & \cdots & & \tilde{g}_{N_2-NU} \end{bmatrix} \tag{12}$$

The control law is implemented using a receding-horizon approach; the current control increment $\Delta u(t)$ is applied to the process and the calculations repeated at the next sampling instant.

3.1.3 Recursion of the Diophantine Equation

For adaptive applications of GPC an efficient method of computing the polynomials G_j and F_j for each horizon j is required. The Diophantine recursion (Clarke et al., 1987a) is a technique for obtaining the polynomials E_{j+1} and F_{j+1} from E_j and F_j. The necessary equations are as follows:

$$e_{j+1,j} = f_{j,0} \tag{13a}$$

$$f_{j+1,i} = f_{j,i+1} - \bar{a}_{i+1} e_{j+1,j} \qquad i=0,\ldots,na \tag{13b}$$

$$E_{j+1} = E_j + e_{j+1,j} q^{-j} \tag{13c}$$

where $\overline{A}=A\Delta$. Note that the first subscript is the prediction horizon and also that $G_{j+1} = E_{j+1}B$.

To initialize the calculations,

$$E_1 = e_{1,0} = C_c(0)P(0) \qquad F_1 = f_{1,0} + f_{1,1}q^{-1} + \ldots = q(C_cP - E_1\overline{A}) \tag{13d}$$

For the general case where $C_c \neq 1$, the identity (6) may be solved for $\tilde{G}_j$ and $\overline{G}_j$ using a different procedure as outlined by Clarke et al. (1987b). A more straightforward method involves using the above Diophantine recursion equations twice. The routine is called with $C_c=1$ giving $\tilde{G}_j'$ (the prime is used to indicate a polynomial cooresponding to $C_c=1$) the coefficients of which are the same as $\tilde{G}_j$ as shown in Appendix A. The routine is then called a second time with the actual C_c polynomial providing G_j. Equation (6) may then be solved for $\overline{G}_j$ directly.

Note that the coefficients of $\tilde{G}_j$ are common for different horizons which is why the first subscript was dropped for the elements in the matrix G_r. For the case where P=1, these elements are step-response coefficients of the process.

3.2 Equivalent General Linear Form of Control Law

The standard form of the control law is suitable for the implementation of GPC in the majority of situations. However, for the purpose of carrying out a closed-loop analysis, the control law can be rearranged (McIntosh, 1988) into the general linear form:

$$T(q^{-1})\Delta u(t) = R(q^{-1})w(t) - S(q^{-1})y(t) \tag{14}$$

Assuming that future setpoints are equal to the current value, $\Delta u(t)$ is computed from equation (11) which may be written:

$$\Delta u(t) = [h_{N_1} \; \ldots \; h_{N_2}] \begin{bmatrix} w(t) - F_{N_1} y^c(t) - \overline{G}_{N_1} \Delta u^c(t-1) \\ \vdots \\ w(t) - F_{N_2} y^c(t) - \overline{G}_{N_2} \Delta u^c(t-1) \end{bmatrix} \tag{15}$$

where the vector **h** is the first row of $\left[G_r^T G_r + \lambda I \right]^{-1} G_r^T$.

This is equivalent to (14) with,

$$T = C_c + q^{-1}\left(\sum_{j=N_1}^{N_2} h_j \overline{G}_j\right), \qquad R = C_c\left(\sum_{j=N_1}^{N_2} h_j\right) \quad \text{and} \quad S = \sum_{j=N_1}^{N_2} h_j F_j \tag{16}$$

The polynomials T,R and S are of degree max(nc,nb), nc and na, respectively.

3.3 Closed Loop Transfer Function

Combining the control law expression (14) with the CARIMA model (1) we obtain the closed-loop transfer function:

$$y(t) = \frac{BRq^{-1}w(t) + TC\xi(t)}{TA\Delta + q^{-1}BS} \tag{17}$$

In Appendix A, it is shown that both the characteristic polynomial and the polynomial R contain C_c only as a direct factor and therefore C_c has no influence on servo control (if there are no process modelling errors). If in addition, the noise model is exact (ie C_c=C) the disturbance predictions are optimal and loads may be rejected in the best possible manner.

3.4 Selection of Overall Tuning Knobs

Generalized Predictive Control represents a flexible approach to the solution of many control problems due to the large number of "tuning knobs" incorporated into the algorithm. The overall closed-loop behavior (servo and regulatory) depends upon the choice of the following parameters:

1) minimum output horizon, N_1
2) maximum output horizon, N_2
3) control horizon, NU
4) control weighting, λ
5) output weighting, $P(q^{-1})$

For ease of application, strategies for assigning proper default values to reduce the number of *active* tuning knobs have been suggested (McIntosh, 1988). One alternative is to disable the weighting terms (ie P=1, λ=0), set N_1 and NU to unity and vary N_2 to achieve the desired speed of response. Large N_2 corresponds to a conservative mean-level controller (Clarke et al., 1987b) while smaller values yield a faster response (N_2 should be kept larger than the maximum expected deadtime of the process).

A second strategy is to increase the value of NU (eg to NU=na+1) giving the controller more degrees of freedom to generate an active response and use $P(q^{-1})$ weighting for necessary detuning. A root locus analysis indicates that closed-loop poles will tend to be 'close' to the roots of P for this "detuned model-following" setup (McIntosh, 1988).

4 POLE PLACEMENT CONTROL

A Pole Placement controller with inherent integral action may also be derived based on the CARIMA process model (Tuffs and Clarke, 1985). In the following development, a design polynomial, C_c, representing prior knowledge of the noise structure will be introduced into the control law as first suggested by Wellstead and Sanoff (1981).

Consider the control law:

$$G(q^{-1})\Delta u(t) = H(q^{-1})w(t) - F(q^{-1})y(t) \tag{18}$$

which when applied to the CARIMA model (1) yields the closed-loop equation:

$$y(t) = \frac{BHq^{-1}w(t) + GC\xi(t)}{GA\Delta + q^{-1}BF} \tag{19}$$

The servo closed-loop poles can be placed at the roots of a user-specified polynomial $P(q^{-1})$ by solving for the controller polynomials F and G from the Diophantine equation,

$$GA\Delta + q^{-1}BF = PC_c \tag{20}$$

and selecting H according to,

$$H = P(1)C_c/B(1) \tag{21}$$

in order to eliminate the noise structure polynomial from the setpoint response and provide zero steady state offset. The resulting closed-loop equation is then:

$$y(t) = \frac{BP(1)}{B(1)P} q^{-1} w(t) + \frac{GC}{PC_c} \xi(t) \tag{22}$$

Again, if the noise model is correct (C_c=C) then optimal disturbance rejection is possible. Note that the open-loop zeros are retained in the closed-loop so that there is no problem with control of nonminimum phase systems.

4.1 Solution of the Diophantine Equation

A well-known potential difficulty with the Pole Placement (PP) technique is the solution of equation (20) when there are near common factors in the estimated process model polynomials $\hat{A}$ and $\hat{B}$ (Clarke, 1982). The only complete solution to this problem is to isolate and cancel process zeros 'near' to process poles but this implies a large increase in computational complexity. A non-theoretical but simple solution is based on the assumption that the true system polynomials are relatively prime (which would normally be the case even if B is overparameterized to handle variable time delays as long as the order of A is selected conservatively small). In this case the controller polynomials F and G are not updated when the system of equations formed from (20) is ill-conditioned. The last numerically robust solution is held until the parameter estimates migrate from transient locations for which common factors in $\hat{A}$ and $\hat{B}$ exist.

5 SIMULATION EXAMPLE

The objective of this section is to demonstrate how the controller design polynomial C_c can be used to tailor the rejection of disturbances and improve robustness of both the GPC and PP controllers in the presence of unmodeled dynamics. The estimator filter polynomial C_e will be shown to be necessary for obtaining a suitable model when the true process order is underestimated.

The third order overdamped stable process,

$$G_P(s) = \frac{1}{(1+5s)(1+3s)(1+s)}$$

when sampled at an interval of 1 time unit has a nonminimum phase discrete time representation with open-loop zeros at -0.18, -2.59 and open-loop poles at 0.37, 0.72 and 0.82.

5.1 Disturbance Rejection

The following difference equation was used for the simulations:

$$Ay(t) = Bu(t-1) + d_u(t) + Ad_y(t) \tag{23}$$

where steps in $d_u(t)$ are similar to nonstationary disturbances acting at the process input and steps in $d_y(t)$ represent nonzero mean disturbances acting directly at the process output. In the following text, these will be referred to loosely as 'input' and 'output' type disturbances.

First, consider Generalized Predictive Control with exact, fixed parameters (ie $\hat{A}=A$ and $\hat{B}=B$). Assume, for convienience, that the process is at steady state with output equal to zero when a unit change in measurement occurs (due to a disturbance of unknown type). Fig. 1 shows the predicted output trajectory f(t+j), j=1,35 for three different settings of C_c (with P=1).

Fig. 2 shows the closed-loop response to setpoint changes and unmeasured disturbances where the overall GPC tuning knobs have been set for detuned model-following. Thirty sampling intervals after each upward-going setpoint change a step in $d_u(t)$ of magnitude 0.1 is applied and removed 40 samples later. This is followed by a step in $d_y(t)$ of magnitude 0.2 applied for 30 iterations.

With $C_c=1$, GPC assumes that disturbances will have an increasingly large effect on the output in the future (Fig. 1) which leads to excellent rejection of the 'input' type loads (Fig. 2, samples 60 to 120). However, these strong predictions are unsuitable for disturbances which have fully manifested themselves at the output and the controller over-reacts strongly in this case (Fig. 2, samples 140 to 180).

With $C_c=A$, the controller is actually making use of the model,

$$y(t) = (B/A)u(t-1) + \xi(t)/\Delta$$

with the assumption that the future effect of disturbances on the output will be the same as the current effect (Fig. 1). The step response equivalent of this model is used in the basic formulation of Dynamic Matrix Control (DMC) (Cutler and Ramaker, 1980). As expected, the rejection of 'input' type disturbances is very sluggish (Fig. 2, samples 250 to 330).

As a compromise, $1/C_c$ may be selected as a low pass filter. For example, with $C_c=1-.8q^{-1}$, the response to the loads (Fig. 2) is intermediate compared to that of the previous cases. Note that while C_c was employed to modify the regulatory response, with an exact process model servo control remained unaffected by this design polynomial. Although not shown, a Pole Placement simulation with $P=1-.8q^{-1}$ (ie a desired closed loop pole at 0.8) is almost indistinquishable from Fig. 2; the effect of C_c on the rejection of disturbances is essentially the same.

Now consider adaptive GPC, where the parameters of the A and B model polynomials are estimated using a RLS routine with variable forgetting factor maintaining a constant trace of the covariance matrix (Sripada and Fisher, 1987). (The on/off criteria discussed in their paper were not employed in this work.) With both the estimator design polynomial C_e and the controller design polynomial C_c set to unity, Fig. 3 indicates that there is excellent rejection of the $d_u(t)$ disturbances entering at samples 160, 200, 240 and 280. However the parameter estimates become very biased when the $d_y(t)$ loads are applied over the period of time from 400 to 520 sampling intervals. This results in a poor response for the setpoint change that follows at sample 560. If C_e and C_c are both selected as $1-.8q^{-1}$ the parameter estimates are slightly biased for both types of disturbances but control is, on average, better (Fig. 4).

5.2 Model-Plant Mismatch

Most real processes are of very high order but, for practical reasons, controllers must be designed using low order models. Parameter estimation must be conducted in the presence of unmodeled dynamics in such a way as to yield a model which closely matches the true process at low frequencies and is therefore useful for making long-range output predictions.

To exaggerate the potential difficulties with unmodeled dynamics, a first order model of the above third order process was identified and subsequently used for control. The identification was carried out in open-loop with a PRBS input (sequence length = 511, amplitude = 1) which excites all modes of the process. Converged or "tuned" parameter estimates were obtained for three different estimator filter polynomials. The unit step responses of the reduced order models are compared with that of the true process in Fig. 5; a reasonable model is identified only for the second order C_e polynomial.

Recall from equation (2) that the regressor is being filtered by Δ/C_e, the frequency response of which is shown in Fig. 6. With $C_e=1$, the differencing operator, Δ, amplifies high frequencies which explains why the steady state gain of the cooresponding identified first order model is grossly in error. On the other hand, for a second order C_e polynomial, Δ/C_e is a band-pass filter leading to a reduced order model with a better low frequency fit. Bandpass filtering of the input/output data has been recommended by several authors including (Sripada and Fisher, 1987; Middleton et al., 1988). The C_e filter is one way to implement this feature.

For successful application, Generalized Predictive or Pole Placement controllers must be insensitive to modelling errors which are inevitably present. Consider control of the third order process using the reduced order model (ROM) identified previously with $C_e=(1-.9q^{-1})^2$. Figures 7 and 8 indicate that without the controller design polynomial (ie $C_c=1$) the closed-loop system is unstable. A first or second order C_c polynomial is necessary to achieve an acceptable response. Note that in the ideal case where the controller receives the exact full order model (FOM) the use of C_c is not required.

6 EXPERIMENTAL APPLICATION

While simulations provide insight into the operation of a control algorithm under known conditions, the true test is application on a real process. In this section, experimental runs with the stirred tank heater, shown schematically in Fig. 9, are presented. The steam flowrate and exit temperature of the water (measured using thermocouple #1 located at the bottom of the tank) were selected as the manipulated (input) and controlled (output) variables, respectively, for the SISO computer controlled temperature loop. Step disturbances were introduced by altering the inlet water flowrate.

Figure 10 shows the open-loop response of the outlet temperature to step changes in the inlet water flowrate (t = 400, 600, 1400 and 1600s) and steam valve position (control signal). During the first half of the run an inlet water deflector was in place; under this circumstance the tank is well-mixed and the response is close to that of an ideal first order process. When the tank is operated without the deflector this is no longer the case; the response is apparently corrupted by a large amount of 'measurement noise'. All of the trials in this paper were conducted without the use of the deflector. The sample time was selected to be 8s - roughly one sixth of the dominant time constant.

6.1 Generalized Predictive Control

Figure 11 contains results for self-tuning Generalized Predictive Control of the stirred tank heater with $C_e=C_c=1$ and a first order model. During startup, four manual step changes in the control signal allowed initial parameter estimation using the variable forgetting factor / constant trace identification algorithm (Sripada and Fisher, 1987). The overall GPC tuning knobs were set according to the first strategy outlined in section 3.4. With $C_e=1$ the identification algorithm fails to come up with a model suitable for long-range predictive control. At 620s the sign of the estimated parameter $\hat{b}_0$ becomes negative and the controller saturates permanently.

When the regressor is band-pass filtered using $C_e=(1-.8q^{-1})^2$, the initial parameter estimates shown in Fig. 12 provide a reasonable low frequency model. However, with $C_c=1$, the controller is active throughout and eventually drives $\hat{b}_0$ to a negative value.

Detuning the (regulatory behavior of the) controller with $C_c=1-.8q^{-1}$ rectifies the problem as demonstrated in Fig. 13. The identified model gives adequate predictions and the controller does not overreact to modelling errors. Step water flowrate disturbances were introduced at times 750 and 900s.

6.2 Pole Placement Control

Results for PP control of the stirred tank heater are similar to those for GPC. With $C_e = C_c = 1$ as in Fig. 14, the closed-loop system is unstable due to the identification of a poor low frequency model (parameter trajectories not shown were similar to those in Fig. 11b). With the estimator and controller design polynomials set to $(1-.8q^{-1})^2$ and $1-.8q^{-1}$, respectively, very good closed-loop behavior is observed in Fig. 15 in spite of the large amount of measurement noise. (Again, parameter trajectories were similar to those of the analagous GPC run in Fig. 13b.)

7 CONCLUSIONS

This paper has demonstrated the importance of using controller and estimator design polynomials, representing prior knowledge of disturbances and model-plant mismatch, to improve the robustness and regulatory performance of adaptive controllers applied to typical chemical processes. A closed-loop transfer function for Generalized Predictive Control was derived to show that the controller design polynomial C_c does not affect the servo response if the process model is exact. This polynomial may be used to modify the prediction of disturbances in GPC and thereby tailor the rejection of disturbances. The inclusion of this design polynomial in a Pole Placement controller has a similar beneficial effect. The estimator design polynomial C_e should be selected to yield bandpass filtering of the incremental regressor to enable the identification of a good low frequency process model.

For the simulations and experimental runs presented, C_c and C_e were selected as first or second order polynomials with roots of 0.8 or 0.9. Since the sample time was selected to be approximately 1/5 to 1/10 of the dominant time constant of the process, the bandwidth of $1/C_c$ and $1/C_e$ corresponded roughly with that of the plant. Experience indicates that these choices are often adequate to ensure closed-loop stability for applications with significant modelling errors and measurement noise. The design polynomials may be adjusted after startup of the adaptive controller in order to fine tune the regulatory response. In particular, as $C_c \to 1$ disturbances are rejected more rapidly with stronger control action. One possibility is to utilize a supervisory system to monitor the closed-loop behavior of the plant and alter C_c and C_e in order to meet user performance specifications (McIntosh, 1988).

8 REFERENCES

Belanger, P.R. (1983). On type 1 systems and the Clarke-Gawthrop regulator. *Automatica*, **19**, (1), 91-94.

Clarke, D.W. (1982). Model following and pole-placement self-tuners. *Opt. Cont. Appl. & Meth.*, **3**, 323-335.

Clarke, D.W. and P.J. Gawthrop (1979). Self-tuning control. *Proc. IEE*, **126**, (6), 633-640.

Clarke, D.W., C. Mohtadi, and P.S. Tuffs (1987a). Generalized predictive control - part I. The basic algorithm. *Automatica*, **23**, (2), 137-148.

Clarke, D.W., C. Mohtadi, and P.S. Tuffs (1987b). Generalized predictive control - part II. Extensions and interpretations. *Automatica*, **23**, (2), 149-160.

Cutler, C.R. and B.L. Ramaker (1980). Dynamic matrix control - a computer control algorithm. *JACC*, San Fransisco.

De Keyser, R.M.C., and A.R. Van Cauwenberghe (1982). Typical application possibilities for self-tuning predictive control. *IFAC Symp. Ident. Syst. Param. Est.*, Washington, D.C., 975-980.

Isermann, R. (1981). *Digital Control Systems*. Springer-Verlag, Berlin.

Ljung, L. and T. Söderström (1983). *Theory and Practice of Recursive Identification*. MIT Press, Cambribge, Mass.

McIntosh, A.R. (1988). *Performance and Tuning of Adaptive Generalized Predictive Control*. M.Sc. Thesis, Dept. of Chem. Eng., Univ. of Alberta.

Middleton, R.H., G.C. Goodwin, D.J. Hill and D.Q. Mayne (1988). Design issues in adaptive control. *IEEE Trans. Automat. Contr.*, **33**, (1) 50-58.

Mosca, E., G. Zappa, and C. Manfredi (1984). Multistep horizon self-tuning controllers. The MUSMAR approach. *IFAC 9th World Congress*, Budapest, **2**, 935-939.

Peterka, V (1984). Predictor-based self-tuning control. *Automatica*, **20**, (1), 39-50.

Sripada, N.R., and D.G. Fisher (1987). Improved least squares identification. *Int. J. Control*, **46**, (6), 1889-1913.

Tuffs, P.S. and D.W. Clarke (1985). Self-tuning control of offset: A unified approach. *Proc. IEE*, **132D**, (3), 100-110.

Wellstead, P.E., D. Prager, and P. Zanker (1979). Pole assignment self-tuning regulator. *Proc. IEE*, **126**, (8), 781-787.

Wellstead, P.E. and S.P. Sanoff (1981). Extended self-tuning algorithm. *Int. J. Control*, **34**, (3), 433-455.

APPENDIX A: INDEPENDENCE OF SERVO CONTROL AND THE C_c POLYNOMIAL

Here it is shown that in the absence of process modelling errors (ie $\hat{A}=A$ and $\hat{B}=B$) the $C_c(q^{-1})$ polynomial incorporated into the GPC control law does not affect the response to setpoint changes.

Using primes to denote polynomials for the case $C_c=1$, the Diophantine equation (3) may be rewritten:

$$P = E_j'A\Delta + q^{-j}F_j' \tag{A1}$$

and equation (6) becomes,

$$G_j' = \tilde{G}_j' + q^{-j}\overline{G}_j' \tag{A2}$$

Multiplying (A1) by C_cB and substituting for $G_j' = E_j'B$ from (A2) gives

$$C_cBP = C_c[\ \tilde{G}_j'A\Delta + q^{-j}(\overline{G}_j'A\Delta + BF_j')\ ] \tag{A3}$$

For the general case ($C_c \neq 1$) multiplying equation (3) by B and substituting for $G_j=E_jB$ from (6) yields:

$$C_cBP = C_c\tilde{G}_jA\Delta + q^{-j}(\overline{G}_jA\Delta + BF_j) \tag{A4}$$

Equating the last two expressions and dividing by $C_cA\Delta$ we find that:

$$\tilde{G}_j + q^{-j}(\overline{G}_jA\Delta + BF_j)/C_cA\Delta = \tilde{G}_j' + q^{-j}(\overline{G}_j'A\Delta + BF_j')/A\Delta \tag{A5}$$

Therefore,

$$\tilde{G}_j = \tilde{G}_j' \tag{A6}$$

and the elements of the matrix G_r involved in the control calculation (11) are independent of C_c.

Continuing on, note also that

$$\overline{G}_jA\Delta + BF_j = C_c(\overline{G}_j'A\Delta + BF_j') \tag{A7}$$

Writing this equation for $j=N_1,\ldots,N_2$, multiplying by the set of real numbers h_j and summing yields:

$$\sum_{j=N_1}^{N_2} h_j\overline{G}_jA\Delta + \sum_{j=N_1}^{N_2} h_jBF_j = C_c\left(\sum_{j=N_1}^{N_2} h_j\overline{G}_j'A\Delta + \sum_{j=N_1}^{N_2} h_jBF_j'\right) \tag{A8}$$

The summation terms are related to the polynomials in the general linear form of the control law. From equation (16) with $C_c=1$:

$$\sum_{j=N_1}^{N_2} h_j\bar{G}_j' = q(T'-1) \quad \text{and} \quad \sum_{j=N_1}^{N_2} h_jF_j' = S' \tag{A9}$$

and with a general C_c:

$$\sum_{j=N_1}^{N_2} h_j\bar{G}_j = q(T-C_c) \quad \text{and} \quad \sum_{j=N_1}^{N_2} h_jF_j = S \tag{A10}$$

So that (A8) becomes, upon substitution:

$$q(T-C_c)A\Delta + BS = C_c[q(T'-1)A\Delta + BS'] \tag{A11}$$

or after rearrangement,

$$TA\Delta + q^{-1}BS = C_c[\, T'A\Delta + q^{-1}BS'] \tag{A12}$$

Thus, the characteristic polynomial for a general C_c is formed as the product of the characteristic polynomial for the case $C_c=1$ and the polynomial C_c. Observing also from (16) that $R = C_cR'$, the closed-loop transfer function given by equation (17) may be written as:

$$y(t) = \frac{C_cBR'q^{-1}w(t) + TC\xi(t)}{C_c[\, T'A\Delta + q^{-1}BS'\,]} \tag{A13}$$

which indicates that C_c appears as a common factor in both the numerator and denominator of the output/setpoint transfer function. Thus, in the ideal case, the servo properties are independent of C_c.

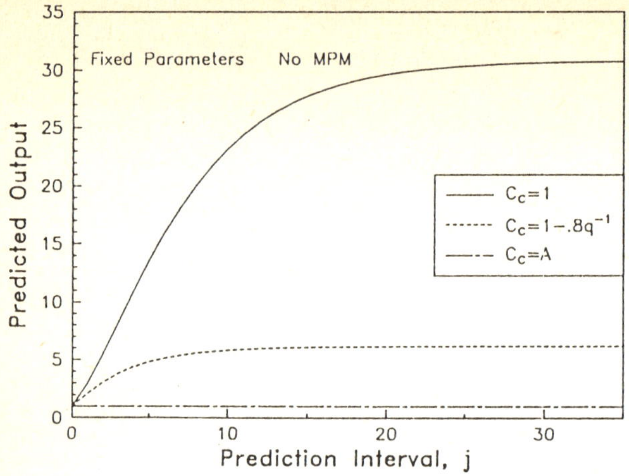

Figure 1 Influence of C_c Design Polynomial on Predicted Output Trajectory for a Unit Disturbance

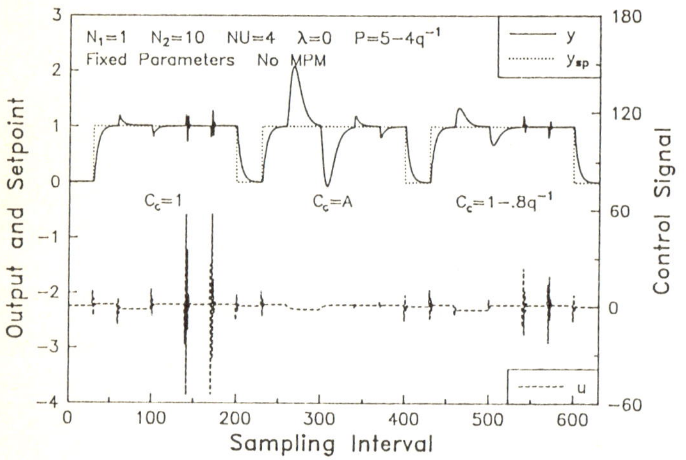

Figure 2 Effect of C_c Design Polynomial on Disturbance Rejection for GPC Control of Process C

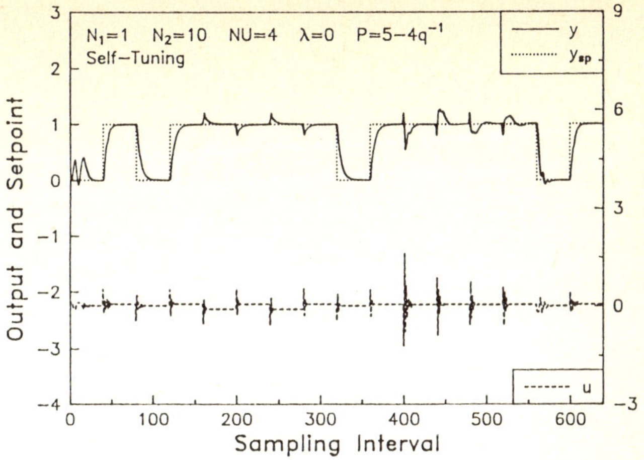

Figure 3 Generalized Predictive Control with $C_c=1$ in the Presence of Various Step Disturbances

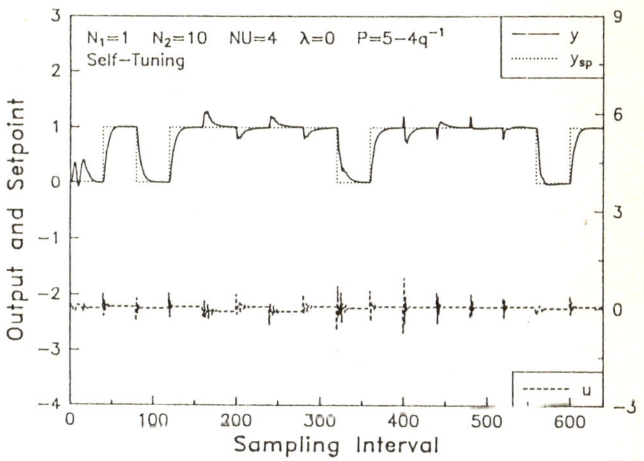

Figure 4 Generalized Predictive Control with $C_c=1-.8q^{-1}$ in the Presence of Various Step Disturbances

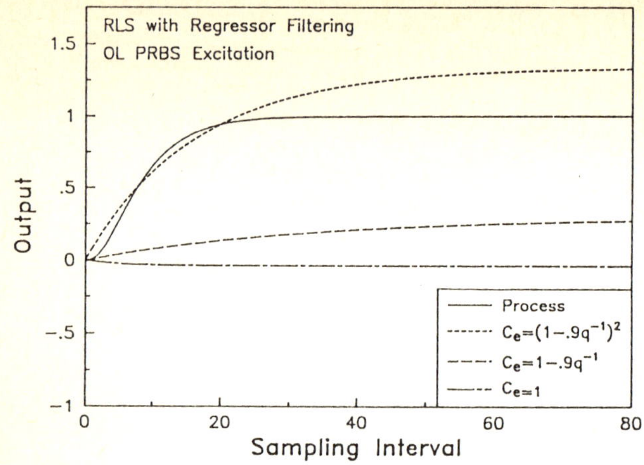

Figure 5 Unit Step Response of Process C and Identified First Order Models

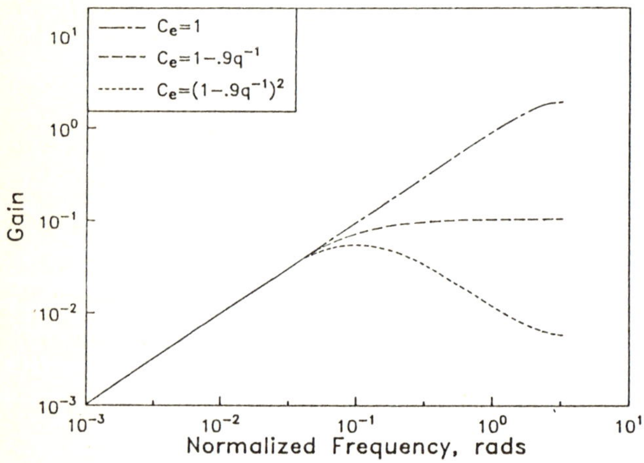

Figure 6 Frequency Response of Filter Δ/C_e

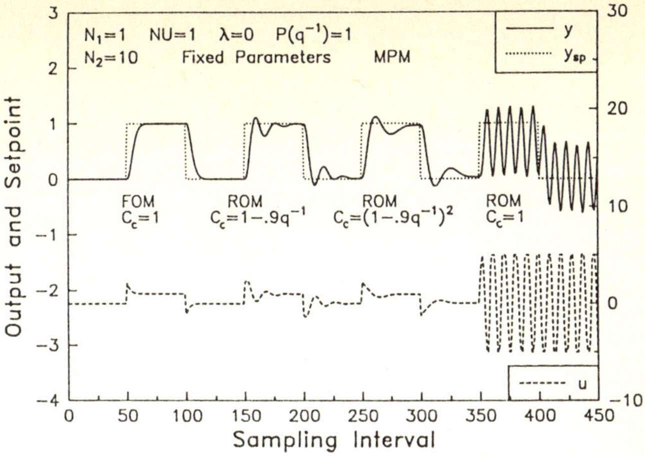

Figure 7 Output Horizon Control of Process C with a First Order Model

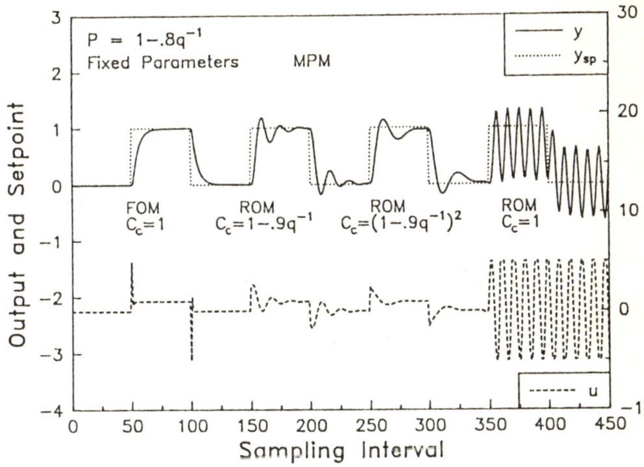

Figure 8 PP Control of Process C with a First Order Model

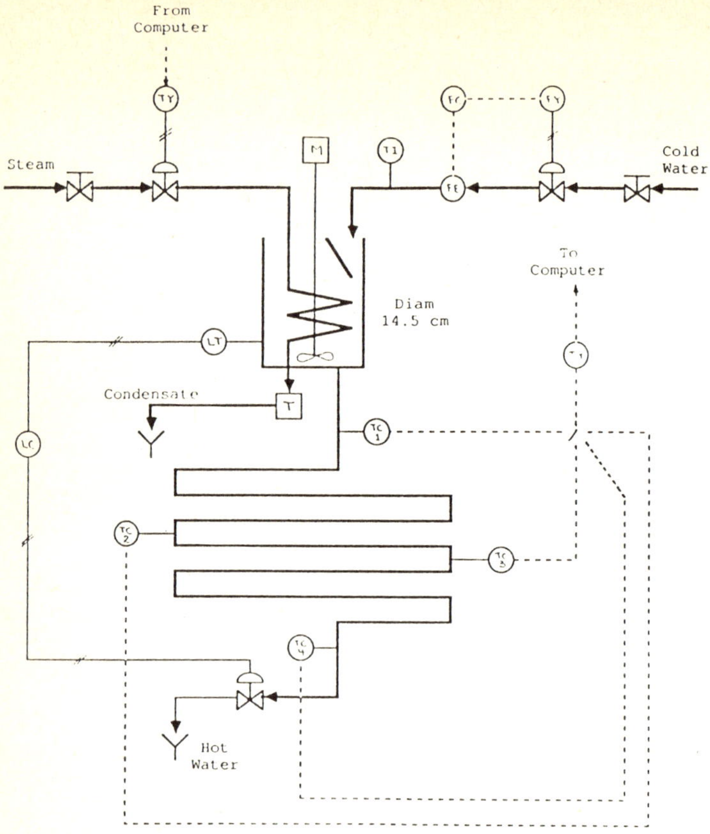

Figure 9 Schematic Diagram of Stirred Tank Heater

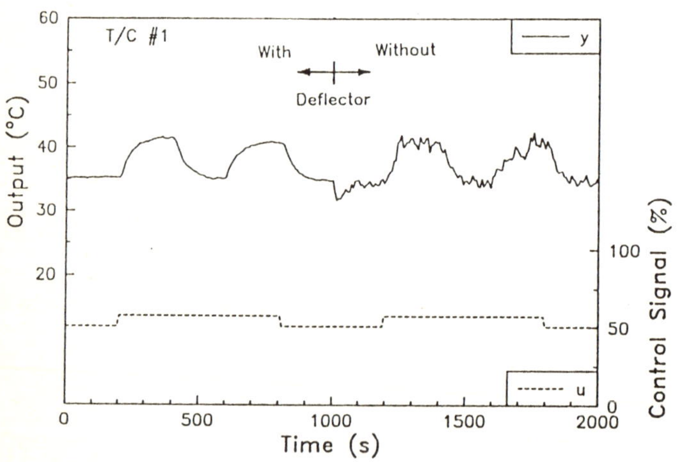

Figure 10 Open Loop Response of Stirred Tank Heater

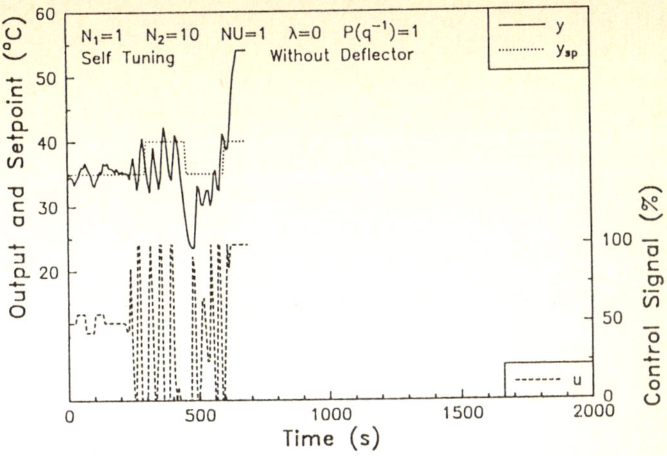

Figure 11a Control of the Stirred Tank Heater using GPC with $C_c=1$

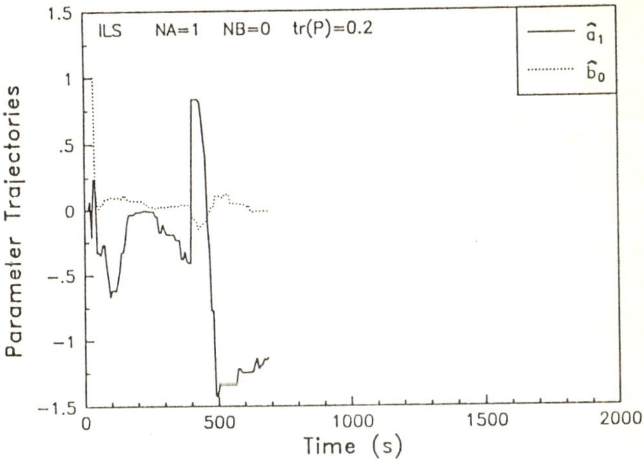

Figure 11b Parameter Estimation with $C_e=1$

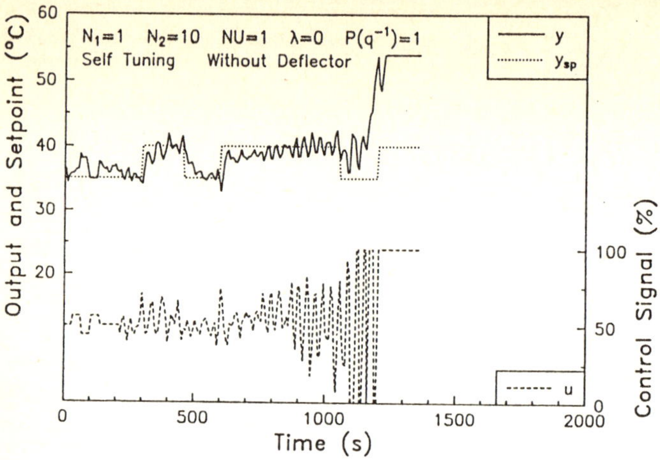

Figure 12a Control of the Stirred Tank Heater using GPC with $C_c=1$

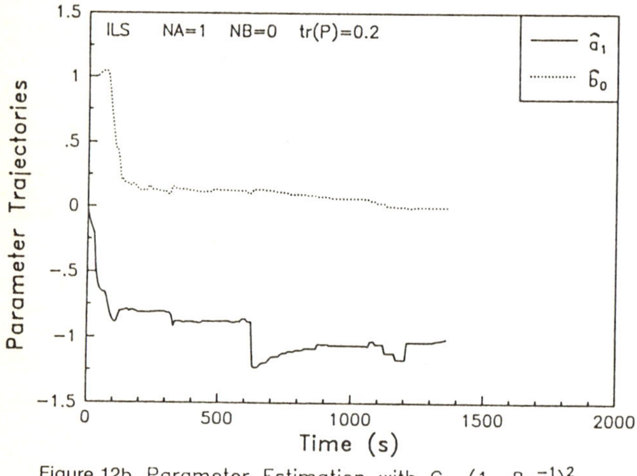

Figure 12b Parameter Estimation with $C_e=(1-.8q^{-1})^2$

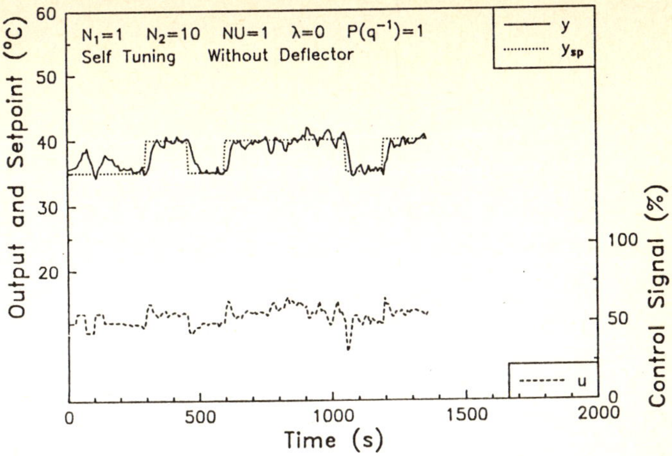

Figure 13a Control of the Stirred Tank Heater using GPC with $C_c=1-.8q^{-1}$

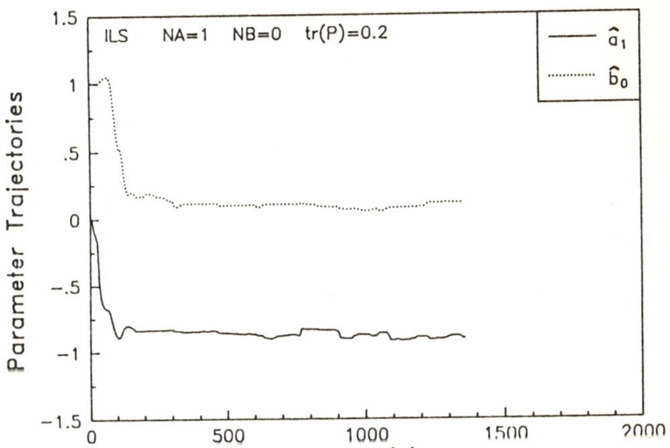

Figure 13b Parameter Estimation with $C_e=(1-.8q^{-1})^2$

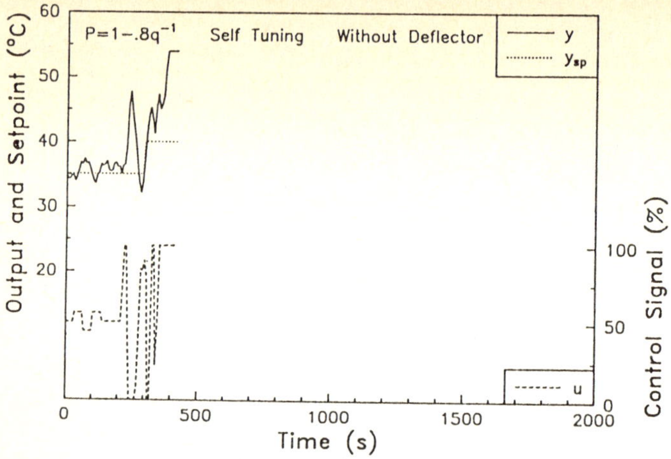

Figure 14 Control of the Stirred Tank Heater using Pole Placement with $C_c=1$

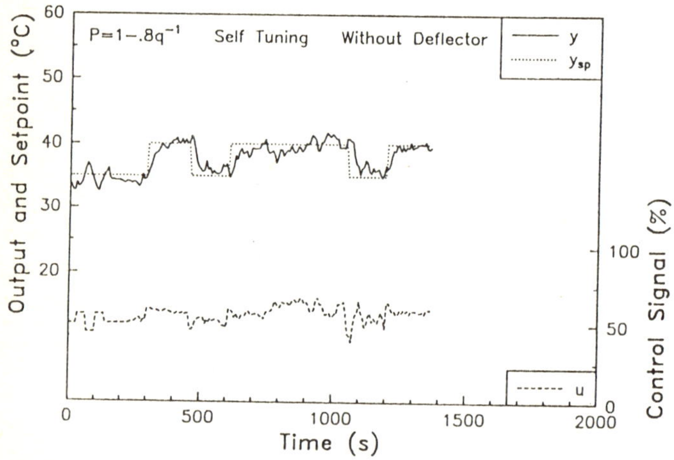

Figure 15 Control of the Stirred Tank Heater using Pole Placement with $C_c=1-.8q^{-1}$

Applicatons
of Adaptive Control

INDUSTRIAL APPLICATION OF AN ADAPTIVE ALGORITHM TO OVERHEAD COMPOSITION CONTROL

Peter J. Vermeer, Brian Roffel, and Patrick A. Chin
Advanced Process Control Group, Polysar Limited
Sarnia, Ontario, Canada N7T 7M2

ABSTRACT

An adaptive control strategy has been successfully applied to overhead composition control of a distillation column in a monomer and solvent recycle area of a rubber production facility. The process is time-varying with a significant time delay. The adaptive algorithm is a simple approach made up of a semi-adaptive PI controller and an adaptive Smith predictor. This paper presents results of the application and documents the experience of the authors in developing this technology.

INTRODUCTION

Research and development in the area of self-tuning and adaptive control technology has been extensive. This renewed interest in adaptive control occurred in the 1970's due to significant theoretical developments such as self-tuning control systems (e.g. Åström and Wittenmark, 1973; Clarke and Gawthrop, 1975) and the widespread availability of inexpensive microprocessor based hardware (Seborg et al. 1986).

This development has led to the commercialization of this technology in products such as the Leeds and Northrup Electromax V, Turnbull Control Systems TCS-6355 and the Foxboro EXACT controllers. Both the development and commercialization of this adaptive control technology has not led to widespread use of adaptive control in industry in general and in particular the process industry.

This paper documents the experience of the authors in developing this technology for application to a selected process. Steps in the development strategy were as follows:

1) Identify a suitable self-tuning control algorithm from the literature.
2) Identify a suitable test application and implement the adaptive control strategy in the field.
3) Design and implement the diagnostic shell of the algorithm which allows it to run unattended on an application.

One of the main purposes of this work was to gain experience in the implementation of adaptive control systems on industrial processes and evaluate the feasibility of adaptive control systems.

PROCESS DESCRIPTION

The test application identified for implementation of an adaptive control strategy was overhead composition control of a distillation column. The tower is the first column in a train of columns in a monomer and solvent recycle area of a rubber production facility. The feed consisting of solvent and monomer is in the vapour state. The control objective of the tower is to control the monomer (heavy key) composition in the overhead product. This objective is realized by manipulating the reflux flowrate.

Conventional control of this overhead loop is degraded for a number of reasons. The first reason is the significant time delay between a reflux flowrate change and its effect on the overhead composition. The time delay consists of the process delay and the cycle time of the gas chromatograph. The second factor is the changing operating conditions of the rubber polymerization reactor system. This includes varying the number of reactors in production at any given time according to production quotas. The third reason for the degradation of control is the changing operating conditions of the tower. Together, these factors result in a process with varying time delay, process gain and process time constant.

The overhead loop is characterized by a significant time delay and a low dynamic gain relating the reflux flow rate to the overhead composition. The effect of load disturbances on the overhead composition is large. Feed flowrate, feed composition, and an occasional change in bottoms temperature have a pronounced effect on the overhead composition. An open-loop step change in the reflux flowrate resulted in the following approximate transfer function model:

$$G(s) = \frac{Y(s)}{U(s)} = \frac{0.022e^{-50s}}{18s + 1} \tag{1}$$

where Y(s) is the deviation variable of the monomer concentration in the overhead stream and U(s) is the reflux flowrate deviation variable. This low order model shows a time delay of 50 minutes and a process time constant of 18 minutes. This transfer function model suggests a low order discrete model will be adequate to represent this system.

The recursive parameter estimation algorithm and the adaptive control algorithm were implemented on the existing control equipment which is comprised of a basic level TDC-2000 distributed control system and a Honeywell 4500 supervisory computer.

PARAMETER ESTIMATION ALGORITHM

Process Representation

The process is represented by a low order linear difference equation of the form:

$$A(z^{-1})y(t) = z^{-k}B(z^{-1})u(t) + C(z^{-1})\xi(t)/\Delta + d \qquad (2)$$

where u(t) and y(t) are the process input and output respectively. For the overhead loop of the distillation column, this corresponds to the reflux flowrate and overhead monomer composition respectively. $\xi(t)$ is a white noise input, k the process time delay including the unit sample delay (k 1), Δ is the backward difference operator $(1-z^{-1})$ and d is the process bias. The polynomials $A(z^{-1})$, $B(z^{-1})$, and $C(z^{-1})$ are low order polynomials in the backward-shift operator z^{-1} as shown below:

$$\begin{aligned} A(z^{-1}) &= 1 + a_1 z^{-1} + a_2 z^{-2} \\ B(z^{-1}) &= b_0 + b_1 z^{-1} \\ C(z^{-1}) &= 1 \end{aligned} \qquad (3)$$

The linear difference equation in Equation (2) postulates a disturbance model of the form:

$$N(t) = \frac{C(z^{-1})}{A(z^{-1})\,\Delta}\,\xi(t) \qquad (4)$$

Disturbances that are usually encountered in the process industries are often well characterized by auto-regressive-integrated-moving-average (ARIMA) time series models in which the disturbance N(t) is the output of Equation (4) with a white noise input (Harris et al., 1982). This model represents nonstationary disturbances with one pole at unity in the z-plane.

This time series model results in the following regressive model for process parameter estimation:

$$\begin{aligned} \Delta y(t) = {} & -a_1 \Delta y(t-1) - a_2 \Delta y(t-1) + b_0 \Delta u(t-3) \\ & + b_1 \Delta u(t-4) + \xi(t) \end{aligned} \qquad (5)$$

The measurement and data vector are zero mean. The data passed to the parameter estimation algorithm are incremental variables.

Recursive Least Squares Algorithm

The regression model of Equation (5) can be expressed in vector notation as:

$$\Delta y(t) = \phi^T(t-1)\theta(t-1) \tag{6}$$

$$\phi(t-1) = [-\Delta y(t-1), -\Delta y(t-2), \Delta u(t-3), \Delta u(t-4)]^T \tag{7}$$

$$\theta(t-1) = [\ a_1,\ a_2,\ b_0,\ b_1\ ]^T \tag{8}$$

The dynamic process parameters are estimated on-line recursively by a recursive least squares algorithm of the form:

$$\theta(t) = \theta(t-1) + K(t)[\Delta y(t) - \phi^T(t-1)\theta(t-1)] \tag{9}$$

$$K(t) = \frac{P(t-1)\phi(t-1)}{\lambda + \phi^T(t-1)P(t-1)\phi(t-1)} \tag{10}$$

$$P(t) = [I - K(t)\phi^T(t-1)]P(t-1)/\lambda \tag{11}$$

This algorithm was implemented by using the U-D factorization method of Bierman (1976) as shown in Åström and Wittenmark (1984).

The variable forgetting factor algorithm of Ydstie et al. (1985) was used to discount old input/output data. The algorithm is as follows:

$$\lambda(t) = [n(t) + (n(t)^2+4w(t))^{.5}]/2 \tag{12}$$

where $\lambda(t)$ is the variable forgetting factor and n(t) and w(t) are intermediate results defined as:

$$n(t) = 1 - w(t) - e(t)^2/\sigma_0 \tag{13}$$

$$w(t) = \phi^T(t-1)P(t-1)\phi(t-1) \tag{14}$$

where σ_0 is a tuning parameter related to the noise variance of the system and asympotic memory length of the estimation algorithm. e(t) is the a priori incremental prediction error.

ADAPTIVE CONTROL ALGORITHM

The control strategy chosen for this loop is an adaptive time delay compensator with a PI controller. This is a simple approach but is consistent with the objective of reducing the complexity of the adaptive control strategy as much as possible.

Time Delay Compensator

The presence of approximately 50 minutes of deadtime in the overhead loop makes the control of the overhead composition difficult without time delay compensation.

The time delay is compensated for by an adaptive Smith predictor (Smith, 1957) as shown in Figure 1 where:

$$G_p(z^{-1}) = \frac{z^{-k}B(z^{-1})}{A(z^{-1})} \qquad (15)$$

$$G_{pr}(z^{-1}) = \frac{z^{-1}B(z^{-1})}{A(z^{-1})} \qquad (16)$$

$$G_d(z^{-1}) = z^{-(k-1)} \qquad (17)$$

The model parameters are estimated by means of the recursive parameter estimation algorithm. The undelayed process output prediction from the adaptive Smith predictor is used in the controller output calculation.

Fig. 1. - Block diagram of Smith predictor

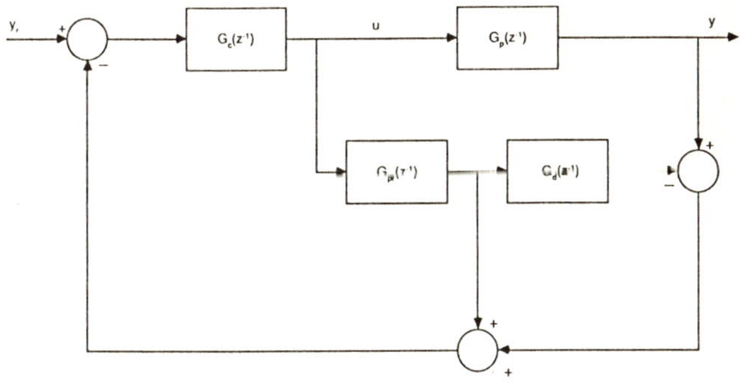

Control Algorithm

A simple PI control law was adopted for the test application of adaptive control. The form of the control algorithm was:

$$\Delta u(t) = K_c[-y(t) + y(t-1) + T_s\varepsilon(t)/T_i] \qquad (18)$$

where $\varepsilon(t) = y_{sp}(t) - y(t)$ and $y_{sp}(t)$ is the process output setpoint. The algorithm is a velocity form with the setpoint on the integral action term only, to avoid the so called "proportional kick" on setpoint changes.

The tuning of the controller parameter K_c and T_i again follows a simple approach. The controller gain is calculated from a rule of thumb as:

$$K_c K_p = \alpha \qquad (19)$$

where K_c is the controller gain, K_p the dynamic process gain, and α is a factor between 0 and 1. As the estimated process gain varies, the controller gain is adjusted appropriately. For this application α was equal to 0.35. The integral time constant was fixed at 30 minutes as this resulted in acceptable closed-loop behaviour.

IMPLEMENTATION ISSUES

A major factor in the success of an adaptive control scheme is the design of a diagnostic shell to ensure its proper operation. The following aspects were found to be useful in building this diagnostic shell.

Numerical Conditioning

Three of the steps taken to minimize this problem were discussed previously. The recursive estimation algorithm was mechanized with the RUD factorization algorithm; a variable forgetting method was adopted with covariance resetting. Another step taken was to scale the input/output data used in the input/output regressor vector. The reflux flowrate changes were 2 orders of magnitude larger than the overhead composition change. Therefore, all reflux flowrate changes were reduced by a factor of ten for parameter estimation. Finally, a low estimated model order was chosen (i.e. DegA=2, DegB=1) to avoid the problems associated with an overparameterized model, namely poor parameter convergence and parameter drifting.

Scheduling

The process measurement in this application was a discrete signal from a gas chromatograph with a cycle time of 20 minutes. This type of process variable requires special treatment to ensure proper execution and sequencing of the adaptive controller upon update of the analyzer. The approach adopted was to synchronize the self-tuner with the first analyser update and then execute at fixed intervals (T_s = 20 min.). If the analyzer update and self-tuner execution were out of phase by more than a pre-defined margin, the execution of the controller was brought into phase with the analyser update.

It is not recommended that the parameter estimation algorithm be continually active. It should be active only when new dynamic information can be obtained about the process, e.g. upon a setpoint change or upon addition of a perturbation signal. Both these criteria were used to initiate identification. The identification of estimation algorithm was initiated periodically with a PRBS contribution added to the reflux flowrate. Identification was stopped when there was no dynamic information about the process. The simple test used to determine this was to test whether a number of consecutive analyzer readings are within a specified deadband of the output setpoint. Other reasons for stopping identification were:

i) control strategy placed in manual mode
ii) the analyzer readings were invalid for an extended period of time
iii) saturation of the manipulated variable.

Some precaution was taken when a major unmeasurable disturbance impacted on the process. A predicted process output was derived based on the estimated dynamic process parameters, the history of the process input and output and a process bias estimate. If the predicted process output differed more than a specified error from the analyzer output, identification was discontinued. Experience has shown that for 80% of load disturbances the identification algorithm may be active.

Reliability of Input/Output Data

Because the process measurement is a discrete analyzer reading which is not always reliable, the following checks on the measurement were made before using it in the identification algorithm:

1) Out of range i.e. outside pre-specified limits.
2) Minimum rate of change (i.e. frozen valve). Even if the process does not change, consecutive analyzer readings will change within a tolerance. If more than a predefined number of readings fall within the tolerance, the analyser reading becomes questionable.
3) Maximum rate of change based on process dynamics and expected disturbance pattern. Reading is unreliable if the rate of change is beyond this limit.

If the analyser reading was unreliable, the last "good" reading was used in the I/O regressor. If the analyser remained "bad" for more than 40 minutes, the application was placed in manual mode.

Parameter Diagnostics

A number of diagnostics on the parameter estimates were made to determine if they were reasonable. The estimated process gain, poles, and zeroes were calculated and checked to see if they were within a minimum and maximum value. If they were not acceptable, the parameters were considered invalid and not used for controller gain calculation. Clamps on the allowable K_c were also imposed.

RESULTS AND DISCUSSION

The adaptive control strategy had been controlling the overhead composition of the distillation column for over a year. The strategy has operated under a variety of operating conditions: normal plant operation, the monomer recovery area in total recycle and the distillation column on total reflux.

Fig. 2. - Adaptive regulatory control of overhead composition to a major feed flowrate disturbance

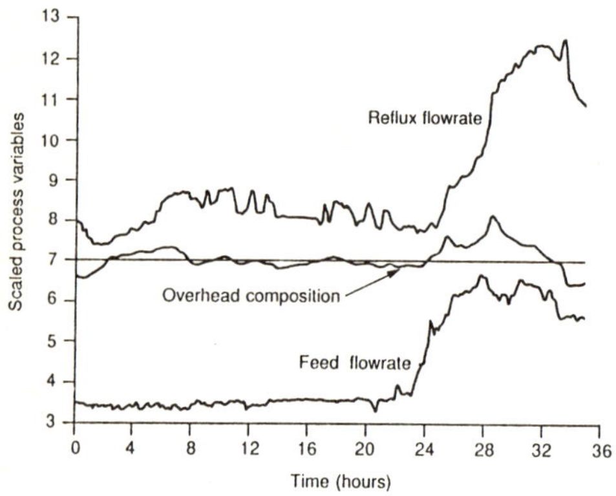

Figure 2 summarizes a regulatory control of the application over a 34 hour period. The major disturbance is a 90% increase in the feed flowrate to the tower. The regulatory control before the feed flowrate is good. After the feed flowrate change the overhead monomer concentration deviates from setpoint but remains within an acceptable deadband of the setpoint given the magnitude of the load disturbance. Note that the reflux flowrate is increased by 73%.

Figure 3 summarizes the servo control performance of the application over a 5 hour period. The reflux flowrate increases immediately after the setpoint change, but the overhead composition does not move until the process time delay has elapsed. The slow servo response is attributed to the form of the PI algorithm (i.e. control error is removed primarily by integral action). The slow response is also a result of feed flowrate and composition changes impacting on the tower after hour 3.

Fig. 3. - Adaptive servo control of overhead composition

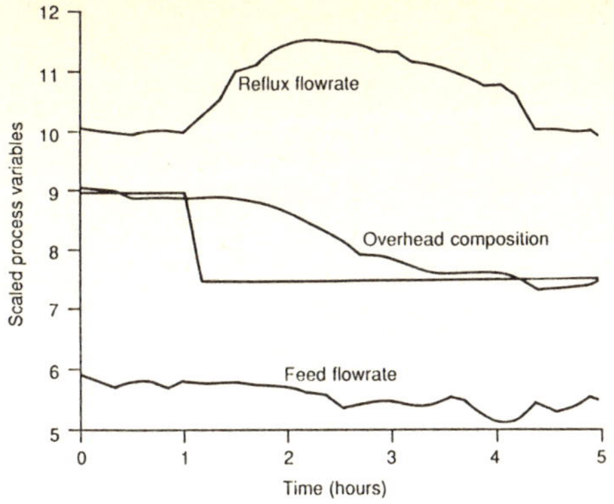

Figure 4 summarizes the combined regulatory and servo control performance of the application over an 8 hour period. The major disturbance during the first 2 hours was caused by a tower bottoms temperature control problem. A setpoint change was made in the second hour. This figure primarily illustrates the recovery of the adaptive application to a process upset and setpoint change.

Fig. 4. - Adaptive servo and regulatory control of overhead composition

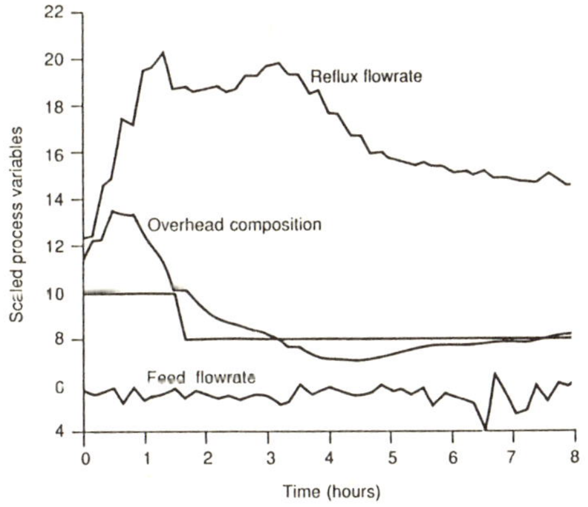

The operation of the recursive parameter estimation algorithm is summarized in Figures 5-7 for a 50 day period. Figure 5 shows the percentage of time the recursive estimation algorithm was actively identifying parameters each day over the 50 day period. The daily average of the estimated a_i parameters is shown in Figure 6. The variation of the trace of the covariance matrix is seen in Figure 7.

Fig. 5. - Service factor of parameter estimation algorithm

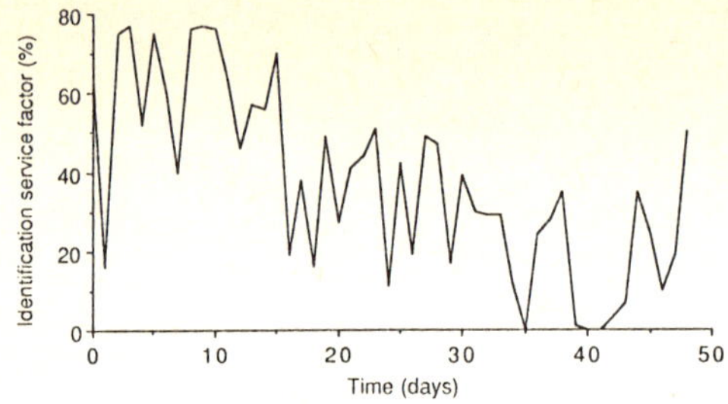

Fig. 6. - Trajectory of ai parameter estimates

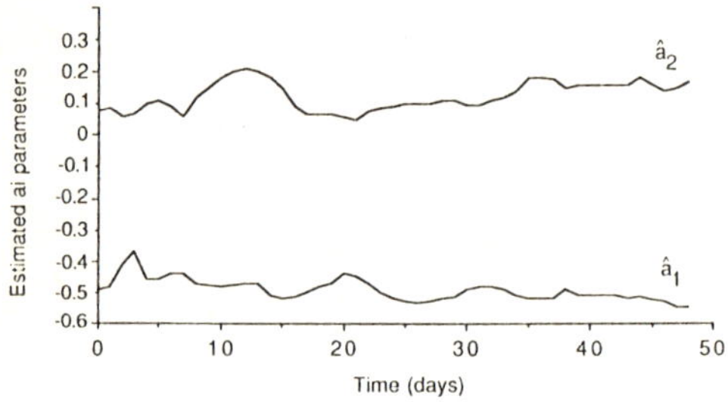

Fig. 7. - Trajectory of covariance matrix trace

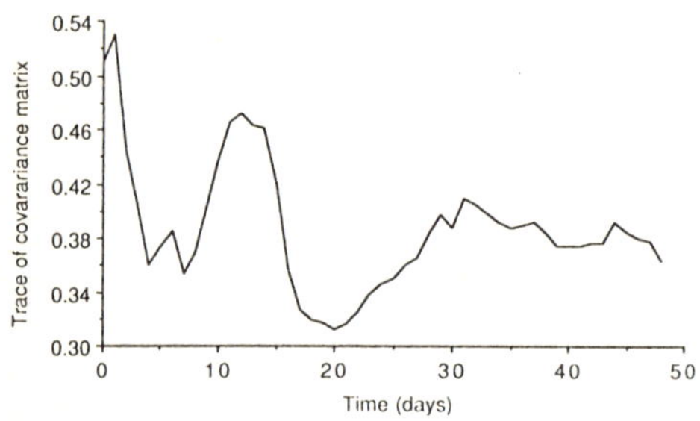

Figure 8 shows the variation of K_c over the same 50 day period. The K_c is clamped at 75 as a safety precaution as seen at Day 12 and at Day 42.

Fig. 8. - Trajectory of adaptive PI controller gain (Kc)

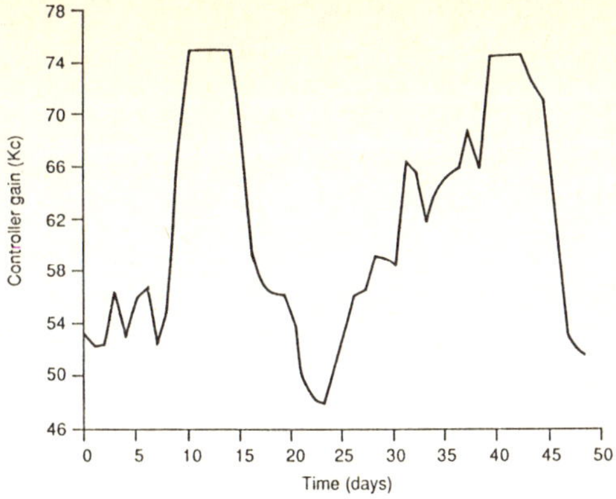

CONCLUSIONS

The evaluation of an adaptive control strategy has been completed with a successful test application on distillation column overhead composition control and has demonstrated the potential of adaptive techniques.

The simplified adaptive algorithm of a Smith predictor and PI controller used for composition control has been shown to be adequate for this application. The design of the diagnostic shell around the adaptive algorithm proved to be very important for the proper functioning of adaptive algorithms (and takes considerable time to develop).

The parameter estimation algorithm uses incremental input/output data, the numerically stable and computationally efficient U D factorization algorithm and the variable forgetting factor method. The industrial application results show the excellent performance of this combination.

REFERENCES

[1] Åström, K.J. and B. Wittenmark (1973). On self-tuning regulators. Automatica, 9, 185-199.

[2] Åström, K.J. and B. Wittenmark (1984). Computer Controlled Systems - Theory and Design. Englewood Cliffs, N.J.: Prentice-Hall Inc.

[3] Bierman, G.J. (1976). Measurement updating using the U-D factorization. Automatica, 12, 401-407.

[4] Clarke, D.W. and P.J. Gawthrop (1975). A self-tuning controller. Proc. IEE, 122, 929-934.

[5] Harris, T.J., J.F. MacGregor and J.D. Wright (1982). An overview of discrete stochastic controllers: Generalized PID algorithms with dead-time compensation. Can. J. Chem. Eng., 60, 425-432.

[6] Seborg, D.E., T.F. Edgar and S.L. Shah (1986). Adaptive control strategies for process control: a survey. AIChE J., 32, 881-913.

[7] Smith, O.J.M. (1957). Closer control of loops with deadtime. Chem. Engr. Progr., 53(5), 217-219.

[8] Ydstie, B.E., L.S. Kershenbaum and R.W.H. Sargent (1985). Theory and application of an extended horizon self-tuning controller. AIChE. J., 31, 1771-1780.

INDUSTRIAL APPLICATIONS OF A NEW ADAPTIVE ESTIMATOR FOR INFERENTIAL CONTROL

A.J. Morris, M.T. Tham and G.A. Montague
Department of Chemical and Process Engineering
University of Newcastle upon Tyne, England.

ABSTRACT

Two methods for inferring process outputs which can only be measured at infrequent intervals of time, from other more rapidly sampled secondary outputs, are presented. The estimators (or soft-sensors) use both a general input-output representation of the plant and a state space representation as starting points for synthesis of the algorithms. The developments are supported by results from some recent industrial scale site tests: on-line determination of product composition in a large industrial distillation tower and in a companion study, biomass concentration in an industrial mycelial fermentation. The technique uses measurements from established instruments such as temperature in the distillation column and off-gas carbon dioxide in the fermenter. The key development is an adaptive estimation relationship for inferring infrequently measured process outputs, from other more rapidly sampled secondary outputs. The structure of the estimators are not application dependent and their parameters can be continuously estimated and updated. As a result, slow variations in plant or disturbance characteristics can be tracked. In contrast to model-based estimators (e.g. Kalman filtering), the methods proposed require minimal effort for estimator design. A significant improvement in overall process control performance is possible using the estimated values of plant output rather than employing direct feedback of the infrequently obtained measurements. Simulation studies and production scale plant tests demonstrate estimator capabilities.

INTRODUCTION

In process control systems, the infrequent measurement of some process outputs, determined by sampling limitations, prevents the early detection of load disturbances. This can result in large deviations from the desired output and long disturbance recovery times. Often, these adverse effects cannot be acceptably overcome by the use of existing advanced control algorithms and can lead to unsatisfactory system performance.

The problem of controlling infrequently measured process outputs has long been studied and publications in this area date back to the early 1970's. There have been essentially two approaches to the problem. One is to design special controllers for the infrequently sampled outputs. For example, Soderstrom [28] formulated a number of minimum variance controllers enabling the manipulated (control) input to be changed between the sampling intervals of the primary process output. These control algorithms were only developed for first order plant models and no comparative results were presented. Parrish and Brosilow [24] also proposed a controller design method based upon the philosophy of reconstructing the effects of disturbance inputs. The controller parameters are determined on-line by heuristic tuning rules. Their simulation results indicated superior performance of their control strategy over that achieved using conventional PID control.

A second approach to deal with the problem of controlling infrequently sampled process outputs is to use the information provided by other easily measurable variables. For example, this information can be used to provide an estimate of the controlled output. The estimated values of the output can then be used for overall control of the plant. Control schemes based on the feedback of estimated outputs are often termed "inferential control schemes". An ideal situation arises when the plant states are completely observable from the secondary outputs. Under such circumstances, Kalman filtering techniques can be employed to estimate plant states using the secondary output measurements. Estimates of the controlled output can then be computed using its relationship with the states. Control of the plant is achieved by feedback of either the state estimates or the output estimates to appropriate controllers. Published literature on the above methods is extensive. The papers by Morari and Stephanopoulos [18, 19] and Morari and Fung [20] are amongst the more recent contributions describing the application issues of these techniques. However, the use of Kalman filtering is confined to situations where the plant is completely observable from the secondary outputs. For most plants such a set of secondary outputs can be difficult to determine or, in some cases, may not even exist. Cwiklinski and Brosilow [3] and Joseph and Brosilow [12] have suggested an estimator design technique using an input-output representation of the plant. The design is approached by obtaining a least squares based static estimator, which can be used to infer the controlled output from secondary measurements at steady state. The estimator is then additionally made applicable to the transient period by incorporating heuristically derived lead-lag elements into its structure. Methods for minimizing the steady state estimation error by appropriate choice of the secondary measurements were also proposed. However, a set of secondary measurements for which this error is satisfactorily small may not exist. In fact, the evaluation work carried out by Patke et al [25] indicated that an inferential control scheme based on the output estimation technique proposed by Brosilow and co-workers can result in significant offsets. Moreover, to apply the technique, it is necessary to know the gains and approximate time constants of the controlled output and all secondary outputs, for all plant disturbances and manipulated inputs.

The control strategies overviewed so far rely either on infrequent measurements of the controlled output or on the use of secondary measurements. It is, however, possible to use both types of measurements. One technique is to set up a parallel cascade control strategy (Luyben [14]). Here measurements of the controlled output are fed to a controller whose output acts as the setpoint to a secondary output controller. Although no offset problems arise with this control configuration, the transient performance can at times be poor (Patke et al [25]). A second technique for using both measurements of the controlled output and secondary output is within an adaptive inferential control framework. In such a scheme, the infrequent measurements of the controlled output are used only during parameter estimation while plant control is achieved using a secondary output at its faster sampling rate. To our knowledge very little investigative work has been carried out in this area, although D'Hulster and Van Cauwenberghe [4] have suggested an algorithm using a similar approach. Their algorithm is, however, restricted to the assumption of a first order plant model, an equal number of disturbance inputs and secondary measurements and the use of a dead beat control law.

There are a number of industrial situations where infrequent sampling of a 'controlled' process output introduces potential operability problems.

<u>Distillation Columns and Chemical Reactors</u>
Here, examples are in product composition control of distillation columns (e.g. Brosilow and co-workers, [3], [12]; Luyben, [14]; Patke et al, [25]) and chemical reactors (e.g. Wright et al, [33]). In these cases, the sampling delay is a direct result of the long cycle time of on-line composition analyzers. Because of potential problems due to infrequent sampling, the control of product quality of many industrial multicomponent or high purity columns is commonly achieved by attempting to maintain an apriori chosen tray temperature at its setpoint value. However, this type of single-temperature feedback control is not always effective since maintaining a constant tray temperature does not necessarily result in constant product composition (Patke et al [25]).

<u>Fermentation Processes:</u>
It is well known that the instrumentation and sensors available for fermentation control do not cover all the desirable (or even necessary) measurements. In fermentation processes, the important internal variables that present measurement difficulties, but

characterise the state and progress of the fermentation, are biomass, substrate and secondary product concentrations. In spite of recent, encouraging, developments in ion selective and enzyme sensors, optical and high frequency based methods, most of the concentration variables in a fermentation liquid phase cannot be measured accurately and reliably on-line. Usually therefore, laboratory analyses/assays are required to support fermentation supervision and control, (Montague et al [17]). There have been a number of successful developments and applications of estimation and advanced fermentation control reported by several authors (e.g. Dochain and Bastin [5]; Halme, [9], [10]; Montague et al [15], [16]; Nihtila et al [23]; San and Stephanopoulos [26]; Stephanopoulos and San [29]; Swiniarski et al [30]). It is interesting that most work has used a mechanistic model-based approach and extended Kalman filtering although this requires the availability of a sufficiently accurate biochemical model of the fermentation being controlled. In contrast the work of Halme [2], and Dochain and Bastin [5], for example, breaks new ground in the development and application of nonlinear methods for biochemical process estimation and control.

An alternative approach is set out in this paper. Two algorithms for estimating a plant controlled output (e.g. product composition of a distillation column, biomass concentration of a fermentation) from a secondary measurement (temperature in a distillation column, off-gas carbon dioxide evolution rate in a fermenter) are presented. The first estimation algorithm is based upon a general input-output representation (Guilandoust et al [8]), whilst the second, summarized for comparison purposes, is derived from a state space representation of the plant (Guilandoust et al [7]). The ideas behind the derivations of the algorithms were initially discussed in Guilandoust and Morris [6]. The self-adaptive Kalman filter approach suggested by Young [39] used a similar model form to that in this work, although a major difference is the multirate aspect addressed in this paper.

ADAPTIVE ESTIMATION USING AN INPUT-OUTPUT PROCESS REPRESENTATION

The process dynamics are assumed represented by the following relationships:

$$y_0(t) = G_1(z^{-1})\, u(t-m_1) + L_1(z^{-1})\, w(t) \tag{1}$$

$$v_0(t) = G_2(z^{-1})\, u(t-m_2) + L_2(z^{-1})\, w(t) \tag{2}$$

$$y(t) = y_0(t-d) + \nu_1(t) \tag{3}$$

$$v(t) = v_0(t) + \nu_2(t) \tag{4}$$

where $y_0(t)$ and $v_0(t)$ are the controlled output (e.g. distillation column product composition, biomass concentration) and secondary output (e.g. column temperature measurement, off-gas CO_2 measurement) at time t. y(t) and v(t) are their corresponding observations contaminated with measurement noises $\nu_1(t)$ and $\nu_2(t)$ respectively, assumed to be zero mean white sequences. d is the analyser (or assay) delay associated with y(t) and is assumed to be an integer multiple of the discretization time step. Therefore, values of y(t) only become available every d time steps.(Extension of the technique to allow process output analyses/assays occurring at irregular intervals of time is straightforward). v(t) is assumed to be measurable at each time step. m_1 and m_2 are the time delays in the responses of the controlled and secondary outputs, respectively, to changes in the manipulated input u(t) (e.g. column reflux flow, bioreactor feed). $\mathbf{w}(t) \in \mathbf{R}^1$ is the vector of stationary random and unmeasurable load disturbances. z^{-1} is the backward shift operator. $G_1(z^{-1})$ and $G_2(z^{-1})$ and all elements of row vectors $\mathbf{L}_1(z^{-1})$ and $\mathbf{L}_2(z^{-1})$ are polynomial ratios with the order of numerator less than or equal to that of the denominator. It is assumed that a change in any one of the components of $\mathbf{w}(t)$ affects both $y_0(t)$ and $v_0(t)$, otherwise $v_0(t)$ will not be suitable for estimation of $y_0(t)$. Consequently an entry of $\mathbf{L}_1(z^{-1})$ cannot be zero, unless the same entry of $\mathbf{L}_2(z^{-1})$ is also zero and vice versa. Eqns. (1) to (4) can be rewritten as:

$$y(t+d) = G_1(z^{-1})u(t-m_1) + \xi(t) + \epsilon(t+d) \qquad (5)$$

$$v(t) = G_2(z^{-1})u(t-m_2) + \zeta(t) \qquad (6)$$

where

$$\epsilon(t+d) = z^d\nu_1(t) - \nu_2(t) \qquad (7)$$

$$\xi(t) = \mathbf{L}_1(z^{-1})\ \mathbf{w}(t) + \nu_2(t) \qquad (8)$$

and

$$\zeta(t) = \mathbf{L}_2(z^{-1})\ \mathbf{w}(t) + \nu_2(t) \qquad (9)$$

The physically non-existing sequence $\epsilon(t)$ is zero mean and white. This is because $z^d\nu_1(t)$ is a zero mean white sequence and so is its algebraic sum with $\nu_2(t)$. Since $\nu_2(t)$ is a zero mean, white sequence and $\mathbf{w}(t)$ is a row vector of stationary signals, Eqns. (8) and (9) may, according to the spectral factorization theorem (Astrom, [2]), be represented as:

$$\xi(t) = H_1(z^{-1})\omega(t) \qquad (10) \qquad \text{and} \qquad \zeta(t) = H_2(z^{-1})\omega(t) \qquad (11)$$

where $H_1(z^{-1})$ and $H_2(z^{-1})$ are stable and proper polynomial ratios and the sequence $\omega(t)$ is zero mean and white. It is noted that both $\xi(t)$ and $\zeta(t)$ are generated by the same white noise sequence $\omega(t)$. If the driving noise for one of the signals was different from the other, then the two signals would clearly be uncorrelated. This cannot be the case, because variations in load disturbances are assumed to affect both controlled and secondary outputs [see Eqns. (5) to (9)]. Moreover neither one of $H_1(z^{-1})$ or $H_2(z^{-1})$ can be zero without the other being zero too. Substituting from Eqns. (10) and (11) into Eqns. (5) and (6) and eliminating $\omega(t)$, results in:

$$y(t+d) = [G_1(z^{-1})z^{-m_1+m_2} - G_2(z^{-1})H_1(z^{-1})/H_2(z^{-1})] * u(t-m_2)$$
$$+ [H_1(z^{-1})/H_2(z^{-1})]v(t) + \epsilon(t+d) \qquad (12)$$

for $m_1 >= m_2$, and for the case where $m_1 =< m_2$:

$$y(t+d) = [G_1(z^{-1}) - z^{-m_2+m_1} G_2(z^{-1})H_1(z^{-1})/H_2(z^{-1})] * u(t-m_1)$$
$$+ [H_1(z^{-1})/H_2(z^{-1})]v(t) + \epsilon(t+d) \qquad (13)$$

Eqns. (12) and (13) relate measurements of the primary output, $y(t)$, to secondary output, $v(t)$, and the values of the manipulated input, $u(t)$. If $G_1(z^{-1})$, $G_2(z^{-1})$, $H_1(z^{-1})$ and $H_2(z^{-1})$ were all known, $u(t)$ and $v(t)$ could be used in Eqn. (12) or (13) to compute:

$$\hat{y}(t+d) = y(t+d) - \epsilon(t+d) \qquad (14)$$

which is an estimate of the controlled output, $y_0(t)$. The variance of the deviations of $\hat{y}(t+d)$ from the controlled output is, by relationship (7), limited to the sum of the variances of $\nu_1(t)$ and $\nu_2(t)$. The variance of the latter noise can, by careful measurement and filtering of the secondary variable, be kept to a minimum. Thus, given a sufficiently accurate plant model, Eqn. (12) or (13) may be used for recursive computation of delay free estimates of the controlled output at the rate $v(t)$ is sampled. The accuracy of the estimates will be close to that of the measurements of the controlled output, $y(t)$. In situations where there are a large number of disturbance inputs or there are complicated or even unknown plant dynamics, the polynomial ratios $G_1(z^{-1})$, $G_2(z^{-1})$, $H_1(z^{-1})$ and $H_2(z^{-1})$ can only be approximately obtained using off-line identification. In this work, however, these polynomial ratios are not assumed to be known apriori. Instead, it is proposed that Eqn. (12) or (13) be approximated on-line.

Based on the definitions of $G_1(z^{-1})$, $G_2(z^{-1})$, $H_1(z^{-1})$ and $H_2(z^{-1})$, it would seem reasonable to choose an approximation for Eqn. (12) or (13) to serve as an estimator for the controlled output:

$$y(t+d) = \frac{B(z^{-1})}{A(z^{-1})}u(t-m) + \frac{C(z^{-1})}{A(z^{-1})}(t) + \epsilon(t+d) \qquad (15)$$

where $m = \mathrm{Min}(m_1,m_2)$ and,

$$A(z^{-1}) = 1 + a_1z^{-1} + a_2z^{-2} +\ldots+ a_nz^{-n}$$
$$B(z^{-1}) = b_1z^{-1} + b_2z^{-2} +\ldots+ b_nz^{-n} \qquad (16)$$
$$C(z^{-1}) = c_o + c_1z^{-1} + c_2z^{-2} +\ldots+ c_nz^{-n}$$

The value of the integer n is chosen to achieve the required accuracy of approximation. It should be pointed out that Eqn. (15) is not of the ARMAX type, although its structure may at first sight suggest otherwise. This is because v(t) is not an unknown signal as in case of an ARMAX relationship. Instead, both the primary, y(t), and secondary, v(t), outputs respond to both the control input and unknown disturbances. The parameters of Eqn.(15), in its present form, cannot be straightforwardly estimated using recursive techniques because y(t) is only available at every d time steps. This problem can, however, be resolved. Using the relationship (14), Eqn. (15) can be written in the form:

$$A(z^{-1})\hat{y}(t+d) = B(z^{-1})u(t-m) + C(z^{-1})v(t) \tag{17}$$

It can be shown that for a given value of n, Eqn. (17) can be reformulated as:

$$(1 + \alpha_d z^{-d} + \ldots + \alpha_{nd} z^{-nd})\hat{y}(t+d) =$$

$$(\beta_1 z^{-1} + \beta_2 z^{-2} + \ldots + \beta_{nd} z^{-nd})u(t-m) + (\gamma_o + \gamma_1 z^{-1} + \gamma_2 z^{-2} + \ldots + \gamma_{nd} z^{-nd})v(t) \tag{18}$$

Since Eqns. (18) and (17) are equivalent, both constitute an estimator for recursive computation of the controlled output. However, Eqn. (18) has the desired form because its parameters, unlike those of Eqn.(11), may be recursively estimated. In order to estimate the parameters of Eqn. (18), it is rewritten as:

$$y(t) =$$

$$- \alpha_d \hat{y}(t-d) - \ldots - \alpha_{nd}\hat{y}(t-nd) + \beta_1 u(t-m-d-1) + \ldots + \beta_{nd} u(t-m-(n+1)d)$$

$$+ \gamma_o v(t-d) + \ldots + \gamma_{nd} v(t-(n+1)d) + e(t) \tag{19}$$

This can be expressed in compact form as $y(t) = \theta^T\phi(t-d) + e(t)$ where:

$$\theta^T = [\ \alpha_d, \ldots, \alpha_{nd}, \beta_1, \ldots, \beta_{nd}, \gamma_o, \ldots, \gamma_{nd}]$$

$$\phi(t-d) = [-\hat{y}(t-d), \ldots, u(t-m-d-1), \ldots, v(t-d), \ldots]$$

and e(t) is the equation error. After updating the parameter estimates, $\hat{\theta}$, at time t, the 'aposteriori' value of $\hat{y}(t) = \theta^T\phi(t-d)$ is computed to update $\phi(t-d)$ at time t+d. Finally, the updated θ is used in the following relationship:

$$\hat{y}(t+d) = \theta^T\phi(t) = -\alpha_d\hat{y}(t) - \ldots - \alpha_{nd}\hat{y}(t-(n-1)d) + \beta_1 u(t-m-1)$$

$$+ \ldots + \beta_{nd} u(t-m-nd) + \gamma_o v(t) + \ldots + \gamma_{nd} v(t-nd) \tag{20a}$$

to compute estimates of the controlled output, $y_o(t)$, between the time interval t+d to t+2d. The parameters of Eqn.(19) are again updated at time t+d when a new measurement of the $y(t)$ becomes available. The expression for estimating $y(t)$ at the faster sampling rate of the secondary output is thus given by Eqn.(20a).

Eqn. (19) has n(2d+1)+1 unknown parameters and d may typically lie in the range 2 to 6. A first order estimator (n=1) may therefore have between 6 and 14 unknown parameters and a second order estimator, between 11 and 27. These figures may, at first sight, give the impression that the tuning in period for larger values of d will be inconveniently long. This is not, however, the case: although the number of unknown parameters increases with d, so does the data being supplied to the estimation algorithm at each sample time of the controlled output [see Eqn.(19)]. Simulation studies have in fact confirmed that the initial tuning in period increases only slightly for larger values of d (the effects of analyzer delays up to 6 sample intervals have been investigated). Indeed, estimates of the controlled output can often be obtained with sufficient accuracy using a first order estimator. The parameter α_d of the first-order estimator may be shown to be related to a_1 by $\alpha_d = (-a_1)^d$. For example, as may be noted from Eqn.(12), $|a_1|$ is normally less then unity (except for the special case where H_2 has a dominant unstable zero). For larger values of d, it therefore follows that $\alpha_d \ll 1$, especially when $|a_1|$ is significantly less than unity. As a result $_d$ becomes negligible and Eqn.(20a) reduces to:

$$\hat{y}(t+d) = \beta_1 u(t-m-1) + \ldots + \beta_d u(t-m-d) + \gamma_o v(t) + \ldots + \gamma_d v(t-d) \tag{20b}$$

An interesting point about the above equation is that an estimator of identical structure can also be derived using a state space representation of the plant (see Eqn. 38b), following an altogether different approach to that adopted here. This agreement between the results of the two different techniques provides additional confidence in the arguments presented.

ADAPTIVE INFERENTIAL CONTROL

An interesting option that now becomes available with the availability of a 'fast' estimate of controlled output (e.g. fermenter biomass concentration) is closed loop 'inferential' control of that variable - Adaptive Inferential Control. The values of $\hat{y}(t+d)$ can be used as a feedback signal for almost any constant parameter or adaptive control algorithm. For example, examination of Eqn. (19) indicates that, unless the order of the denominator of the controller transfer function is zero, problems of correlated data should not arise while estimating the parameters of that equation. For a time invariant process, the estimated parameters should converge to final values after an acceptable number of sample intervals. At this stage, parameter estimation could be stopped if desired, with estimates of the controlled output still being available without loss of accuracy. An analyser could then be made available say for maintenance or indeed any other duties. In the case of a slowly time varying process, however, it may be preferable to continue with the parameter estimation at all times. If such a decision were made, it may also be sensible to use an adaptive control algorithm in order to cope with plant variations. As with all self-tuning algorithms, adaptive regulation can be realised in either an explicit or an implicit manner. The corresponding closed-loop identifiability conditions for the plant model in the explicit case and for the modified plant model in the implicit case also apply here. This is because the values of $\hat{y}(t+d)$ are generally uncorrelated (see Eqn.(18)). Issues of closed-loop identifiability are discussed in detail in Isermann [11].

ADAPTIVE ESTIMATION BASED ON A STATE SPACE PROCESS REPRESENTATION

For comparison with the input-output derivation above, a state space based analysis is reviewed here.
The process dynamics are represented by the following state space model:

$$\tilde{\mathbf{x}}(t+1) = \mathbf{A}'\tilde{\mathbf{x}}(t) + \mathbf{B}'u(t-m) + \mathbf{L}'\mathbf{w}(t) \tag{21}$$

$$y(t) = \mathbf{D}'\tilde{\mathbf{x}}(t-d) + \nu_1(t) \tag{22}$$

$$v(t) = \mathbf{H}'\tilde{\mathbf{x}}(t) + \nu_2(t) \tag{23}$$

The aim is to estimate the system states from observations of v(t) and then obtain an estimate of y(t) using Eqn. (22).

A note on disturbances and their representation
Morari and Stephanopoulos [19] observed that disturbances affecting process plants are characterized by infrequently occurring random changes, lasting for periods of the order of at least the process response time. To represent this behaviour, disturbances have been modelled as Brownian motion processes (Joseph and Brosilow [12]; Morari and Stephanopoulos [18], [19]; Tuffs and Clarke [31]). This random process was chosen because it can be pictured as a sequence of step changes with independent random amplitudes, occurring at times described by a Poisson distribution. Although this model of process disturbances may well represent the physical situation its use, however, poses an observability problem. The Brownian motion process possesses a pole on the unit circle and hence is nonstationary. This necessitates that, for the plant to be observable from any set of measurements, the number of measurements should exceed the number of disturbance inputs (Morari and Stephanopoulos [19]) - a situation rarely met in process control. The random step-like behaviour of disturbances can alternatively be modelled by the stationary Exponentially Correlated

Noise process. Such a representation does not introduce the observability problem associated with the Brownian motion process. In the subsequent analysis, all elements of $\mathbf{w}(t)$ are assumed stationary random sequences so that each, according to the spectral factorization and representation theorems (Astrom [2]), may be represented as the response of a stable filter to a white input sequence.

Now, the poles of $\mathbf{w}(t)$ in Eqn.(21) can be augmented into the system matrix such that the resulting plant model becomes:

$$\mathbf{x}(t+1) = \mathbf{A}\mathbf{x}(t) + \mathbf{B}u(t-m) + \mathbf{L}\omega(t) \tag{24}$$

$$y(t) = \mathbf{D}\mathbf{x}(t-d) + \nu_1(t) \tag{25}$$

$$v(t) = \mathbf{H}\mathbf{x}(t) + \nu_2(t) \tag{26}$$

where $y(t)$ and $v(t)$ are measurements of the controlled primary and secondary outputs respectively, and $\nu_1(t)$ and $\nu_2(t)$ are the corresponding measurement noises. The measurement noises are assumed to be independent random Gaussian sequences with zero mean and finite variances. $\mathbf{x}(t)\in\mathbf{R}^n$ is the state vector with initial value $\mathbf{x}_o$ having properties $E\{\mathbf{x}(0)\}=\mathbf{x}(0)$ and $E\{[\mathbf{x}(0)-\mathbf{x}_o][\mathbf{x}(0)-\mathbf{x}_o]^T\}=\mathbf{P}_o$. d is the analyzer delay associated with $y(t)$. The secondary measurement, $v(t)$, is available at each sample instant. m is the smallest of the time delays in the responses of the controlled output and secondary output to changes in the manipulated input $u(t)$. Any difference between the two delays can be included in the model by extending the state vector. $\omega(t)\in\mathbf{R}^l$ is the vector of random and unmeasurable load disturbances.

If matrices **A**, **B**, **L**, **H** and **D**, and also the noise covariances, were known then the state estimates, $\hat{\mathbf{x}}(t)$, could be computed from $v(t)$ by Kalman filtering and substituted into the expression:

$$\hat{y}(t+d) = \mathbf{D}\hat{\mathbf{x}}(t) \tag{27}$$

to obtain estimates of the controlled output, $\hat{y}(t+d)$. The Kalman filter for Eqns.(4) and (5) is the recursive relationship (Anderson and Moore [1]):

$$\hat{\mathbf{x}}(t+1) = \mathbf{A}\hat{\mathbf{x}}(t) + \mathbf{B}u(t-m) + \mathbf{K}(t)\epsilon(t) \tag{28}$$

where $\mathbf{x}(0) = \mathbf{x}_o$; $\epsilon(t) = v(t) - \hat{v}(t)$ and $\hat{v}(t) = \mathbf{H}\hat{\mathbf{x}}(t)$. Here $\hat{v}(t)$ is the filtered value of $v(t)$, $\mathbf{K}(t)$ is the Kalman gain determined from the well known Riccati equations and $\epsilon(t)$ is the innovations sequence which is white and Gaussian. This method is one of the more popular approaches adopted for estimating controlled outputs. The need for detailed information about the process model and disturbance statistics can, however, be avoided if the estimation is carried out adaptively, on-line. Consequently, a suitable low order innovations model is fitted to the process as a replacement for Eqns.(24) to (26):

$$\hat{\mathbf{x}}(t+1) = \mathbf{A}\hat{\mathbf{x}}(t) + \mathbf{B}u(t-m) + \mathbf{K}\epsilon(t) \tag{29}$$

$$y(t) = \mathbf{D}\hat{\mathbf{x}}(t-d) + \nu_1(t) \tag{30}$$

$$v(t) = \mathbf{H}\hat{\mathbf{x}}(t) + \epsilon(t) \tag{31}$$

Here $\mathbf{K}$ is the time invariant (limiting) Kalman gain, while $\hat{\mathbf{x}}$ is the optimal estimate of the reduced order model states, $\mathbf{x}$, using v.

Since Eqns.(29) and (31) are equivalent to Eqns.(24) and (26) (Anderson and Moore [1]), either representation may be employed. It is, however, more convenient to use the innovations representation - Eqns.(29) and (31). The filtering process is thus avoided as the state estimates can be computed directly from Eqns.(29) and (31). The matrices of the model can be parameterized in various ways. Since no prior knowledge about the model structure is assumed, it is reasonable that the parameterization be made in a manner that the identified parameters would directly correspond to an input-output model (Ljung and Soderstrom,1983). This is best achieved by choosing an observer canonical structure for the model. Having estimated the states, Eqn.(27) could then be used to obtain $\hat{y}(t+d)$ if the row vector $\mathbf{D}$ was known. In principle, $\mathbf{D}$ could be estimated using values of y(t) and Eqn.(25), if the state estimates were not correlated. [As v(t) is assumed to be obtainable rapidly and with little measurement noise, the state estimates are considered accurate, hence justifying the replacement of $\mathbf{x}(t)$ by $\hat{\mathbf{x}}(t)$.] However, this is not the case as may be noted from Eqn.(29). To overcome this problem we first note that in the observer canonical structure, Eqns.(29) and (31) are of the form:

$$\hat{\mathbf{x}}(t+1) = \begin{bmatrix} -a_1 & & \\ \cdot & I_{n-1} & \\ \cdot & & \\ \cdot & & \\ -a_n & \cdots & 0 \end{bmatrix} \hat{\mathbf{x}}(t) + \begin{bmatrix} b_1 \\ \cdot \\ \cdot \\ \cdot \\ b_n \end{bmatrix} u(t\ m) + \begin{bmatrix} k_1 \\ \cdot \\ \cdot \\ \cdot \\ k_n \end{bmatrix} \epsilon(t) \tag{32}$$

$$v(t) = [\ 1,0,\ldots\ldots,0\ ]\hat{\mathbf{x}}(t) + \epsilon(t) \tag{33}$$

From the definition, $\epsilon(t) = v(t) - \hat{v}(t)$, Eqn.(33) becomes:

$$\hat{v}(t) = \hat{x}_1(t) \tag{34}$$

The state estimates $\hat{x}_1(t)$ to $\hat{x}_n(t)$ can be eliminated from the RHS of Eqn.(32) by successive substitution to yield:

$$\hat{x}_1(t+1) = \hat{v}(t+1)$$

$$\hat{x}_2(t+1) = - a_2\hat{v}(t) - .. - a_n\hat{v}(t-n+2) + b_2u(t-m) + .. + b_nu(t-m-n+2)$$
$$+ k_2\epsilon(t) + .. + k_n\epsilon(t-n+2) \qquad (35)$$

$$\vdots$$

$$\hat{x}_n(t+1) = - a_n\hat{v}(t) + b_nu(t-m) + k_n\epsilon(t)$$

If the above equations for the state estimates are now used in Eqn.(6), the result is of the form:

$$y(t) = \beta_1u(t-m-d-1) + .. + \beta_{n-1}u(t-m-n-d+1)$$
$$+ \gamma_o\hat{v}(t-d) + .. + \gamma_{n-1}\hat{v}(t-n-d+1)$$
$$+ \delta_1\epsilon(t-d-1) + .. + \delta_{n-1}\epsilon(t-n-d+1) + \nu_1(t) \qquad (36)$$

Since this expression does not depend on the state estimates, there should not be any difficulty in estimating its parameters. A more convenient form is:

$$y(t) = \theta^T\phi(t-d) + e(t) \qquad (37)$$

where: $\theta^T = [\ \beta_1, .. , \beta_{n-1}, \gamma_o, .. ,\gamma_{n-1}, \delta_1, .. , \delta_{n-1}\ ]$

$$\phi(t-d)^T = [\ u(t-m-d-1), .. , \hat{v}(t-d), .. ,\epsilon(t-d-1), ..\ ]$$

and $e(t)$ is the equation error. The parameters of Eqn.(37) can then be estimated. After updating the parameters $\hat{\theta}$ at time t, the value $\hat{y}(t)=\hat{\theta}^T\phi(t-d)$ is computed to update $\phi(t-d)^T$ and the updated θ used in:

$$\hat{y}(t+d) = \theta^T\phi(t)$$
$$= \beta_1u(t-m-1) + .. + \beta_{n-1}u(t-m-n+1) + \gamma_o\hat{v}(t) + .. + \gamma_{n-1}\hat{v}(t-n+1)$$
$$+ \delta_1\epsilon(t-1) + .. + \delta_{n-1}\epsilon(t-n+1) \qquad (38a)$$

to obtain estimates of the controlled output - see Eqn.(27). The parameters of Eqns.(36) and (38a) are again updated at time t+d when a new measurement of the controlled output becomes available. Equation (38a) thus serves as an estimator for computing the values of the controlled output at the faster secondary output sample rate. The number of parameters to be estimated in Eqn.(36) is 3n-2. This implies that, if the process can be adequately represented by a first order model, then the two outputs are proportionally related. Our experience indicates that a value of, n=2 or 3 is normally sufficient. The values of $\epsilon(t)$ in Eqns.(36) and (38a) can be computed from Eqns.(12) and (13) after the parameters of Eqn.(32) have been updated (Tsay and Shieh [32]). Often, secondary output measurements, v(t), can be obtained with low noise levels and load disturbances only occur infrequently. Under these circumstances, the ϵ terms tend to be very small and may be ignored. Computation of $\hat{y}(t+d)$ then becomes very much simplified as it is no longer necessary to identify a model for the secondary output. This results in a significant reduction in the number of unknown parameters in Eqn.(36), and Eqn.(38a) simplifies to:

$$\hat{y}(t+d) = \beta_1 u(t-m-1) + .. + \beta_{n-1} u(t-m-n+1) + \gamma_0 v(t) + .. + \gamma_{n-1} v(t-n+1) \tag{38b}$$

The discussion above was based upon the assumption of complete observability of the plant from v(t). The question arises, however, as to what happens if this is not the case. As is evident from Eqns.(24) to (26), y(t) will then share only some of its poles with v(t). The zeros of y(t) and v(t) are, in general, different from each other. Loosely speaking, substitution of the state estimates into Eqn.(25) may be regarded as assigning the poles of v(t) to y(t). The zeros assigned to y(t) are then determined by the process of estimating the parameters of Eqn.(36). As may be expected, this operation places the zeros such that the effect of the difference between the poles of v(t) and y(t) is minimal. It is also possible that some of the poles of v(t) are effectively cancelled. Simulation studies carried out on systems with widely different controlled and secondary output dynamics indicate satisfactory estimation despite the differences in dynamics.

INDUSTRIAL EVALUATIONS AND SIMULATION STUDIES

In the application of the adaptive estimators several important implementation aspects arose. In particular the need for careful data conditioning prior to parameter estimation. Data scaling, in that the input / output data magnitude ranges were of the same order, is particularly important. Two additional techniques of data and algorithm conditioning for successful robust on-line application were also adopted. Firstly, whenever a primary output analysis (biomass, column product composition) becomes available it is used in the algorithm to replace the estimated value. Secondly, all data is conditioned to represent deviations about a mean process operating level. The appropriate mean values are continuously determined using tiered exponential filters. The techniques adopted here reflect the experiences with our work on adaptive and self-tuning control (e.g. Montague et al [15], [16]; Morris et al [21]).

Industrial fermentation validation of both the input-output and state space based adaptive estimation schemes was undertaken on the site of one of our industrial collaborators. A continuous fermenter, operated in stirred tank and air lift fermenter configurations, produces a fungal mycelium. Following downstream processing the mycelia are the final process product. However, specifications placed upon its quality dictate certain fermenter operating conditions. In particular, biomass concentration has been found to be a variable which requires 'tight' regulation. Present industrial biomass regulation is based upon the analysis of four hourly samples and the dilution rate adjusted upon the return of the assay information. Process analysis reveals that a sampling interval of one hour would be more suitable for the maintenance of a constant biomass concentration. It is, however, not possible to sample at this frequency. Indeed present policy is based around increasing the sampling frequency so as to reduce the demand for 24 hour supervision. A move to 8 or 12 hourly analysis would therefore be desirable. This has important consequences on the control policy in that an alternative technique to off-line analysis must be found in order to regulate biomass effectively. The adaptive estimators, discussed in the previous sections, provides a means by which frequent estimates of biomass can be obtained using measurements of the carbon dioxide evolution rate. Off-gas analysis can be obtained at a fast rate, with little measurement delay, providing an indication of changes in growth rate. Hence an inferential estimator supplied with hourly CO_2 and dilution rate measurements, 'updated' by infrequent and irregular biomass concentration assays, will enable hourly dry weight predictions. Industrial scale trials have been promising, with data obtained from both stirred tank and air-lift fermentations providing verification. Results are presented for a stirred tank bioreactor.

Figure 1 shows the carbon dioxide and dilution rate measurements over a 470 hour period of continuous operation. For confidentiality reasons, in this and following figures,the ordinate scales have been removed. (It is noted, however, that the ordinate scale does not start at zero). Figure 2 shows the biomass (dry weight) assay results (step-like response) over the fermentation period and how they compare with inferred biomass concentration. Here the estimator is supplied with filtered CO_2 measurements and dry weight assays every four hours. It can be seen that the ability of the estimator to predict biomass transient behaviour is very acceptable. Moving from four to eight hourly dry weight samples (figure 3) can be seen to have little effect, with the accuracy of prediction being maintained.

Distillation columns can also exhibit controllability problems, especially those with high purity products, which can be exacerbated by the delays introduced by on-line product analysers. In one such column, an industrial demethaniser, the aim of the control strategy is to regulate the concentration of ethylene in the methane product by varying reflux flow. Here, a column temperature is measured every five minutes whilst a product composition analyser measurement becomes available every twenty minutes. The task here is to use the frequent temperature measurement to infer product concentration at five minute intervals. Figure 4 shows logged data over 10 1/2 days of a 25 day test run on the column with responses for measured temperature (plot a), feed flow (plot b) and reflux flow (plot c). The process variables in these and the following figures are plotted as deviations about a nominal steady state. For reasons of commercial confidentiality, units have been omitted from the ordinate scales. A comparative plot of actual product concentration with the estimated concentration superimposed shown in figure 5. The latter part of this run (2500 minutes), is expanded and shown in fig. 6. As can be seen, the estimates of top product composition are very close to its actual measurement with the errors being within the tolerances demanded by the plant operating personnel. The next stage of the project will be to incorporate the estimator into the plant supervisory control strategy.

In implementing the adaptive estimators a prime concern was the ability of the algorithm to cope with particular process plant problems that proved to be most demanding. For instance, periodic recalibration of the infra-red carbon dioxide analyser (or on-line chromatograph in the case of the distillation column, unless accounted for, can cause significant problems. Due to the adaptive nature of the algorithm, however, it is capable of dealing with the slow drift in calibration usually experienced. Difficulties can also arise if the raw plant data is not filtered adequately, or even correctly. Care must be taken to ensure that the measured signals are filtered for noise rejection, anti-aliasing and sudden changes or spikes.

The adaptive estimation scheme described can be incorporated into a feed back control system to provide an inferential control strategy. This possibility has been studied by simulation of the control of bottom product composition of an 8 tray binary distillation column. The column simulation package (Simonsmier [27]) uses a comprehensive set of equations describing the dynamics of a ten stage pilot plant column separating a 50-50 wt.% methanol-water feed entering at stage 5. Feedrate is 18.23 g/s with the bottom and top products being regulated to 5 wt.% and 95 wt.% respectively. The pilot plant is installed at the University of Alberta, Canada. Both the simulation package and the pilot plant have been widely used by many investigators to study different control schemes (e.g. Morris et al [21], [22]). The manipulated input for the control of bottom product composition was the steam flow rate the composition is measured by an on-line analyzer with a cycle time of 3 minutes. The temperature of the liquid on various trays was chosen as the secondary output and was sampled every half-minute. The disturbances considered were step changes in feed flow rate. In the studies, w represents feed flow disturbances, v - stage 4 (tray 3) temperature and y and u the bottom product composition and steam flow rate respectively. The controllers used were of the fixed parameter PI type. The PI settings were obtained by Ziegler-Nichols ultimate sensitivity tuning rules followed by subsequent fine tuning to obtain responses with low integral of absolute error. Estimation of the controlled output has been investigated using up to third order estimators, although in most cases lower order estimators performed equally as well. A recursive least squares algorithm based on the UD factorization technique (e.g. Ljung and Soderstrom [13]) was used for parameter estimation. In all the composition plots the data shown corresponds to actual bottoms composition, $y_0(t)$ rather than its analyzer delayed value, y(t).

The open loop responses of bottom product composition and tray 3 liquid temperature to $\pm$10 per cent step changes in steam and feed flows are shown in Figures 7 and 8 respectively. It can be seen that both bottom product composition and tray 3 temperature exhibit different gains and time constants for both positive and negative steps in steam and feed flow rates. The linear estimator, therefore, has to cope with the resultant nonlinearities of all these responses. This is an important factor when assessing the adaptation properties of the estimator to changing plant and disturbance characteristics. The nonlinearity of the column dynamics is very high, especially for large (> $\pm$20%) changes in either reflux, steam or feed flow rates.

Comparative performances of PI feedback control and adaptive inferential control (AIC) under major disturbance conditions was assessed by subjecting the column to $\pm$25% feed flow changes. Figure 9 shows the PI feedback control performance, using the product composition analyser as the measured signal, to be compared with that of a first-order adaptive estimator (AIC) feedback control scheme shown in Fig. 10. Deviations of actual composition from its steady state value (full line) are plotted along with the estimated composition (broken line). Clearly an overall improved transient behaviour is observed. (Particular note should be made of the difference in ordinate scales in figures 9 and 10). Although the estimator tracks the actual composition response closely during the first part of the transient following the +25% increase in feed rate, the actual composition response exhibits slow offset removal as the estimator parameters tune-in under conditions of minimal excitation of the estimator. This is particularly visible during the two major disturbing steps occurring at times 20 and 160 mins. respectively. The conventional PI control strategy settings were K=-60 and T_I=20 min. whilst the settings used for the inferential scheme were K=-175 and T_I=2.5 min. Degaining the fixed parameter PI controller and adjustment of its integral time to take account of the delay in the bottom loop actually resulted in a much more sluggish response as expected, at the expense of an increase in peak overshoot of about 0.6 wt.% (response not shown). However, fine tuning of the PI settings in the adaptive inferential control scheme could well improve those responses even further. In addition a continuously repeating train of disturbances, such as used in this study, would not be expected to occur in practice and a longer, and more excited, tuning period would be available.

Other comparative studies have been carried out using the multi-component distillation column model described by Patke et al [25]. A much improved, offset free control performance, has been reported using the estimators described (Guilandoust et al [6], [7], [8]). The ability to anticipate disturbance effects results in a significantly improved control behaviour. This has been achieved here despite a poor choice of secondary measurement location. The studies also hopefully indicate that the use of these estimators in multicomponent column control should enable improved product quality regulation. It is worth noting, however, that in practice it is not likely that all available secondary outputs will posses very nonlinear dynamics. Neither would the disturbances affecting the plant change as sharply as the steps used in the present study. Hence the ability of the adaptive algorithm to track slow changes in plant dynamics should be adequate. As has been shown, the adaptive inferential control technique described above appears to possess such capabilities.

CONCLUSIONS

In this paper, two adaptive estimators for inferring infrequently measured plant outputs, using a more frequently measured secondary outputs, have been derived. In one approach a general input-output representation of the plant formed the starting point for the algorithm and this was compared with one developed from a state space concepts. The algorithms are termed 'soft-sensors' since they are software based rather than being hardware based. Both estimators approximate the complex relationship between the two output variables. Estimator parameters can be identified on-line using values of the infrequently sampled plant output and the relatively faster sampled secondary outputs and inputs. The technique enables the controlled output to be inferred at the same rate that the secondary output is being sampled.

The algorithms have been applied to the estimation of biomass concentration, using off-gas CO_2 and fermenter dilution rate measurements in an industrial bioreactor system, and to the estimation of top product composition on an industrial depropaniser column using column temperature and reflux flow measurements. The results highlight potential for the improvement of two 'difficult' industrial control problems using the adaptive scheme proposed.
With respect to closed loop 'inferential' control, early detection of the effects of load disturbances and hence significantly faster disturbance rejection can also be achieved. Consequently, it is expected that control schemes using the estimated output in conjunction with fixed parameter PI or PID controllers could outperform other conventional feedback PI/PID schemes, or indeed self-tuning control schemes, which rely solely on infrequent output measurements. The potential for inferential control was demonstrated by the distillation column simulation studies.

ACKNOWLEDGEMENTS

The authors would like acknowledge support from ICI, RHM, the UK SERC, and the University of Newcastle Upon Tyne.

REFERENCES

[1] Anderson, B.D.O. and Moore, J.L. (1979), "Optimal Filtering", Prentice Hall.

[2] Astrom, K.J. (1970), "Introduction to Stochastic Control Theory", Academic Press.

[3] Cwiklinski, R.R. and Brosilow, C.B. (1977), "Inferential Control of An Industrial Depropanizer", Proc. JACC, pp 1530-1536.

[4] D'Hulster, F.M. and van Cauwenberghe, A.R. (1981), "Applications of Parameter Adaptive Inferential Control to Non-Continuously Measurable Quality Variables", Proc. 2nd World Congress on Chemical Engineering, Montreal, pp 378-380.

[5] Dochain D. and Bastin G. (1984), "Adaptive identification and control algorithms for nonlinear bacterial growth systems." Automatica, 20, pp 621-634.

[6] Guilandoust, M.T. and Morris, A.J. (1985), "Adaptive Inferential Control of Processes with Slow Measurement Rates." Proc. of 35th Canadian Chemical Eng. Conf., Calgary, Canada. pp 213-217.

[7] Guilandoust, M.T., Morris, A.J. and Tham, M.T. (1987), "Adaptive Inferential Control". IEE Proc. Part D, 134, 3, pp 171-179.

[8] Guilandoust, M.T., Morris, A.J. and Tham, M.T. (1988), "An Adaptive Estimation Algorithm for Inferential Control", Ind. Eng. Chem. Res., 27, pp 1658-1664.

[9] Halme, A. and Selkainaho, A. (1982), "Application of a non-linear filter to multivariable parameter adaptive control in a distributed micro computer." 6th. IFAC Symp. on Identification and System Parameter Estimation, Washington, USA.

[10] Halme, A. (1985), "A method to consider delayed laboratory analysis in state and parameter estimation of bioreactors." 1st. IFAC Symp. on Modelling and Control of Biotechnological Processes, Noordwijkerhout, Holland.

[11] Isermann, R. (1981), "Digital Control Systems", Springer-Verlag.

[12] Joseph, B., and Brosilow, C.B. (1978), "Inferential Control of Processes", Parts I and III, AIChE Journal, Vol.24, No.3, pp 485-491 and pp 500-509.

[13] Ljung, L. and Soderstrom, T. (1983), "Theory and Practice of Recursive Identification", MIT Press.

[14] Luyben, W.L. (1973), "Parallel Cascade Control", Ind. Eng. Chem. Fundam., Vol 12, No 4, pp 463-467.

[15] Montague, G. A., Morris, A. J., Wright, A.R., Aynsley, M. and Ward, A. C. (1986-a),"On-line estimation and adaptive control of penicillin fermentation". Proc. IEE Part D, 133, 5, pp 240-246.

[16] Montague, G. A., Morris, A. J., Wright, A. R., Aynsley, M. and Ward, A. C. (1986-b), "Modelling and adaptive control of fed batch penicillin fermentation". Can. Journal Chem. Eng., 64, pp 567-580.

[17] Montague, G. A., Morris, A. J. and Bush, J. R. (1988), "Considerations in control scheme development for fermentation process control." IEEE Control Systems Mag.

[18] Morari, M. and Stephanopoulos, G. (1980-a), "Optimal Selection of Secondary Measurements within the Framework of State Estimation in the Presence of Persistent Unknown Disturbances", AIChE Journal, Vol.26, No.2, pp 247-260.

[19] Morari, M. and Stephanopoulos, G. (1980-b), "Minimizing Unobservaibility in Inferential Control Schemes", Int. J. Control, Vol.31, No.2, pp 367-377.

[20] Morari, M. and Fung, K.W. (1982), "Non-linear Inferential Control", Computers and Chemical Engineering, Vol.6, No.4, pp 271-281.

[21] Morris, A.J., Nazer, Y., Wood, R.K. and Lieuson, H. (1981), "Evaluation of Self-tuning Controllers for Distillation Column Control", IFAC Conf. on Digital Computer App. to Process Control, Ed. Isermann and Haltenecker, Dusseldorf, Germany, pp 345-354.

[22] Morris, A.J., Nazer, Y. and Wood, R.K. (1982), "Multivariate Self-tuning Process Control". Optimal Control and Applications, 3, pp363-387.

[23] Nihtila, M., Harmo, P. and Perttula, M. (1984), "Real-time growth estimation in batch fermentation". 9th. World Congress of IFAC, Budapest, Hungary.

[24] Parrish, J.R. and Brosilow, C.B. (1985), "Inferential Control Applications". Automatica, 21, 5, pp 527-538.

[25] Patke, N.G., Deshpande, P.B. and Chou, A.C. (1982), "Evaluation of Inferential and Parallel Cascade Schemes for Distillation Control", Ind. Eng. Chem. Process Des. Dev., 21, pp 266-272.

[26] San, K-Y. and Stephanopoulos, G. (1984), "Studies on on-line bioreactor identification. II, Numerical and experimental results". Biotech. & Bioeng., 26, pp 1189-1197.

[27] Simonsmier, U. (1977), "Non-linear Binary Distillation Column Models", M.Sc. Thesis, Univ. of Alberta, Canada.

[28] Soderstrom, T. (1980), "On Some Adaptive Controllers for Stochastic Systems with Slow Output Sampling", in Unbehaven, H. "Methods and Applications in Adaptive Control", Springer-Verlag.

[29] Stephanopoulos, G. and San, K-Y. (1984), "Studies on on-line bioreactor identification, I, Theory". Biotechnol. & Bioeng., 26, pp 1176-1188.

[30] Swiniarski,R., Lesniewski, A., Dewski, M. A. M., Ng, M. H. and Leigh, J. R. (1982), "Progress towards estimation of biomass in a batch fermentation process". 1st. IFAC Workshop on Modelling and Control of Biotechnical Processes, Espoo, Finland.

[31] Tuffs P.S. and Clarke, D.W. (1985), "Self-tuning Control of Offset: a unified approach", Proc. IEE, 128, Pt. D, 3, pp 93-101.

[32] Tsay, Y.T. and Shieh, L.S. (1981), "State-space approach for self-tuning feedback control with pole assignment", IEE Proc. D, Control Theory & Appl., 128, 3, pp 93-101.

[33] Wright, J.D., MacGregor, J.F., Jutan, A., Tremblay, J.P. and Wong, A. (1977), "Inferential Control of an Exothermic Packed Bed Tubular Reactor", Proc. JACC, pp 1516-1522.

[34] Young, P.C., (1979), "Self-Adaptive Kalman Filter", Electronics Letters, Vol 15, No 12, pp 358-360.

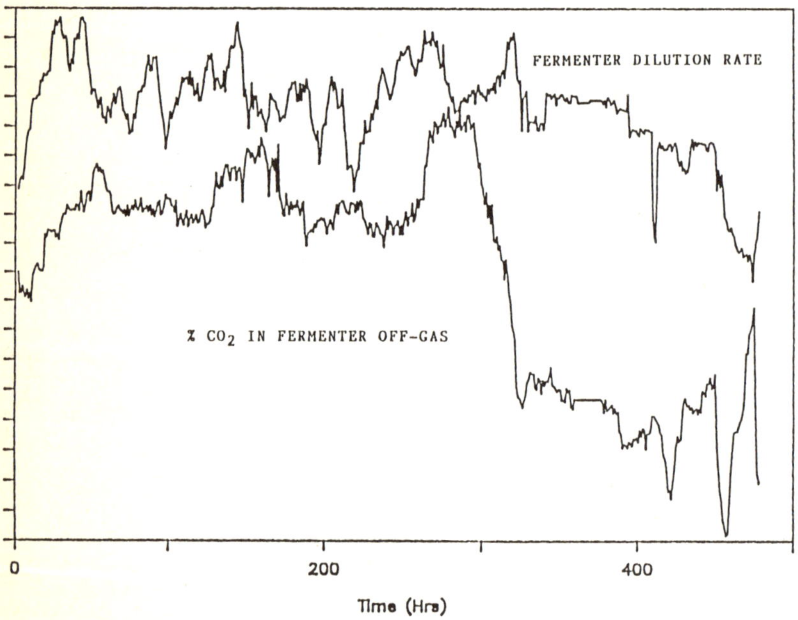

FIG. 1. FERMENTER CO_2 EVOLUTION RATE and DILUTION RATE

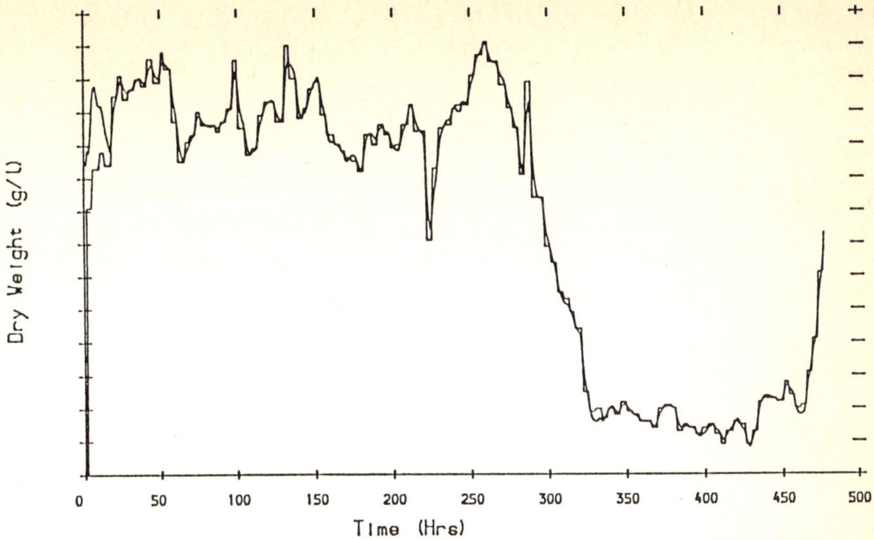

FIG. 2. FERMENTER DRY WEIGHT ASSAYED EVERY 4 HOURS compared with ESTIMATED BIOMASS

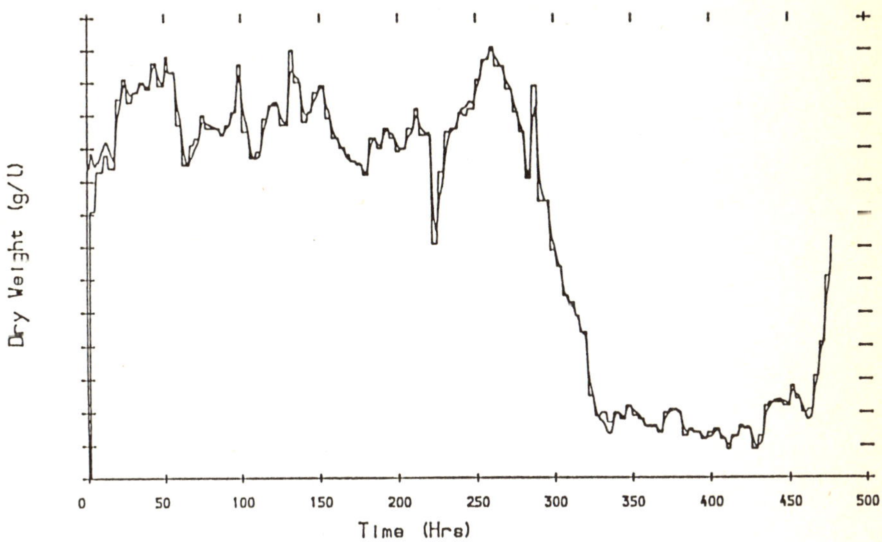

FIG. 3. FERMENTER DRY WEIGHT ASSAYED EVERY 8 HOURS compared with ESTIMATED BIOMASS

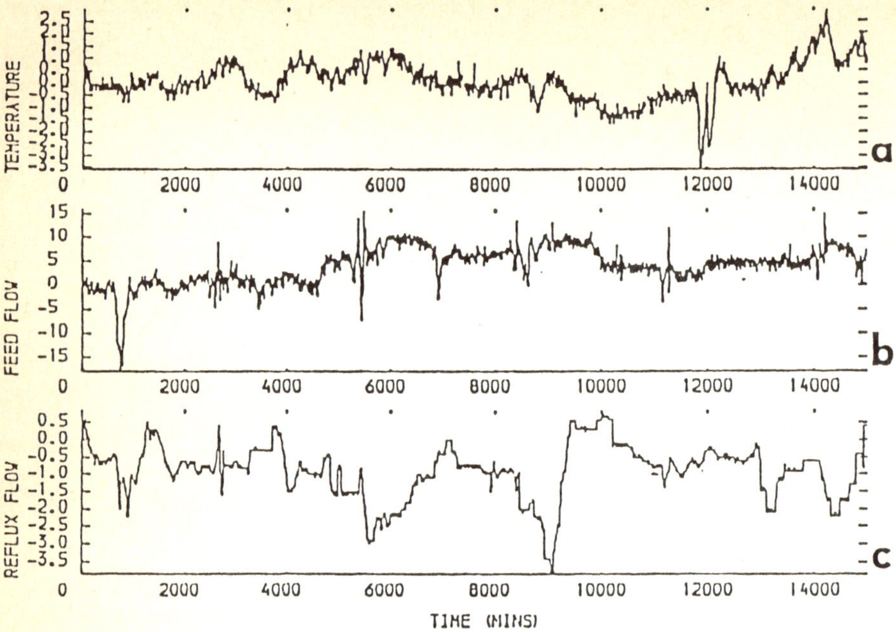

FIG. 4. DEMETHANISER COLUMN TEMPERATURE, FEED FLOW and REFLUX FLOW RESPONSES

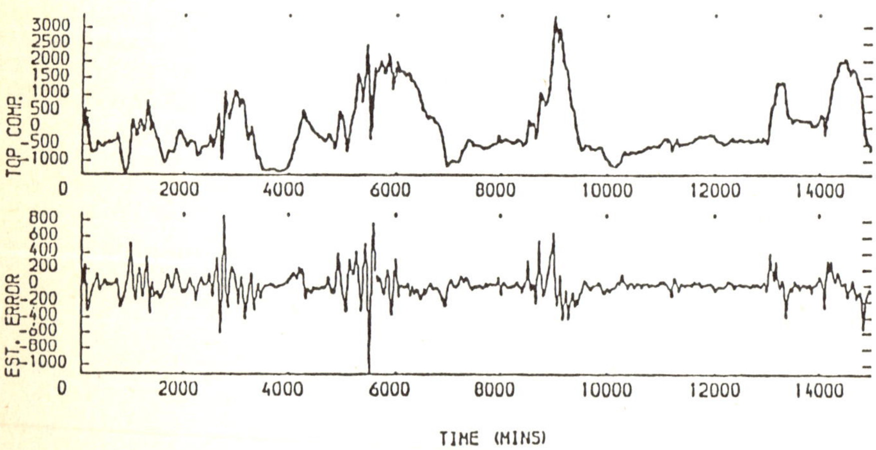

FIG. 5. DEMETHANISER COLUMN PRODUCT COMPOSITION ANALYSIS compared with ESTIMATED COMPOSITION OVER 15120 MINUTES

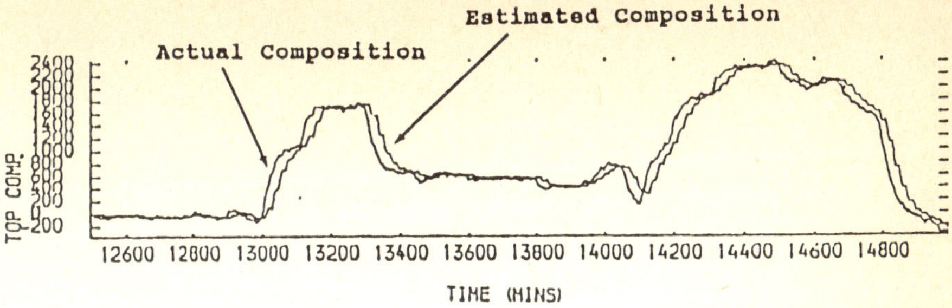

FIG. 6. DEMETHANISER COLUMN PRODUCT COMPOSITION ANALYSIS compared with ESTIMATED COMPOSITION OVER 2500 MINUTES

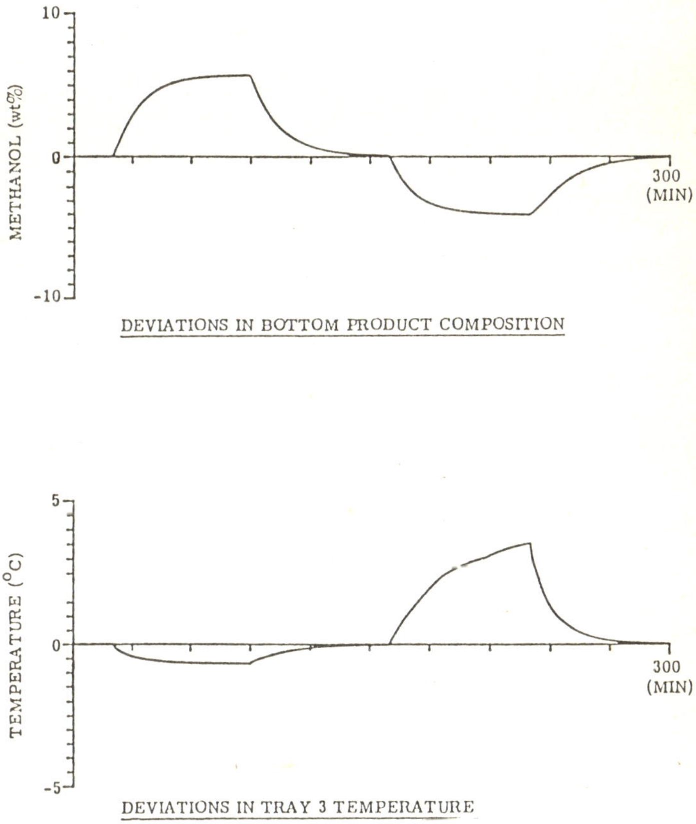

FIG. 7. METHANOL COLUMN OPEN LOOP RESPONSES ($\pm$10% feed flow changes from steady state)

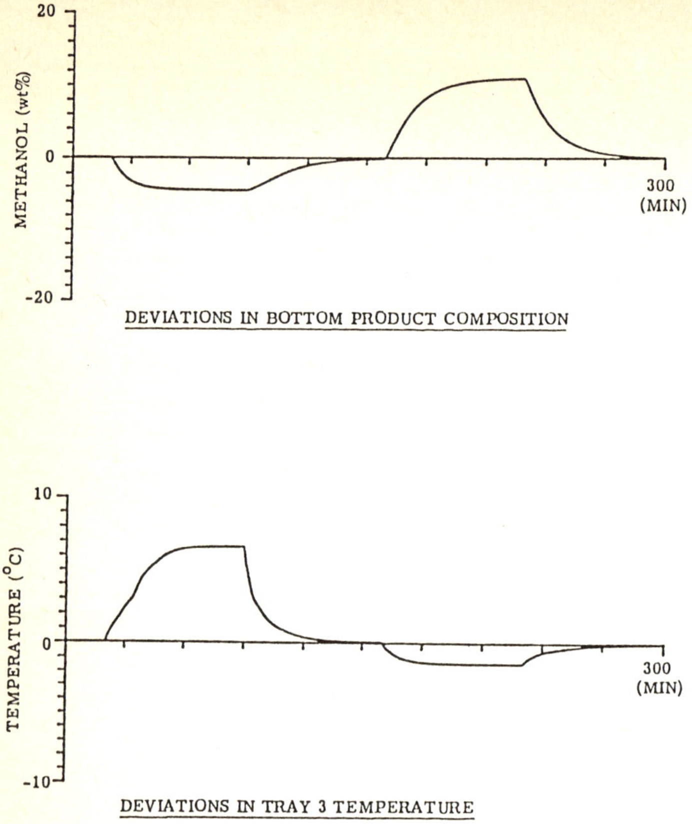

FIG. 8. METHANOL COLUMN OPEN LOOP RESPONSES
($\pm$10% steam flow changes from steady state)

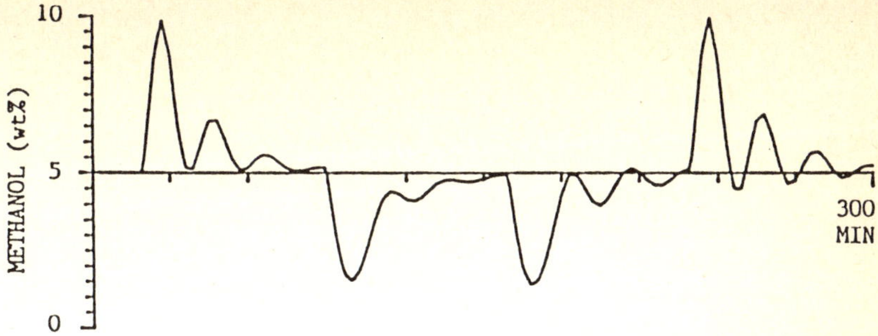

FIG. 9. METHANOL COLUMN PRODUCT COMPOSITION
(PI Control) ±25% feed flow disturbances

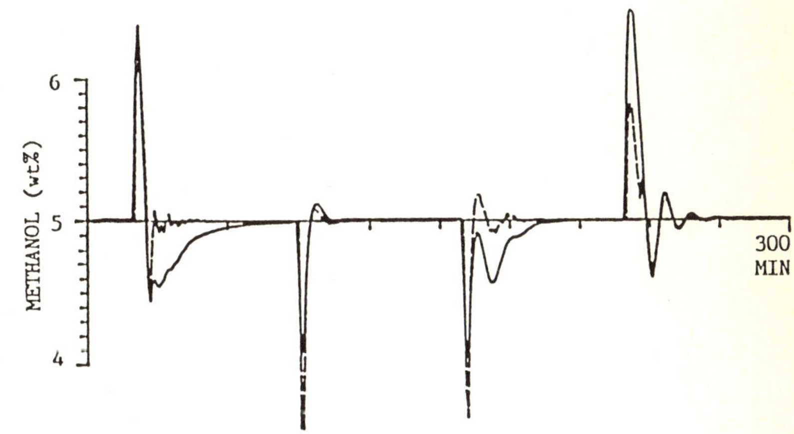

FIG. 10. METHANOL COLUMN BOTTOM PRODUCT COMPOSITION
(Adaptive Inferential Control)
±25% feed flow disturbances
(——Actual Composition ---- Estimated Composition)

ADAPTIVE ESTIMATION AND CONTROL OF BIOTECHNOLOGICAL PROCESSES

D. Dochain+, G. Bastin*, A. Rozzi and A. Pauss++**

+ : Ecole Polytechnique de Montréal, Campus de l'Université de Montréal
Case Postale 6079, Succursale "A", Montréal H3C 3A7, CANADA
* : Laboratoire d'Automatique, Dynamique et Analyse des Systèmes,
Université Catholique de Louvain, Bâtiment Maxwell, Place du Levant, 3
1348 Louvain-la-Neuve, BELGIUM
** : Instituto di Ingeneria Sanitaria, Politecnico di Milano,
Piazza Leonardo da Vinci,32 , 20133 Milano, ITALY
++ : Institut de Recherches en Biotechnologie, CNRC,
6100 avenue Royalmount, Montréal H4P 2R2 , CANADA

1. Introduction

For more than a decade, fermentation processes have found a larger and larger place in industrial applications. Generally speaking, the problems arising from their implementation are similar to those of more classical industrial processes, and the need for efficient control systems is obvious. However the automatization of biotechnological processes is developing slowly. Two main reasons can be put forward to explain this phenomenon :

1) The internal working and dynamics of fermentation are as yet badly grasped. It appears very difficult to develop models which take into account the numerous factors influencing the parameters which characterize the micro-organism growth. Moreover, because they involve living organisms, their dynamics are strongly non linear and non stationary.

2) There is a lack, in many instances, of a cheap and reliable instrumentation to measure on-line the important process variables, i.e. the biological variables (biomass, substrate, synthesis product).

The classical control methods do not prove very efficient to tide over these basic difficulties. Therefore there is a need to develop appropriate estimation and control algorithms which take into account the nonlinearities and the non stationary features of the processes.

The objective in this paper is to illustrate with a particular fermentation process, the anaerobic digestion, how to take advantage of :

- the nonlinear structure of the process
- the few available on-line measurements

to implement simple efficient adaptive controllers. It is also worth noting that the proposed algorithms do not require any analytical expression for the fermentation parameters (like the specific growth rates).

The paper is organized as follows. The anaerobic digestion and its basic dynamics are described in section 2. The following sections present three different control examples.

In anaerobic digestion, it is important to control the substrate concentration. In section 3, a nonlinear adaptive control algorithm for regulating the substrate concentration is developed and experimental results are presented. However in practice, it may be difficult to have at one's disposal on-line measurements of the substrate concentration which are at the same time reliable and at a fair cost. Therefore there is a clear incentive to look for other possible controlled variables whose on-line measurement is easier and which gives an appropriate representation of the substrate concentration and its dynamics. In such a perspective, alternative control solutions are presented in sections 4 and 5. Section 4 is dedicated to the adaptive control of the bicarbonate concentration. And in section 5, it is suggested how to use the hydrogen measurement to control anaerobic digestion processes. The control algorithms presented in sections 4 and 5 are based on ideas similar to those developed in section 3. They are illustrated by realistic simulation results.

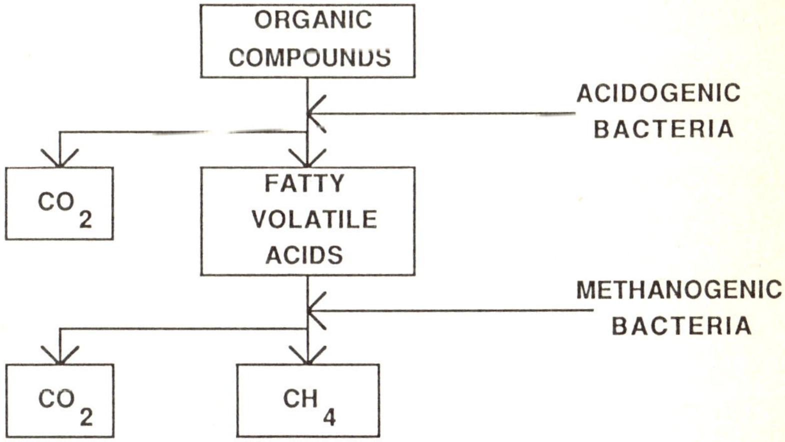

Fig.1. Basic scheme of the anaerobic digestion

2. The Anaerobic Digestion : description of the process

The anaerobic digestion is a fermentation process used for the biological treatment of organic wastes, and in which methane gas is produced. Basically, it is a two-step fermentation process (fig.1)(e.g. Mc Carty, 1964; Hill & Barth, 1977). In the first step, the organic compounds are fermented into volatile fatty acids and carbon dioxide (CO_2) by a group of acidogenic bacteria. And in the second step, a group of methanogenic bacteria convert the acids into methane gas (CH_4) and carbon dioxide (CO_2).

Basic Dynamical Model

By using mass balances considerations, the anaerobic digestion (as described above) in a CSTR (continuous stirred tank reactor) can be described by the following state-space equations (e.g. Hill & Barth, 1977) :

$$\frac{dX_1}{dt} = \mu_1 X_1 - DX_1 \qquad (1.a)$$

$$\frac{dS_1}{dt} = -k_1\mu_1 X_1 + DS_{in} - DS_1 \qquad (1.b)$$

$$\frac{dX_2}{dt} = \mu_2 X_2 - DX_2 \qquad (1.c)$$

$$\frac{dS_2}{dt} = -k_2\mu_2 X_2 + k_3\mu_1 X_1 - DS_2 \qquad (1.d)$$

$$Q_{CH4} = k_4\mu_2 X_2 \qquad (1.e)$$

$$Q_{CO2} = k_5\mu_1 X_1 + k_6\mu_2 X_2 \qquad (1.f)$$

with X_1 : acidogenic bacteria concentration

S_1 : organic matter concentration

X_2 : methanogenic bacteria concentration

S_2 : volatile acid concentration

S_{in} : influent organic matter concentration

D : dilution rate (i.e. hydraulic flow rate)

Q_{CH4} : methane gas flow rate

Q_{CO2} : carbon dioxide flow rate

μ_1 : acidogenic bacteria specific growth rate

μ_2 : methanogenic bacteria specific growth rate

k_1 , k_2 , k_3 , k_4 , k_5 , k_6 : yield coefficients

In these equations, the specific growth rates μ_1 and μ_2 are known to be a complex functions of many physico-chemical and biological factors like the biomass concentrations , the substrate concentrations , the pH or the temperature.

Many different analytical structures have been proposed in the literature to express the dependence between the specific growth rate and the most influential variables : more than fifty different growth rate expressions are listed in Dochain (1986). The most popular is certainly the "Monod law" (Monod, 1942) :

$$\mu = \frac{\mu_{max} S}{K_M + S}$$

where μ_{max} is the maximum specific growth rate, K_M is the Michaelis-Menten parameter and S the substrate concentration.

In an automatic control context, two important difficulties are arising when dealing with the microbial specific growth rates :

1) How to choose the appropriate analytical expression to describe the dependence of the specific growth rates to the process variables?

2) When you try to calibrate the parameters of the chosen expression (e.g. μ_{max} and K_m in the above Monod law), you have to face intricate identifiability difficulties due to the lack of accuracy of the measurements and to their limited number and frequency (see Holmberg (1982); Vialas et al (1985) for a more detailed discussion on the subject).

In the following, in order to avoid these choice and identifiability difficulties, we shall consider the specific growth rates as a time-varying parameters which will then be estimated on-line (section 5). This is in line with a number of recent works on estimation and control of fermentation processes (see e.g. Holmberg and Ranta, 1982, Dochain and Bastin, 1984; Stephanopoulos and San, 1984; Bastin and Dochain, 1986) .

Another typical problem arising in anaerobic digestion processes is that the bacterial populations X_1 and X_2 are not accessible from any direct measurement, even from off-line analyses.

Statement of the control objective

An important feature of anaerobic digestion is that it may become very unstable when its input variables (i.e. the influent flow rate D and the influent substrate concentration S_{in}) are modified. This phenomenon takes place under the form of accumulation of volatile fatty acids (acetate, proprionate); it may destabilize the process and lead it to a wash-out (i.e. a state where the bacterial life has completely disappeared). In such a case, it is obvious that efficient feedback control is needed to stabilize the process.

Furthermore, if the anaerobic digestion is used for waste treatment purposes, the control objective clearly consists of regulating the output pollution at a prescribed level despite the fluctuations of the input pollution by acting on the hydraulic flow rate.

Throughout sections 3, 4 and 5, we assume that the dilution rate D (i.e. the hydraulic influent flow rate) is the control input.

3. Adaptive Control of the Substrate Concentration

Statement of the control algorithm

It is clear that from equations (1), we can define the pollution level as the total substrate concentration S :

$$S \triangleq S_1 + S_2$$

Assume that :

1) the substrate concentration S, the influent substrate concentration S_{in}, the gaseous flow rates Q_{CH4} and Q_{CO2} are available for on-line measurement;
2) the yield coefficients k_i (i = 1 to 6) are unknown.

By combining equations (1.b)(1.d)(1.e)(1.f), we obtain the following dynamical equation for S:

$$\frac{dS}{dt} = -\alpha_1 Q_{CH4} + \alpha_2 Q_{CO2} + DS_{in} - DS \qquad (2)$$

with $\alpha_1 \triangleq \frac{1}{k_4}\left[k_2 + \frac{k_3 - k_1}{k_5}\right]$

$$\alpha_2 \triangleq \frac{k_3 - k_1}{k_5}$$

Note that in equation (2), we have substituted the specific growth rates μ_1 and μ_2, and the biomass concentrations X_1 and X_2 for on-line measurable variables (Q_{CH4}, Q_{CO2}).

By using a first order Euler approximation for dS/dt, equation (2) can be written in discrete-time as follows :

$$S_{t+1} = S_t - \alpha_1 TQ_{CH4t} + \alpha_2 TQ_{CO2t} + TD_t S_{int} - TD_t S_t \tag{3}$$

where T is the sampling period and the index t is the time index (t = 0, 1, 2,...).

Let us now state the control algorithm.

Define S* as the desired substrate concentration. If we wish that the regulation error :

$$\tilde{S} \triangleq S^* - S_t \tag{4}$$

has a closed-loop exponentially stable dynamical behaviour, i.e. :

$$\tilde{S}_{t+1} = c\tilde{S}_t \qquad 0 < c < 1 \tag{5}$$

we can calculate the command input D_t as follows :

$$D_t = \frac{(1-c)(S^* - S_t) + \alpha_1 TQ_{CH4t} - \alpha_2 TQ_{CO2t}}{T(S_{int} - S_t)} \tag{6}$$

Since we have assumed that the parameters α_1 and α_2 are unknown, they should be replaced in equation (6) by on-line estimates :

$$D_t = \frac{(1-c)(S^* - S_t) + \hat{\alpha}_{1t} TQ_{CH4t} - \hat{\alpha}_{2t} TQ_{CO2t}}{T(S_{int} - S_t)} \tag{7}$$

Since equation (3) is linear in the parameters α_1 and α_2, these can be estimated on-line via some parameter adaptation algorithm like, for instance, a recursive least square algorithm with a forgetting factor :

$$\hat{\theta}_{t+1} = \hat{\theta}_t + P_t \Phi_t e_{t+1} \tag{8.a}$$

$$P_t = \frac{P_{t-1}}{\gamma}\left[I_2 - \frac{\Phi_t \Phi_t^T P_{t-1}}{\gamma + \Phi_t^T P_{t-1} \Phi_t} \right] \qquad P_0 = a I_2 \quad I_2 = \begin{bmatrix} 1 & 0 \\ 0 & 1 \end{bmatrix} \quad a > 0 \tag{8.b}$$

with $\hat{\theta}_t = [\hat{\alpha}_{1t}, \hat{\alpha}_{2t}]^T$

$\Phi_t = [-TQ_{CH4t}, TQ_{CO2t}]^T$

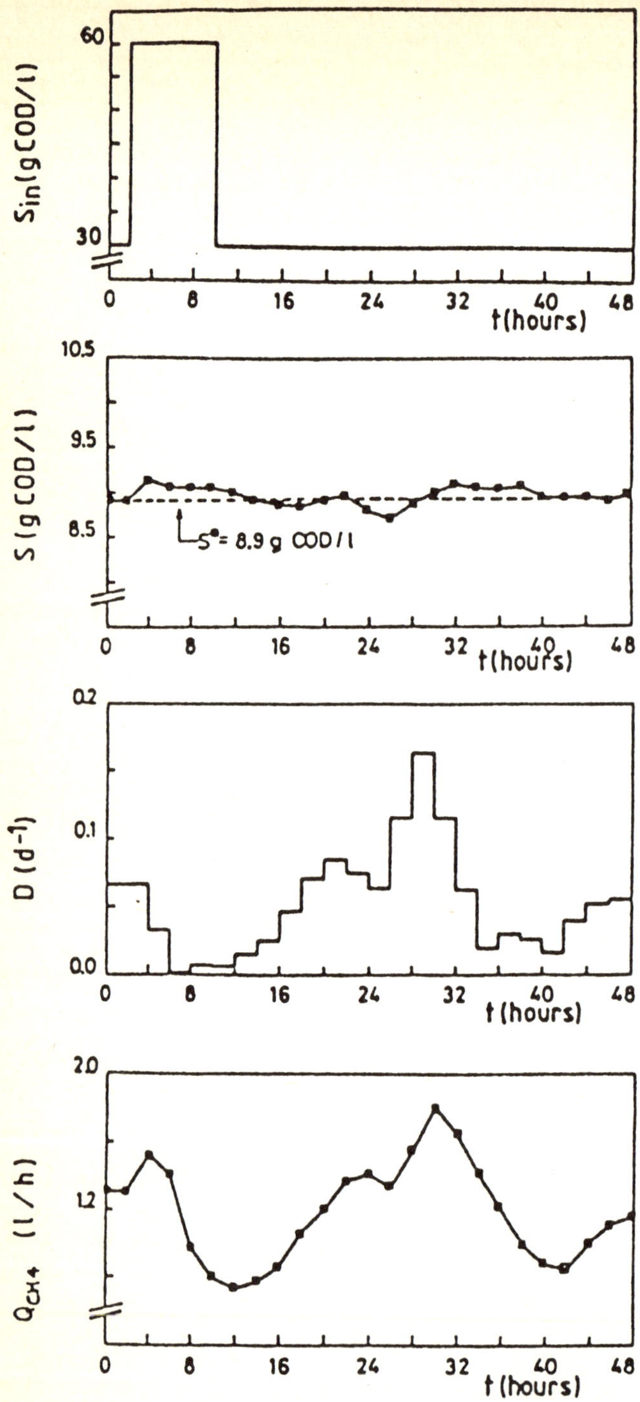

Fig.2. Adaptive control of the substrate concentration

$$e_{t+1} = S_{t+1} - S_t + \hat{\alpha}_{1t} TQ_{CH4t} - \hat{\alpha}_{2t} TQ_{CO2t} - TD_t S_{int} + TD_t S_t$$

P_t : gain of the estimation algorithm

γ : forgetting factor ($0 < \gamma \leq 1$)

A forgetting factor γ has been introduced in the estimation algorithm to account for possible variations of the parameters α_1 and α_2.

Moreover, in practice, the dilution rate D must be positive and upper bounded :

$$0 \leq D_t \leq D_{max} \tag{9}$$

with D_{max} : the upper bound on D

Equations (7)-(9) constitute the nonlinear adaptive control algorithm. Its stability and convergence properties are analyzed in Dochain & Bastin (1988a).

Real-life implementation of the control algorithm

The above control algorithm (7)-(9) has been implemented under the (realistic) assumption that $k_1 = k_3$ (i.e. $\alpha_2 = 0$) on a pilot bioreactor. In this case, the control equation (7) reduces to :

$$D_t = \frac{(1 - c)(S^* - S_t) + \hat{\alpha}_{1t} TQ_{CH4t}}{T(S_{int} - S_t)} \tag{7.bis}$$

One control experiment is presented in fig.2. In this experiment, the substrate concentration was measured via off-line COD (Chemical Oxygen Demand) analysis every two hours (which corresponds to the measuring duration). Despite the high variation (100%) of the influent substrate concentration (which could have led the process to a wash-out in open loop), the process remained stable and the output substrate concentration S has undergone small variations (less than 3%). This control experiment is described in details in Renard et al (1988).

4. Adaptive Control of the Bicarbonate Concentration

The difficulty, in practice, of having at one's disposal reliable and cheap on-line measurements of the substrate concentration (e.g. via a COD analysis) may limit the applicability of the control algorithm (7)-(9). This section presents an alternative control solution, based on ideas similar to those developed in the preceding section and using the bicarbonate concentration which is easier to measure on-line (Rozzi, 1984a) (Section 5 will present another alternative based on the hydrogen measurement).

It has been shown (Rozzi, 1984a, 1984b; Houba & Dehem, 1983) that in anaerobic digestion processes, the bicarbonate concentration B can be related to the fatty volatile acid concentration S_2 as follows :

$$B = Z - \rho S_2 \qquad (10)$$

Where Z is the net cation concentration (Z represents the ion concentration available for the bicarbonates and fatty volatile acids) and ρ is a conversion factor. The dynamical behaviour of Z can be described by the following equation (Houba & Dehem, 1983) :

$$\frac{dZ}{dt} = -DZ + DZ_{in} \qquad (11)$$

with Z_{in} : influent net cation concentration.

Our conjecture is that regulating the bicarbonate concentration B will help to avoid fatty volatile acid accumulation and process destabilization. We shall now introduce a bicarbonate concentration regulation scheme which do not require any on-line substrate concentration measurement.

After some manipulations on equations (1)(11) (Van Breusegem et al, 1988), we can deduce the following dynamical equation for the bicarbonate concentration B :

$$\frac{dB}{dt} = \beta_1 Q_{CH4} + \beta_2 Q_{CO2} + DB_{in} - DB \qquad (12)$$

with β_1, β_2 : yield coefficients.
B_{in} : influent bicarbonate concentration

Assume that :

1) the bicarbonate concentrations B and B_{in}, and the gaseous flow rates Q_{CH4} and Q_{CO2} are available for on-line measurement;
2) the parameters β_1 and β_2 are unknown;

By using arguments similar to those of section 3, we can write the following control equation :

$$D_t = \frac{(1-c)(B^* - B_t) + \hat{\beta}_{1t} T Q_{CH4t} + \hat{\beta}_{2t} T Q_{CO2t}}{T(B_{int} - B_t)} \qquad (13)$$

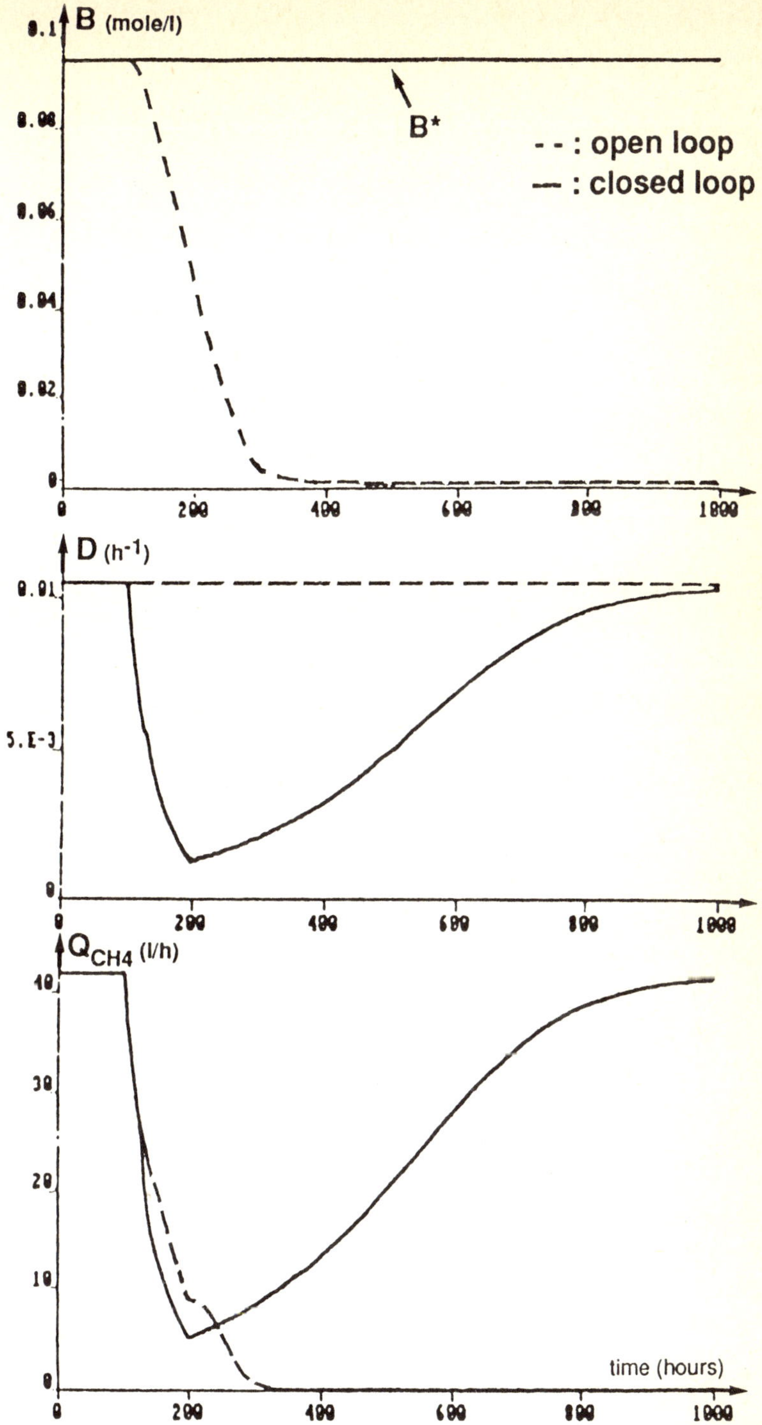

Fig.3. Adaptive control of the bicarbonate concentration

with B^* : desired bicarbonate concentration

$\hat{\beta}_{1t}, \hat{\beta}_{2t}$: on-line estimates of β_1 and β_2

($\hat{\beta}_{1t}$ and $\hat{\beta}_{2t}$ are updated via, for instance, the recursive least square algorithm (8) with $e_{t+1} = B_{t+1} - B_t + TD_tB_t - TD_tB_{int} - \hat{\beta}_{1t}TQ_{CH4t} - \hat{\beta}_{2t}TQ_{CO2t}$ and $\Phi_t^T = [TQ_{CH4t}, TQ_{CO2t}]$).

Simulation result

Fig.3 describes a simulation experiment in which a pulse of toxic matter is introduced in the bioreactor after 100 hours (see Van Breusegem et al, 1988). In open loop (with a constant dilution rate D), this disturbance destabilizes the process and leads it to a wash-out, while in closed loop, with the adaptive controller (13)(8), the bioreactor remains stable, close to the desired bicarbonate concentration B^*.

5. Adaptive Control of the Hydrogen Concentration

Quite recently, the role of hydrogen in the dynamics of anaerobic digestion processes has been emphasized (Mosey, 1983). Basically, the influence of hydrogen can be summarized as follows. The methanogenic step (see fig.1) can be divided into two parallel metabolic phases. In the first one (as already mentioned in section 2), fatty volatile acids are transformed into methane gas and carbon dioxide by (acetoclastic) methanogenic bacteria. And in the second one, hydrogen (also produced in the acidogenic step) is combined with CO_2 to produce methane gas under the action of (hydrogenophilic) methanogenic bacteria.

In an automatic control context, the use of hydrogen appears to be very promising for the following three reasons :

1) In some instances (e.g. in the presence of glucose-type substrate (like wastewater from sugar industry)), up to 30% of the methane gas may be produced from hydrogen.

2) The two methanogenic phases are, as far as we know, assumed to be completely parallel, but the hydrogen kinetics are faster than the corresponding acid dynamics (see Dewan et al, 1984). Under this assumption, it is reasonable to believe that acting directly on the hydrogen will prevent any process instability more quickly.

3) It is easier to measure on-line hydrogen than fatty volatile acids. In fact, under the (realistic) assumption that the hydrogen in the liquid phase is in equilibrium with the hydrogen in the gas phases, it only requires to determine the hydrogen content of the output gas (for further details, see Pauss et al, 1988).

It is obvious that here again, the introduction of the new variable, the hydrogen, makes the dynamical model of the process more complex. In particular, the dynamical behaviour of the hydrogen concentration can described by the following equation (Dochain & Perrier, 1989) :

$$\frac{dX_3}{dt} = \mu_3 X_3 - DX_3 \tag{14.a}$$

$$\frac{dS_3}{dt} = -DS_3 - Q_{H2} + k_7\mu_1 X_1 - k_8\mu_3 X_3 \tag{14.b}$$

with S_3 : hydrogen concentration (in the liquid phase)

Q_{H2} : gaseous hydrogen flow rate

X_3 : hydrogenophilic methanogenic bacterial concentration

μ_3 : specific growth rate of X_3

k_7, k_8 : yield coefficients.

For control purposes, we can rewrite equation (14.b) as follows.

By using singular perturbation arguments (e.g. Kokotovic et al, 1986) about the substrate dynamics in the acidogenic step (equation (1.b)), we write :

$$k_1\mu_1 X_1 = DS_{in} \tag{15}$$

Define the auxiliary variable Z and the parameter γ_1 :

$$Z \triangleq S_3 + k_8 X_3 \tag{16}$$

$$\mu_3 \triangleq \gamma_1 S_3 \tag{17}$$

Note that equation (17) does not introduce any restriction on the structure of the specific growth rate μ_3. It simply implies that $\mu_3 = 0$ when $S_3 = 0$, in accordance with the physical reality.

By using (15)(16)(17), equation (14.b) becomes :

$$\frac{dS_3}{dt} = -DS_3 - Q_{H2} + \gamma_2 DS_{in} - \gamma_1 S_3(Z - S_3) \tag{18}$$

where $\gamma_2 = k_7/k_1$ and Z obeys to the following dynamical equation :

$$\frac{dZ}{dt} = -DZ + \gamma_2 DS_{in} - Q_{H2} \tag{19}$$

If we assume that :

1) S_3, Q_{H2} are available for on-line measurement
2) the parameter γ_2 is known (it is simply a stoechiometric coefficient between glucose and hydrogen !)
3) the parameter γ_1 is unknown

we can deduce from equation (18) a nonlinear adaptive control algorithm similar to (7)-(9), i.e. :

$$D_t = \frac{(1-c)(S_3^* - S_{3t}) + TQ_{H2t} + \hat{\gamma}_{1t} T S_{3t}(\hat{Z} - S_{3t})}{T(\gamma_2 S_{int} - S_{3t})} \tag{20}$$

with S_3^* : desired hydrogen concentration

$\hat{\gamma}_{1t}$, $\hat{Z}$:on-line estimates of γ_1 and Z.

γ_1 can be estimated by using (as before) a recursive least square estimation algorithm, while Z is updated via equations (19), i.e. in discrete-time :

$$\hat{Z}_{t+1} = \hat{Z}_t - TD\hat{Z}_t + T\gamma_2 DS_{int} - TQ_{H2t} \tag{21}$$

Simulation result

A crucial problem in operating anaerobic digestion plants is the reactor start-up, since substrate overloading effects may easily appear and then lead the process to a wash-out. Fig.4 shows start-up simulation experiments, both in open loop and in closed loop. In open loop, the dilution rate was kept to a value equal to 0.1 and the process was led to a wash-out. On the other hand, the adaptive controller proved successful in compensating the overloading effects and in leading the process to a stable steady-state.

6. Conclusions

In this paper, we have illustrated on a particular fermentation process how to take advantage of the nonlinear structure of the system and of the few available on-line measurements to implement simple adaptive controllers. Their efficiency have been illustrated by real-life and simulation results. Ideas similar to those developed in this paper can be used for other control or estimation situations (e.g. Dochain and Bastin, 1984, 1988;Bastin and Dochain, 1986; Dochain, 1986).

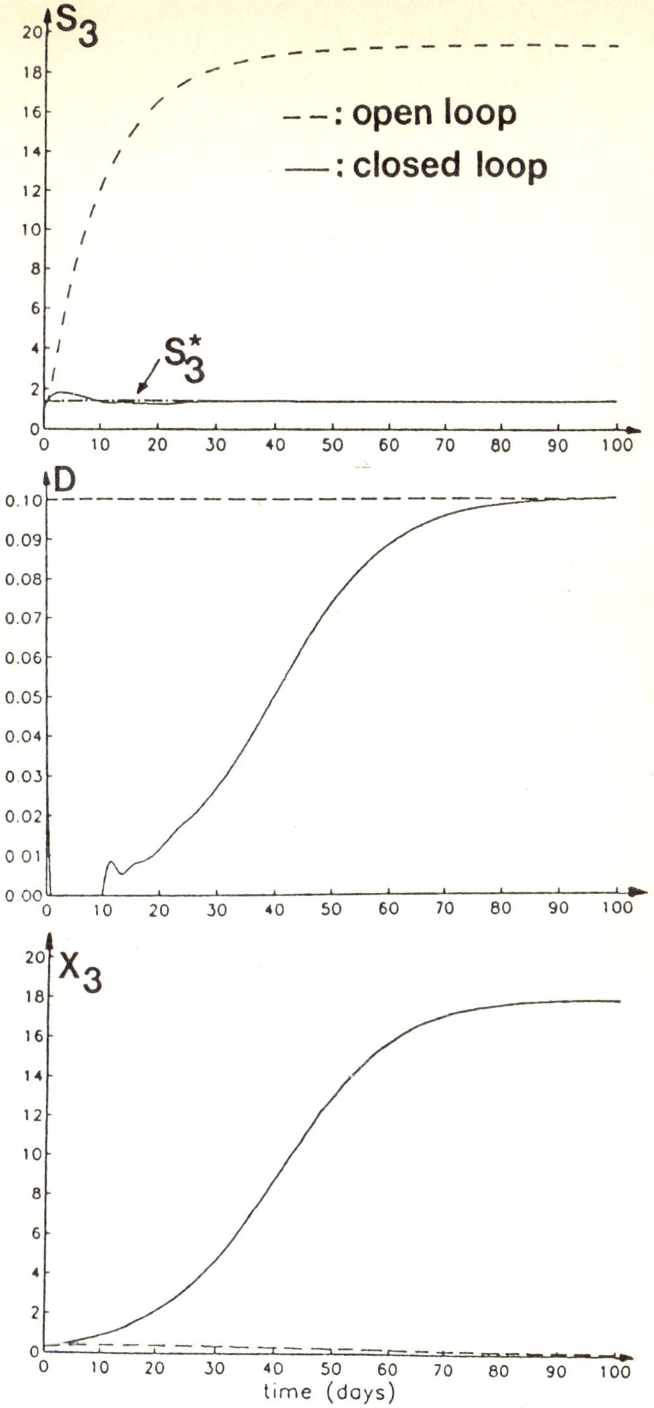

Fig.4. Adaptive control of the hydrogen concentration

References

BASTIN G. and D. DOCHAIN (1986). On-line estimation of microbial specific growth rates. *Automatica*, vol.22 (6), 707-710.

DEWAN M., D. PONCELET, D. DOCHAIN and M. HAEST (1984).Présentation d'un système d'acquisition de données et de contrôle automatique de bioréacteurs. Internal report, *Lab. Aut. Dyn. Anal. Syst.*, University of Louvain, Belgium.

DOCHAIN D. and G. BASTIN (1984). Adaptive identification and control algorithms for non linear bacterial growth systems. *Automatica*, Special Issue on Adaptive Control, 20, 621-634.

DOCHAIN D. (1986). On-line Parameter Estimation, Adaptive State Estimation and Adaptive Control of Fermentation Processes. Ph. D. Thesis, University of Louvain, Belgium.

DOCHAIN D. and G. BASTIN (1988). Nonlinear Adaptive Control Algorithms for Fermentation Processes. *Proc.1988 American Control Conference*, Invited Session on Advances in Control of Biotechnical Processes, Atlanta, 1124-1128.

DOCHAIN D. and M. PERRIER (1989). Adaptive control of the hydrogen concentration in anaerobic digestion. Submitted for publication.

HILL D.T. and C.L. BARTH (1977). A dynamic model for simulation of animal waste digestion. *J. WPCF*, 2119-2143.

HOLMBERG A. and J. RANTA (1982). Procedures for parameter and state estimation of microbial growth process models. *Automatica*, 18, 181-193.

HOLMBERG A. (1982). On the practical identifiability of microbial growth models incorporating Michaelis-Menten nonlinearities. *Math. Biosc.*, 62, 23-43.

HOUBA M. and F. DEHEM (1983). Contribution à la modélisation et à la régulation d'un procédé biotechnologique: adjonction des équations chimiques. Internal Report, *Lab. Aut. Dyn. Anal. Syst.*, University of Louvain, Belgium.

KOKOTOVIC P., H.K. KHALIL and J. O'REILLY (1986). Singular Perturbation Methods in Control : Analysis and Design. Academic Press, London.

MC CARTY P.L. (1964). Anaerobic waste treatment fundamentals. *Public Works*, Sept. (107-112), Oct. 123-126), Nov. (91-94), Dec. (95-96).

MONOD J. (1942). Recherches sur la Croissance des Cultures Bactériennes. Hermann, Paris.

MOSEY F.E. (1983). Mathematical modelling of the anaerobic digestion process : Regulatory mechanisms for the formation of short-chain volatile acids from glucose. *Wat. Sci. Tech.*, vol.15, 209-232.

PAUSS A., S. GUIOT and R. SAMSON (1988). Sensors for anaerobic digestion management. *5th International Symposium on Anaerobic Digestion*, Bologna, Italy, 22-26 May.

RENARD P. D. DOCHAIN, G. BASTIN, H. NAVEAU and E.J. NYNS (1988). Adaptive control of anaerobic digestion processes. A pilot-scale application. *Biotechnol. & Bioeng.*, vol.31, 287-294.

ROZZI A. (1984a). Direct bicarbonate determination in anaerobic digester liquors by measurement of the pressure of carbon dioxide. *Process Biochem.*, 19(6), 201-203.

ROZZI A. (1984b) Modelling and control of anaerobic digestion processes. *Trans. Inst. Meas. & Control*, vol.6, n^{0}3, 153-159.

STEPHANOPOULOS G. and K-Y SAN (1984). Studies on on-line bioreactor identification. *Biotech. & Bioeng.*, vol.26, 1176-1188.

VAN BREUSEGEM V., G. BASTIN and A. ROZZI (1988). Feedback control of anaerobic digestion processes through adaptive bicarbonate regulation. *5th International Symposium on Anaerobic Digestion*, Bologna, Italy, 22-26 May.

VIALAS C., A. CHERUY and S. GENTIL (1985). An experimental approach to improve the Monod model identification. In A. Johnson (Ed.), *Modelling and Control of Biotechnological Processes*, Pergamon, 175-180.

BLOOD PRESSURE POSTOPERATIVE TREATMENT: MODEL REFERENCE ADAPTIVE CONTROL WITH CONSTRAINTS

G. Pajunen, M. Steinmetz, R. Shankar
Florida Atlantic University
Boca Raton, Florida 33431

Abstract

This paper presents an approach for adaptive control of blood pressure using sodium nitroprusside. A modified stochastic model reference adaptive control algorithm with a time varying reference model is developed and used for this purpose. In order to optimize the performance of the closed loop system while meeting medical constraints imposed on the infusion rate and the mean arterial pressure an automatic adjustment of the reference model is proposed. Extensive computer simulations show the robustness of the proposed controller in the presence of relatively high noise levels, over the full range of plant parameters and for large parameter variations.

1. Introduction

Immediately following heart surgery a patient's blood pressure may be at a high level. To decrease the chances of complication it is necessary to reduce the pressure to a lower level as soon as possible. One method commonly used to treat this condition is the continuous infusion of the fast-acting drug sodium nitroprusside. Clinical experiments reported in the literature have shown that the automation of this procedure is safe, effective and often superior to manual methods (Reves, 1978; Slate, Sheppard, 1982). The transfer function relating the reduction in pressure to the infusion rate was developed by Slate (1980). This transfer function includes two initially unknown time delays. There can be as much as 30:1 range in the gain of this function over a typical patient population. Moreover, the parameters of the transfer function associated with a particular patient are exponentially varying. These characteristics of the blood pressure process are the reasons why a constant parameter controller works unsatisfactorily. There were two groups of methods proposed recently in the literature to deal with this problem: the multiple model adaptive approach and the single model adaptive approach. The multiple model adaptive controller (MMAC) assumes that the plant at any given time may be represented by one of a finite number of sub models (He et al., 1986, Martin et al., 1987). The design process entails the selection of these models, designing a controller for each model, and devising an algorithm to select the appropriate model at each time. The sub models of the model bank considered in the above mentioned work only account for a time-varying drug sensitivity. As sub models are included for all other plant parameters, the design of the MMAC becomes increasingly complex. This is the major drawback of the MMAC design.

The single model adaptive controller, e.g., minimum variance, pole placement or model reference adaptive controller is based on the on-line estimation of the process parameters. The stochastic adaptive controller based on minimum variance control and least-squares parameter estimation is applied by Stern et al. (1981), Walker et al. (1982) and Arnsparger et al (1983). This type of adaptive controller is characterized by large amplitude transients which are not acceptable for blood pressure control for medical reasons. Model reference adaptive control (MRAC) of drug infusion rate was proposed by Kaufman et al, (1984). Using this control method smoother transients can be obtained. The limitation of this work is that no stochastic disturbance is considered and only the compensation for initially unknown gain is provided. Moreover, all the above mentioned adaptive control system are based on the assumption that the time delay of the process is known a priori. This assumption cannot be accepted for new patients. An adaptive algorithm which can handle uncertain time-delay is based on pole-placement control. This algorithm is considered by Serna et al. (1983) for adaptive control of multiple drug infusion.

The general limitation of practical applications of adaptive control theory is that it largely ignores the presence of realistic input-output constraints. In the blood pressure control system one has to consider medical limitations on minimum blood pressure, its maximum rate of change and maximum infusion rate. The adaptive control algorithms with saturated input and output signals have been implemented, but no theory exists that considers the stability of such systems. Recently a modified MRAC system with constraints was proposed by Pajunen and Steinmetz (1987). It includes the application of a time-varying reference model instead of a constant prespecified one. This adaptive controller is applied here to control the drug infusion rate. The plant is considered initially unknown with exponentially time-varying parameters and corrupted by stochastic noise. The simulation results are presented and the comparison of the effectiveness of the proposed MRAC, and MMAC used by Martin et al. (1984) is given.

2. The Problem Formulation

A continuous-time model describing the relationship between the change in the blood pressure and drug infusion rate as developed by Slate (1980) is as follows:

$$\Delta P(s) = \frac{K e^{-T_i s}(1+\alpha e^{-T_c s})}{1+\tau s} I(s) \quad . \tag{1}$$

In Eq. (1) ΔP is the change in blood pressure, I is the infusion rate, K is drug sensitivity, α is the recirculation constant, T_i is the initial transport delay, T_c is the recirculation time constant, and τ is the system time constant.

The corresponding discrete-time model for this process can be given as follows (Steinmetz, 1987):

$$\Delta P(t) = \frac{q^{-d}(b_0 + b_m q^{-m})}{1 - a_1 q^{-1}} I(t) \tag{2}$$

where q^{-1} denotes a unit delay operator.

A range of values for the parameters in the model (1) for different patients is given in Slate (1980). Using these values and the sampling time T equal to 15 seconds, a range of values for the parameters in model (2) can be computed. It is given in Table 1.

One can see that there is a considerable difference in the parameter values, including the pure time delay, for different patients. For a given patient time delay is unknown, but it can be assumed constant. The parameters b_o , b_m and a_1 , however, drift in an exponential manner during the infusion procedure. Thus, the controller should be tuned for a particular patient and should be able to handle drifting parameters and initially unknown time delay.

The following physical and medical constraints have to be introduced for this system (Slate, 1980):

- On the infusion rate

(C1) $\quad 0 \leq I(t) \leq 180$ ml/hr

- On the maximum rate of change of the mean arterial pressure (MAP) per 15 seconds sampling interval

Table 1. Range of Values for Parameters of the Discrete-time Plant Model for Sampling Time 15 Seconds.

Parameter	Minimum	Maximum	Nominal
b_o	0.053	3.546	0.187
b_m	0	1.418	0.075
a_1	0.606	0.779	0.741
d	2	5	3
m	2	5	3

(C2) $\quad \Delta MAP = MAP(t-1) - MAP(t) \leq 15$ mm Hg

- On the minimum value of the pressure which should not drop more than 20 mm Hg below the set point. In this paper desired pressure is 100 mm HG, so

(C3) $\quad MAP(t) \geq 80$ mm Hg

The objective of this paper is to develop a blood pressure control system taking into consideration initially unknown and exponentially time-varying process parameters, initially unknown pure time delay, and input-output constraints (C1)-(C3). This system should be robust with respect to stochastic as well as deterministic disturbances.

3. Control System Design

(a) Unknown and Exponentially Convergent Parameters

The transfer function given by Eq. (2) is invertibly stable for all ranges of values for parameters given in Table 1; therefore a model reference approach can be used. Taking into consideration significant stochastic noise a direct stochastic model reference controller was implemented (Goodwin, Sin 1984). Based on the paper by Goodwin et al. (1986), it is easy to prove that this algorithm developed originally for time-invariant plants is also globally stable when parameters are exponentially convergent (Pajunen, Steinmetz, 1987). This means that a direct stochastic model reference controller can be used without modification for exponentially convergent parameters.

The extended least-squares algorithm with covariance resetting was chosen for the adaptation algorithm. The stochastic MARAC algorithm is given as follows:

$$\hat{\theta}(t) = \hat{\theta}(t-1) + \frac{P(t-1)\,\phi(t-d)}{1+\phi(t-d)^T P(t-1)\,\phi(t-d)}\,\varepsilon(t) \tag{3a}$$

where

$$\varepsilon(t) = y_F(t) - \phi(t-d)^T\,\hat{\theta}(t-1) \tag{3b}$$

$$\phi(t)^T = \left[\Delta P(t)\, I(t) \ldots I(t-d-m) \; - y_F(t+d-1) \ldots - y_F(t+d-r)\right] \tag{3c}$$

$$y_F(t) = \phi(t-d)^T\,\hat{\theta}(t-d)\ . \tag{3d}$$

Covariance Resetting:

$$P(t) = P(t-1) - \frac{P(t-1)\,\phi(t-d)\,\phi(t-d)^T P(t-1)}{1+\phi(t-d)^T P(t-1)\,\phi(t-d)} \tag{4a}$$

$$P(t) = \rho I \;\; \textit{iff} \;\; P_{ii}(t-1) < P_t\ ,\quad i = 1,2,\ldots,n+m+r \quad \text{and} \quad \frac{1}{6}\sum_{i=1}^{6} \varepsilon(t-i) \geq SD \tag{4b}$$

Finally, the control input is generated from

$$\phi(t)^T\,\hat{\theta}(t) = 1 + e(t)\,q^{-1})\,\Delta P^*(t) \tag{5}$$

The values for P_t and SD in Eq. (4b) are determined based on the noise level. The value of ρ chosen for the simulations was 0.01 times the initial P(0).

(b) Input and Output Constraints

The limitation of this algorithm is that it cannot handle input-output constraints; hence a modification of a stochastic MRAC algorithm is proposed (Steinmetz, 1987; Pajunen, Steinmetz, 1987). It includes the application of a time-varying reference model instead of a constant, prespecified one. In this case the following first order reference model is used.

$$\Delta P^*(t) = \frac{q^{-d}(1-e(t))}{1-e(t)\,q^{-1}}\,\Delta P_d(t) \tag{6}$$

where ΔP_d represents the desired change in pressure and ΔP^* is the reference model output. The automatic adjustment of the parameter e(t) is done based on a so-called long-range prediction principle. It means that the control input and the process output are predicted over a time horizon h greater than the impulse response of the system plus pure time delay. The constraints are checked for all present and future values of input and output sequences. If *all* the constraints are satisfied and in addition all the values are sufficiently far from the limits, then the reference model is speeded up. That means e(t) is decreased. If, however, *any* of the constraints are violated or if some of the values are too close to the limit, the model is slowed down. This means that e(t) is increased. In order to assure a global stability of the closed-loop system, it is important that the reference model remains stable, so e(t) was constrained to

$$|e(t)| < 1$$

In this case the following algorithm for modification of the reference model was used.

- *Slow down* the reference model (increase e(t)) *iff* for any i, $i = 1, 2, \ldots, h$

  $I(t+i) > 165$ ml/hr

  or $I(t+i) < 0$

  or $MAP(t+i) < 95$ mm Hg

  or $\Delta MAP(t+i) > 12$ mm Hg

- *Speed up* the reference model (decrease e(t)) *iff* for any i, $i = 1, 2, \ldots, h$

  $I(t+i) < 125$ml/hr

  and $MAP(t+i) > 115$ mm Hg

  and $\Delta MAP(t+i) < 5$ mm Hg

- Leave the reference model unchanged for all other conditions.

(c) Estimation of Pure Time Delay

In order to design a MRAC algorithm one needs to know the pure time delay of the system. For first-time patients it is, however, unknown. (See Table 1). That is why the model reference blood pressure adaptive controller proposed here includes a period of an open-loop operation during which the system delay and, at the same time, the initial values for system parameters were identified. During this period, a pseudo-random binary signal (PRBS) with an amplitude dependent on the noise was used to excite the system. An offset was added to the PRBS equal to its amplitude to prevent negative infusion rate. A total number of 10 parameters was estimated and based on them the delay was computed. If no noise was present, then 15 samples were enough for reliable estimation of the delay. When the variance of the noise was set to 4 mm Hg^2, then a minimum of 75 sampling periods were required for accurate estimates (Steinmetz, 1987). Once the delay was determined, the loop was closed and the controller parameters were updated directly.

(d) System Block Diagram

A block diagram of the complete control system proposed here is shown in Figure 1.

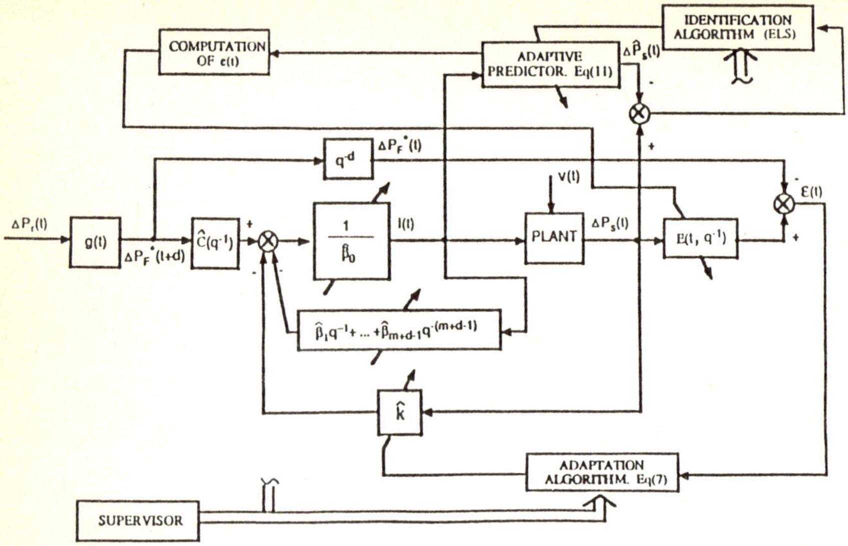

Figure 1. A block diagram of the model reference adaptive blood pressure control system

The system is comprised of three main sections. The standard MRAC block which consists of the process, controller and adaptation algorithm (Eqs. 4-6). The second part consists of an additional identification algorithm and the subroutine for updating the reference model. The last section is the supervisor block. This block determines when to switch from open-loop to closed-loop operation. It also generates the PRBS and sets its amplitude for different operating modes and noise levels.

4. Simulation Results

The objective of the simulation was to demonstrate the robustness of the controller under those conditions which most closely resemble the actual operating environments. All of the simulation results presented here assumed that the initial pressure was 150 mm Hg and the objective was to reduce this to 100 mm Hg and maintain this level to within ± 15 mm Hg . The system was first run open loop for 50 samples, then the loop was closed. The parameters start to change exponentially, in the simulated system, at sample interval 100. The system was corrupted by white or colored noise. The highest noise level had a variance 4 mm Hg^2 . This sequence had peak-to-peak noise variations of 12 mm Hg . This is considered to be a moderate to high noise level in the physiological system (Slate, 1980).

To establish a reference a simulation was run without the adaptive controller. A fixed controller was used for the plant with average parameters and the drug sensitivity was allowed to change during the run. As the results in Fig. 2 show, the blood pressure drops to 60 mm Hg after the plant changes. This is 40 mm Hg below the desired level. Obviously, some form of adaptive control is required.

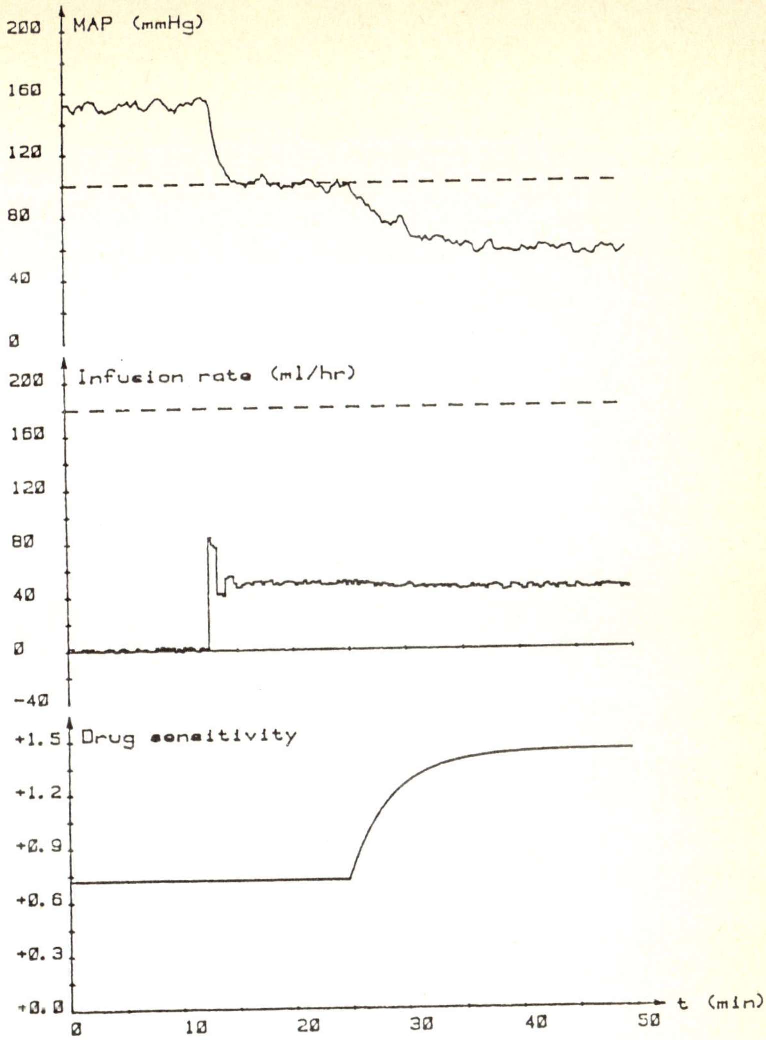

Figure 2. Results for time varying drug sensitivity and fixed controller.

Then the same process was simulated and the usual model reference adaptive control algorithm with a fixed reference model was used. The parameter e(t) was set to 0.1. The results are shown in Fig. 3a. One can see that all the constraints are violated. The results with clipping input and output signals are shown in Fig. 3b. Finally the results with the time-varying reference proposed here are shown in Fig. 3c. The maximum and minimum infusion rates, the maximum change of pressure per sampling interval, and the minimum value of the pressure for three of the above mentioned cases are summarized in Table 2.

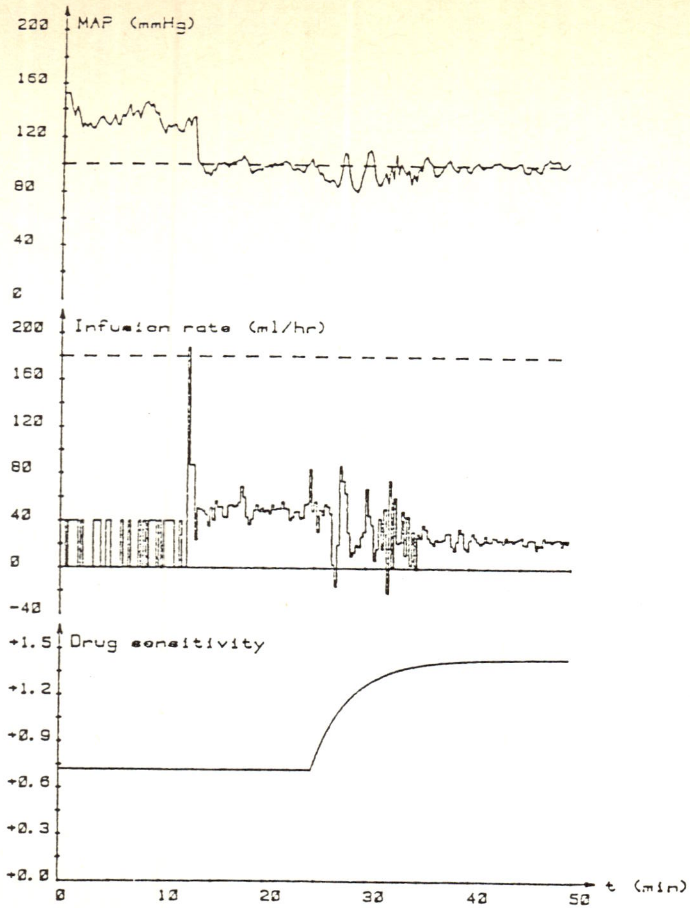

Figure 3a: Results for Time-varying Drug Sensitivity MRAC with Constant Reference Model

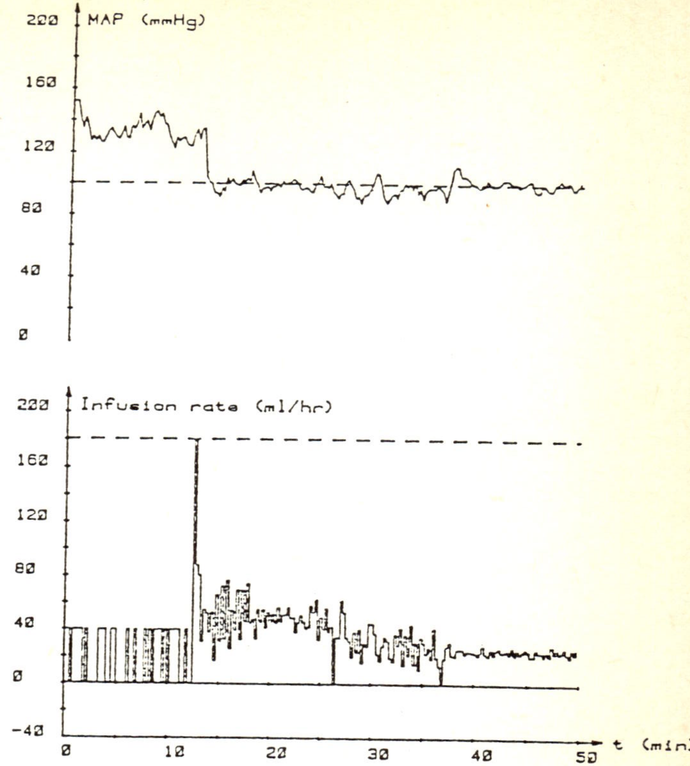

Figure 3b: Results for Time-Varying Drug Sensitivity MRAC with Clipping

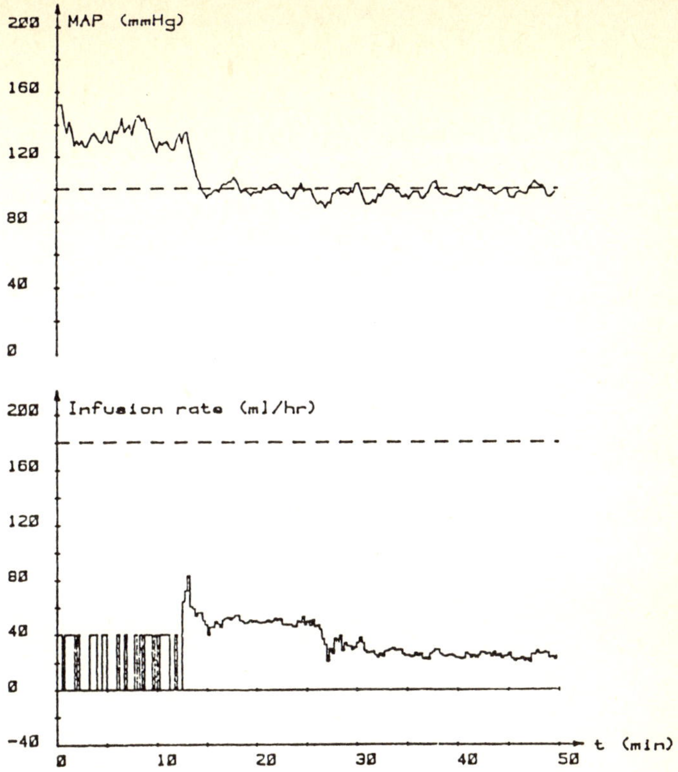

Figure 3c: Results for Time-Varying Drug Sensitivity MRAC with Time-varying Reference Model

Table 2. Summary of Results for Simulations With and Without Constraints

Fig. 3	Infusion Rate		Maximum ΔMAP	Minimum ΔMAP
	Min.	Max		
No constraints (a)	−21.7	187.4	32.5	60.8
Clipping (b)	0	180	31.1	87.6
Time–varying reference model (c)	0	83.1	9.5	88.8

The previous simulations were run with the initial plant parameters at their average values. Simulations were also done with the parameters set to the minimum and maximum values given by Slate (1980). Each of the parameters in turn was set to one of its extreme values while the other parameters were set to the average values. In all cases the drug sensitivity was varying except for $K = 9$ where α was varied instead. The time-varying reference model was used in all cases. The maximum and minimum infusion rates, the maximum rate of change of pressure per sampling interval, the minimum value of the pressure and the setting time are listed in Table 3.

Table 3. Simulation Results Using Minimum and Maximum Parameter Values. K was allowed to vary in all cases except for K = 9, where α was varied instead. The parameters not given in the left column were set to their average values. A time varying reference model was used in all cases.

Parameter	Infusion rate ml/hr1		ΔMAP mm Hg max	MAP mm Hg min	Settling Time sec
	min	max			
K = .25	0.0	162.4	7.88	83.8	180
K = 9.0	0.0	7.0	10.6	84.6	90
α = 0.0	0.0	93.4	8.5	10.6	195
τ = 60	0.0	80.9	10.9	22.3	135
τ = 30	0.0	64.0	10.9	14.3	150
T_i = 15	0.0	75.8	10.2	12.5	120
T_i = 45	0.0	92.6	8.5	13.2	120
T_c = 30	0.0	84.8	9.7	12.3	135
T_c = 60	0.0	85.6	8.0	11.7	120

Based on Table 3 one can see that quite a short settling time is achieved and all signals remain within constraints.

5. Comparison of MMAC and MRAC With Time-varying Reference Model

The MMAC developed by Martin et al (1987) seems to be the most universal, efficient and best tested adaptive controller developed so far. The limitation of this controller is its complexity. Using the MRAC technique one does not need to build a priori a bank of process models. The adaptive algorithm will find the actual model of a particular process and adjust the controller appropriately. Similar simulation tests as reported by Martin et al (1987) were performed for our controller. The results of these simulations are summarized in Table 4 and they can be compared with the results in Table 3 in the above mentioned paper. Comparing the results reported in the paper by Martin et al (1987) with our results, we found:

- The MRAC is simpler and more general than the MMAC.
- Both methods result in a good control of blood pressure in the presence of noise.
- Our method results in a shorter settling time, but for both methods it is quite short.
- The MMAC has a lower undershoot, but the undershoot for our method is also acceptable.
- Both methods include the estimation of the time delay a priori.
- The MMAC algorithm was designed for time-varying gain and good control with this respect was obtained. It was not checked for time-varying recirculation constant or system time constant. It provides adaptation for time-varying time delay. This is, however, not necessary

since time delay is only patient dependent and it remains constant during drug infusion procedures. Using the MRAC algorithm all parameters may be time-varying except for the time delay and very good control results were obtained in all cases.

- Both methods include include the input and output constraints. In the MMAC algorithm the constraints are realized by saturation of input and output signals. The MRAC algorithm generates the long-range prediction of input and output sequences and the constraints are checked for all of the predicted values as explained in the paper.

6. Conclusions

A modified MRAC algorithm with a time varying reference model has been designed and applied to control blood pressure using sodium nitroprusside. The time-varying reference model is automatically tuned so that the physical and medical constraints posed on the system are met and at the same time close to optimal performance is achieved. This ensures that the proposed controller is effective and safe. Extensive simulation of the controller has shown that it has a short settling time, and it is capable of maintaining the blood pressure within ± 15 mm Hg of the desired set point. Moreover, the infusion rate, MAP and its derivative are within constraints. This is true even in the presence of relatively high noise levels and over the full range of plant parameters. The pressure is held to within ± 15 mm Hg even when the plant undergoes large changes in its parameters.

Table 4. Simulation Results for Different Plant Gains and Time Delays.
Values are for set point changes after the loop was closed.
Each simulation was run with white Gaussian noise (variance of 4 mm Hg^2).

Plant Gain mm Hg ml/hr	Time Delay (sec)	(ΔMAP/T) max (mm Hg)	MAP min (mm Hg)	Undershoot (min Hg)	Settling Time (sec)
	15	7.73	80.6	7.41	60
0.25	30	7.88	83.8	6.01	180
	60	6.5	81.6	0.95	105
	15	7.3	85.6	5.61	60
2.1	30	13.7	80.1	2.68	180
	60	9.05	80.9	13.7	90
	15	8.98	82.8	9.68	60
9.0	30	10.6	84.6	8.44	90
	60	9.43	86.7	1.94	105

Although not presented here, the ability of the controller to handle deterministic disturbances was investigated. Most of these simulations involved fairly low noise levels and produced reasonable results. More work in this area is needed.

The MRAC system described here is simpler and more general than the multi-model adaptive algorithm proposed in the literature. Simulation tests show that both control systems are effective, meet all the medical requirements, and can be used safely for automatic control of blood pressure. The experimental evaluation of the MRAC is going to be performed on dogs in the near future.

References

Arnspager, J.M., B.C. McInnis, J.R. Glover, Jr., and N.A. Normann, "Adaptive control of blood pressure", *IEEE Trans. Biomed. Eng.*, BME-30 (3), pp. 168-176 (1983).

Goodwin, G.C. and K.S. Sin, *Adaptive Filtering, Prediction and Control*, Prentice-Hall, New Jersey (1984).

Goodwin, G.C., D.J. Hill and X. Xianya, "Stochastic adaptive control for exponentially convergent time-varying systems", *SIAM J. Control and Optimization*, 24(4), pp. 589-603 (1986).

He, W.G., H. Kaufman and R. Roy, "Multiple model adaptive control procedure for blood pressure control", *IEEE Trans. Biomed. Eng.* BME-33(1), pp. 10-19 (1986).

Kaufman, H., R. Roy and X. Xu, "Model reference adaptive control of drug infusion rate", *Automatica* 20(2), pp. 205-209 (1984).

Martin, J.F., A.M. Schneider and N.T. Smith "Multiple-model adaptive control of blood pressure using sodium nitroprusside", *IEEE Trans. Biomed. Eng.* BME-34(8), pp. 603-611 (1987).

Pajunen, G.A. and M. Steinmetz, "Model reference adaptive control for exponentially convergent systems with input and output constraints", *Proc. IEEE Conf. Decision and Control, Los Angeles, CA*, pp. 831-836 (1987).

Reves, J.G., L.C. Sheppard, R. Wallach and W.A. Lell, "Therapeutic uses of sodium nitroprusside and an automated method of administration", *Int. Anesth. Clin.* 16(2), pp. 51-88 (1978).

Serna, V., R. Roy and H. Kaufman, "Adaptive control of multiple drug infusion", *Proc. American Control Conf., San Francisco, CA* pp. 22-26 (1983).

Slate, J.B., "Model-based design of a controller for infusing sodium nitroprusside during post-surgical hypertension", Ph.D. Thesis, University of Wisconsin - Madison, 1980 (No. 8028208, University Microfilms International, Ann Arbor, Michigan).

Slate, J.B. and L.C. Sheppard, "A model-based adaptive blood pressure controller", *Proc. IFAC Identification and System Parameter Estimation, Washington, D.C.*, pp. 1437-1441 (1982).

Steinmetz, M., "Model reference adaptive control of blood pressure", M.S. Thesis, Florida Atlantic University, Boca Raton, Florida (1987).

Stern, K.S., B.K. Walker, and P.G. Katona, "Automated blood pressure control using a self-tuning regulator", *Proc. IEEE 1981 Frontiers in Engineering in Health Care Conference, Houston, TX*, pp. 255-258 (1981).

POWER SYSTEM DAMPING: A SIMULATION PROGRAM AND ENHANCED LQ SELF-TUNING STRATEGIES

Donald A. Pierre
Department of Electrical Engineering
Montana State University
Bozeman, Montana 59717 USA

1. Introduction

Electrical power systems contain large synchronous machines interconnected with electrical loads in power networks. When changes in the network or in the generation pattern occur, the electrical machines tend to oscillate around new equilibrium values. As power systems become more complex, there is an increasing awareness of the need for robust damping controls (Hauer, 1988). There are several methods (Mittelstadt, 1968) of adding damping to low-frequency oscillations in power systems: first, power system stabilizer loops (PSS loops) can be included in the generator exciter circuits; second, if dc transmission lines are included in the system, dc power modulation schemes can be used to counter the oscillations; third, electrical brakes (resistors) can be switched into action to absorb power surges when large loads are disconnected from the system; and fourth, reactive devices can be controlled to modulate the power flowing in a transmission line for damping purposes. All of these damping methods can be employed in a large electrical network, and various adaptive control techniques have been proposed to enhance power system damping (Pierre, 1987b).

In this paper, the focus is on volt-ampere reactive devices (VAR devices) which can be used to modulate real power flow, while not dissipating real power in themselves, to add damping to low-frequency oscillations in power systems. In power system terminology, when a nonrotating VAR device is controlled, it is called a static (nonrotating) VAR system or an SVS (Piwko, 1987).

The objectives of the paper are: 1) to describe a power system simulation program which efficiently solves algebraic network equations along with synchronous machine dynamic equations and in which control strategies for controlling one VAR unit are readily incorporated; 2) to describe an enhanced linear-quadratic self-tuning adaptive controller (an ELQ adaptive controller); 3) to present preliminary simulation results that illustrate effects of using ELQ adaptive control on an SVS unit, to add damping to power systems; and 4) to suggest improved strategies for damping low-frequency oscillations in power systems on the basis of locally measured signals. This paper represents a status report; much remains to be done before we can conclude that adaptive self-tuning of SVS units will work reliably for damping oscillations in large power systems.

2. Power System Swing Equations

The swing equation representation of power systems consists of second-order synchronous machine models and an interconnection network (e.g., Elgerd, 1982). The electrical angular velocity of every machine in the system is assumed to remain in the neighborhood of the system synchronous frequency, so that electrical power outputs are assumed to be characterized by values obtainable by phasor methods. The RMS phasor voltage of a given machine is assumed to be of fixed amplitude E_i but to have a varying angle called the machine angle δ_i.

The time derivative of each machine angle, when added to the system synchronous frequency, results in the instantaneous frequency of the given machine. The second derivative of the i-th machine angle is governed by:

$$\ddot{\delta}_i = c_i(F_i - P_i) - c_{di}\dot{\delta}_i \quad , \quad i = 1, 2, \ldots, M \tag{1}$$

where c_i and c_{di} are proportionality constants, F_i is the mechanical power supplied to the machine, P_i is the electrical power produced by the machine, and $c_{di}\dot{\delta}_i$ is a natural damping term. Each such second-order differential equation is called a swing equation, and a set of these coupled equations forms a basic characterization of a power system for studies involving low-frequency oscillations.

The electrical power P_i from the i-th machine can be shown (Elgerd, 1982, or Anderson and Fouad, 1977) to be of the form

$$P_i = E_i \sum_{j=1}^{M} E_j q_{ij}(\beta) \sin[\delta_i - \delta_j + \Theta_{ij}(\beta)], \quad i = 1, 2, \ldots, M \tag{2}$$

in which $q_{ij}(\beta)$ and $\Theta_{ij}(\beta)$ generally are nonlinear functions of the vector β of network parameters (e.g., network conductances and susceptances). When time-varying network elements are included in the system, the algebraic network equations need to be solved repetitively along with the system differential equations in order to generate the varying electrical power terms.

The network equations are simplified considerably if the network consists of all reactive elements. This is assumed in this paper. Under this assumption the network is lossless, and the mechanical powers delivered to the synchronous machines must sum to zero in order for an equilibrium to exist. Thus some of the synchronous machines are generators delivering power to the network while other synchronous machines are motors receiving power from the network. Also the amount of power that can be delivered through a network is limited by the network configuration and parameter values.

3. System Simulation Features

Because the electrical power delivered from each machine is a function of network connections and network parameter values, changes in network parameters must be accompanied by the solution of a set of linear algebraic network equations at each integration step. However, of the elements in the network in this study, only the controlled network element changes at each time step, in general. Because of this, the network calculations are partitioned so that most of the network solution effort is done at the start of the simulation (Pierre, et al., 1987a).

The simulation allows for an arbitrary number M of synchronous machines to be tied together in a network consisting of reactive elements. Each machine is modeled by a fixed voltage amplitude, with voltage angle determined by machine angle, which is characterized by a second-order differential equation as described in the previous section. The SVS being controlled is tied from the (M + MV)'th network bus (node) to ground, in the network which consists of M synchronous machine buses and MV nonmachine buses. The user of the program selects machine parameters and the network susceptances. Buses in the network are of two classes: machine buses and nonmachine buses. Between each machine bus and ground there is a machine phasor voltage in series with a susceptance, and susceptance connections between all buses are assigned by the user.

The organizing subroutine in the simulation program is subroutine SYSTEM which is called four times during each integration time step from a general fourth-order Runge-Kutta simulation package. During the first call to SYSTEM, data is input relative to the network configuration, the network and machine parameters, and the parameters used in the SVS control strategy. Then NETCALC is called from SYSTEM to perform a substantial part of the network matrix inversion by use of an appropriate matrix partitioning scheme. Other parameters of interest to the simulation also are initialized, prior to the point where all subsequent calls from the Runge-Kutta package enter SYSTEM.

The first activity in SYSTEM that is performed every time that SYSTEM is called is a call to subroutine NETSOLVE. NETSOLVE requires nominal matrix multiply and add operations to obtain: 1) the real and reactive parts of the voltages at the machine buses; 2) the real and reactive parts of the machine currents; 3) the electrical power generated by each machine; 4) the phasor voltage at the SVS bus; and 5) the real power flowing in the line between the number 1 bus and the SVS bus.

Because of the digital nature of the control, the control sample period (typically 0.1 second) differs from the simulation time step (typically 0.01

second). Control changes are allowed only at the start of a Runge-Kutta iteration, and only if the elapsed time since the last control update equals or exceeds the control period. Prior to the call to CONTROL, past values of the system output are updated, and the current output YC is assigned. Control updates are obtained by calls to subroutine CONTROL which in turn calls other subroutines to do all the calculations necessary for the adaptive control described in Section 5. Subroutine CONTROL returns an appropriate value for the SVS susceptance. After storing appropriate past values of control actions, NETSOLVE is called to return the new values of electrical machine powers and the real power in the line between machine number 1 and the SVS bus. In the final assignment that is performed each time SYSTEM is called, the right-hand sides of the state equations are updated appropriately.

4. Local Signals Used for Damping Control

It is assumed that the SVS bus is located on a line between a major generation bus (bus 1) and a major load bus (bus 2). Other tranmission lines in the system connect bus 1 and bus 2 to additional generation and load buses. Between the SVS bus and ground is a controlled susceptance with a design center value and a controlled value that is dictated by the adaptive controller.

The damping control examined in this paper is based on outputs of the power system that are measured locally at the SVS bus. Two signals are of particular interest: 1) an incremental change in power flow in the line from bus 1 to the SVS bus; and 2) the frequency deviation from 60 Hz. Because the simulation uses phasor methods, however, instantaneous local frequency is not directly available in the simulation. In place of local frequency, an alternative signal is formed on the basis of the phase angle of the voltage at the SVS bus. The time rate of change of this phase angle is approximated as follows. The SVS bus voltage (real part and imaginary part) is saved at the start of a control period, immediately after a change in SVS susceptance. At the end of the control period, but before the controlled susceptance is changed, both the current bus voltage and the previously saved value are used in an appropriate finite difference calculation to estimate the time derivative of the SVS bus voltage phase angle, scaled to have units of cycles per second. For lack of a better name, this signal is called the SVS bus frequency deviation. Note that this signal does not include the direct effect of instantaneous phase change caused by a step change in the SVS susceptance at a sample instant.

In the simulation, step changes in susceptance also generally result in instantaneous changes in real power flow. In this study, we exclude such instantaneous changes in deriving a local signal based on power, as follows.

Immediately after a change in SVS susceptance, the value of the real power flowing in the line from bus 1 to the SVS bus is saved. At the end of the control period, but before the controlled susceptance is changed, the new value of real power in the line is recorded, and the previously saved value of real power is subtraced from the current value to form an incremental change in real power flow.

In this paper, the above signals are combined to form one system output y, as follows. First, the signals are assigned weighting factors, w_p for the power-based signal and w_f for the frequency-based signal; and a linear combination of these signals is used as the input to a bandpass filter. The filter serves two purposes. It has an upper corner frequency of 5 Hz as an antialiasing filter for the 10 Hz sampling frequency of the controller, and it has a 0.05 Hz lower corner frequency to eliminate dc and very-low-frequency components of the input signal. The passband gain of the filter is unity. In power system terminology, the filter is a low-pass/wash-out filter. An additional digital filter stage operates on the output v of the low-pass/wash-out filter. This simple digital filter accumulates a "steady-state" value by adding 5/1000 times the current v to 995/1000 times the current steady-state value. The sampled system output y is then v minus the steady-state value. The purpose of this last operation is to allow the power system to be ramped at a slow rate from one power level to another or from one frequency to another without having the damping controller react to the scheduled change.

5. An Enhanced LQ Self-Tuning Controller

Fig. 1 displays an enhanced LQ self-tuning digital controller which consists of three main features: 1) an identifier implemented with the UD version of recursive least squares (Bierman and Thornton, 1978, and Clarke, 1980); 2) an observer using an observable-canonical state-variable form; and 3) an enhanced linear-quadratic optimal controller (an ELQ controller) which incorporates penalty weights both on the control magnitude squared and on the squared rate of change in control (Pierre, 1987c). This adaptive controller is related to the LQ adaptive controllers of Samson (1982), Clarke, et al. (1985a&b), and Landau (M'Saad, et al., 1986). The new feature that we have added to this adaptive controller is the additional penalty term in the performance measure. By penalizing both control magnitude squared and the squared rate of change in control, smoother control often can be achieved. Continuous versions of this type of optimal control date back to the 1960's (Moore and Anderson, 1967, Johnson, 1968, Pearson and Ding, 1969). When unmodeled dynamics are present in a system being controlled, the use of adaptive control may result in a form of erratic control known as bursting. The ELQ controller tends to reduce the impact of bursting and erratic control (Pierre, 1987c, Smith, et al., 1988, 1989).

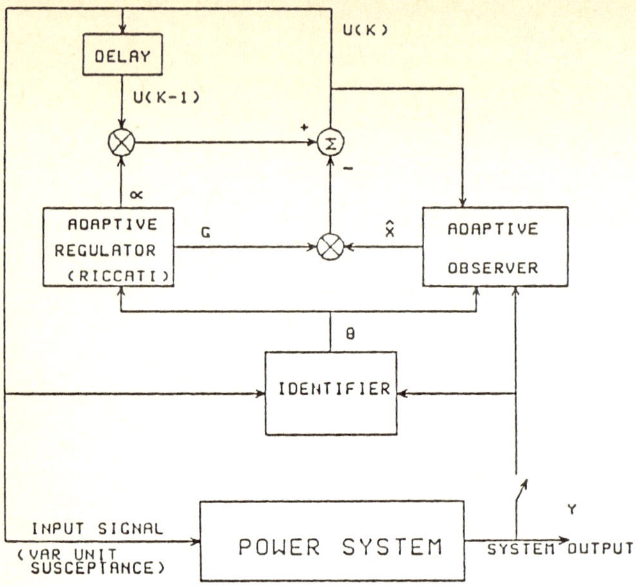

FIG. 1. An enhanced LQ self-tuning control system.

Parameter Identifier

The system being controlled has an output y and controlled input u. A reduced-order model for the system is represented by a linear discrete-time transfer function:

$$y \cong \frac{z^{-1}\bar{B}(z^{-1})}{1 + z^{-1}\bar{A}(z^{-1})} \, u \tag{3}$$

where

$$\bar{A}(z^{-1}) = a_1 + a_2 z^{-1} + \ . \ . \ . \ + a_n z^{-n+1} \tag{4}$$

$$\bar{B}(z^{-1}) = b_1 + b_2 z^{-1} + \ . \ . \ . \ + b_n z^{-n+1} \tag{5}$$

and

$$z^{-1}y(k) = y(k-1) \tag{6}$$

The sampled output is y(k) at integer sample point k. The sampling period used in this power system damping study is 0.1 second.

For parameter identification, equation (3) is written in the form

$$y(k) \cong \phi^T(k-1)\Theta \tag{7}$$

where

$$\Theta^T = [a_1\ a_2\ \ .\ .\ .\ a_n\ b_1\ \ b_2\ \ .\ .\ .\ b_n\ ] \tag{8}$$

and

$$\phi^T(k-1) = [-y(k-1)\ .\ .\ .\ -y(k-n)\ u(k-1)\ .\ .\ .\ u(k-n)] \tag{9}$$

To identify the parameter vector Θ, recursive identification in the form of the UD algorithm (Bierman and Thornton, 1978, Clarke, 1980) is used.

Observer

The purpose of the observer is to generate the state variables of the reduced-order model of the system, so that they can be used for state variable feedback control. Because of the reduced-order model, the observer states do not have a one-to-one correspondence to system states. However, if the unmodeled dynamics of the system are not dominant, it is expected that control based on the observer states will be reliable. The observer states are denoted by $\hat{x}(k)$, and the observer output by $\hat{y}(k)$. An observable canonical form is useful in implementing the observer, the equations for which are as follows:

$$\hat{x}(k+1) = A\hat{x}(k) + Bu(k) + K(k)(y(k) - \hat{y}(k)) \tag{10}$$

and

$$\hat{y}(k) = C\hat{x}(k) \tag{11}$$

where

$$A = \begin{bmatrix} -a_1 & 1 & 0 & \cdots & 0 \\ -a_2 & 0 & 1 & & 0 \\ & & & \cdot & \cdot \\ & & & \quad\cdot & \cdot \\ \cdot & & & \quad\ \cdot & \cdot \\ \cdot & & & & 1 \\ -a_n & 0 & \cdot & \cdot\ \ \cdot & 0 \end{bmatrix} \tag{12}$$

$$B = [b_1\ b_2\ \ .\ .\ .\ \ b_n]^T \tag{13}$$

$$C = [\,1 \;\; 0 \; . \; . \; . \; 0\,] \tag{14}$$

and, for the case where all observer poles are at the origin of the z plane,

$$K^T = [\,-a_1 \; -a_2 \; . \; . \; . \; -a_n\,] \tag{15}$$

The parameter values used in the observer are those obtained from the identifier and generally are time varying. Using this observer, the error between the system output y(k) and the observer output $\hat{y}(k)$ is driven to zero, provided the system identification is sufficiently accurate.

ELQ Controller

The performance measure used in the control is

$$J = \sum_k \{y^2(k) + w_1 u^2(k) + w_2[u(k) - u(k-1)]^2\} \tag{16}$$

in which the summation is over future sampling instants. Weighting factor $w_1 > 0$ is included to penalize control magnitude, and $w_2 > 0$ is used to penalize rapid or erratic swings in control. The control which minimizes J can be shown (Pierre, 1987c) to be

$$u(k) = \alpha u(k-1) - G^T x(k) \tag{17}$$

where the scalar α and the nx1 vector G are computed iteratively based on appropriate Riccati equations. Riccati terms of interest are an nxn matrix R, an nx1 vector Q, and a scalar s, with initial values of

$$R = C^T C \tag{18}$$

$$s = w_1 w_2/(w_1 + w_2) \tag{19}$$

and initial Q being a zero vector. Using these initial values, the optimal α and G result when the following equations are iterated until steady-state solutions are obtained. An auxiliary vector V and an auxiliary scalar β are included in the sequenced equations:

$$V = RB + Q \tag{20}$$

$$\beta = w_1 + w_2 + s + B^T(V + Q) \tag{21}$$

$$G = A^T V/\beta \tag{22}$$

$$\alpha = w_2/\beta \tag{23}$$

$$R(k+1) = C^TC + A^TR(k)A - \beta GG^T \tag{24}$$

$$s = (1 - \alpha)w_2 \tag{25}$$

and

$$Q = w_2G \tag{26}$$

The matrices A, B, and C are the same as those used in the observer. Because the parameters used are from a reduced-order model, they generally are time varying. The way in which the above equations are used in adaptive control must therefore differ from the ideal steady-state optimal controller. Thus, in the adaptive controller, the above Riccati equations are iterated just once during each control period, and the resulting α and G are used in (17) along with $\hat{x}(k)$ in place of $x(k)$. Over a sufficiently long time interval, if the identified parameters approach true steady-state values, the controller gains also will approach their optimal steady-state values.

Like the LQ adaptive controller (Samson, 1982), the ELQ adaptive controller is robust with respect to the following: 1) nonminimum phase zeros and poles of the plant; 2) common roots in the numerator and the denominator of the identified plant transfer function; and 3) zero values of leading numerator coefficients in the identified transfer function. Erratic control behaviour can be tempered by adjusting the second penalty weight in the ELQ adaptive controller. As shown in the following simulation results, the controller tolerates some degree of unmodeled dynamics. Before it is applied in an actual power system, however, it should be tested on a detailed simulation of that system.

6. Simulation Results

Simulation results are given for a two-machine system, a three-machine system, and a five-machine system. Data for the systems are given in the appendix. The data used is intended to illustrate the performance of the controller under conditions of little or no natural system damping, and to produce typical modes of power system oscillations in the 0.1 to 1.5 Hz range. A one-line diagram for the five-machine system is shown in Fig. 2. In the two-machine system, only buses 1, 2, and 6 are retained. In the three-machine system, buses 1, 2, 3, and 6 are retained, and a reactance directly between buses one and three is inserted.

All the simulations start with the machines at their equilibrium angles for the first 0.5 seconds. At $t = 0.5$ s, a step change in mechanical power inputs is made, causing the system to seek new equilibrium angles. At $t = 1.0$ s, the SVS system is activated, and for the next 3.5 seconds an open loop, pseudo random, rectangular wave varies the SVS susceptance by plus or minus 0.02 per unit; this probing therefore is from $t = 1.0$ s to $t = 4.5$ s. At $t = 1.6$ s, the parameter identifier is activated. At $t = 3.5$ s, both the observer and the Riccati equation solver are activated. And at $t = 4.5$ s, the adaptive feedback control begins.

A fifth-order model is identified in all of the simulation results shown. The effects of using other model orders are discussed.

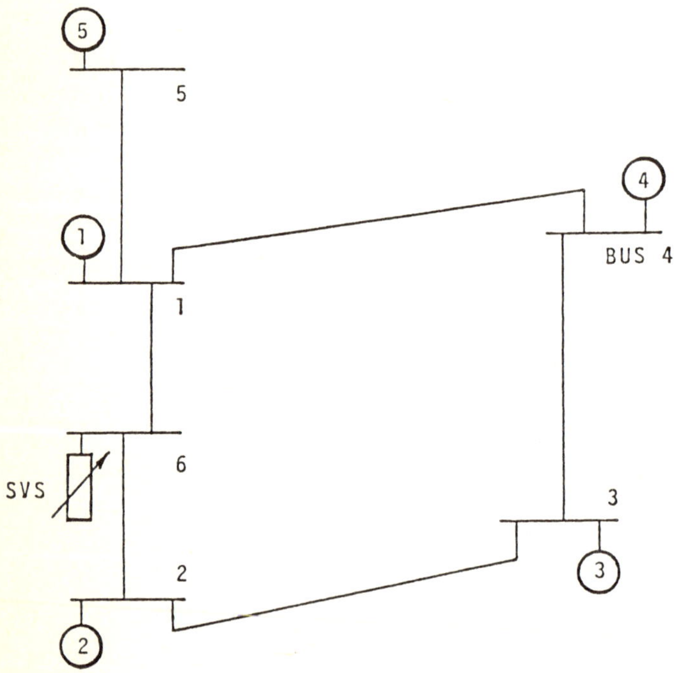

FIG. 2. One-line diagram of a five-machine power system.

Two-Machine System

The two-machine system, including the output signal filtering and conditioning, is characterized by six continuous state variables and one discrete-time state variable, for a total of seven state variables in the plant. The damping coefficient on both machines is set at zero. At t = 0.5 s, the mechanical power delivered to machine 1 is increased by 10%, as also is mechanical power taken from machine 2.

The effect of the w_2 weighting factor on the performance is illustrated in Figs. 3 and 4. The adaptive control action in Fig. 3 is much more erratic than that in Fig. 4, with the only difference being that w_2 is a factor of 8 larger for the results of Fig. 4. The machine angles resulting from the control are shown in Fig. 5.

Using a third-order model in the adaptive controller instead of the fifth-order model, the two-machine system was simulated under the same conditions as above. The results were similar, except that the adaptive control action was more erratic for the third-order model cases.

Three-Machine System

The three-machine system, including the output signal filtering and conditioning, is characterized by nine state variables. Natural damping on all

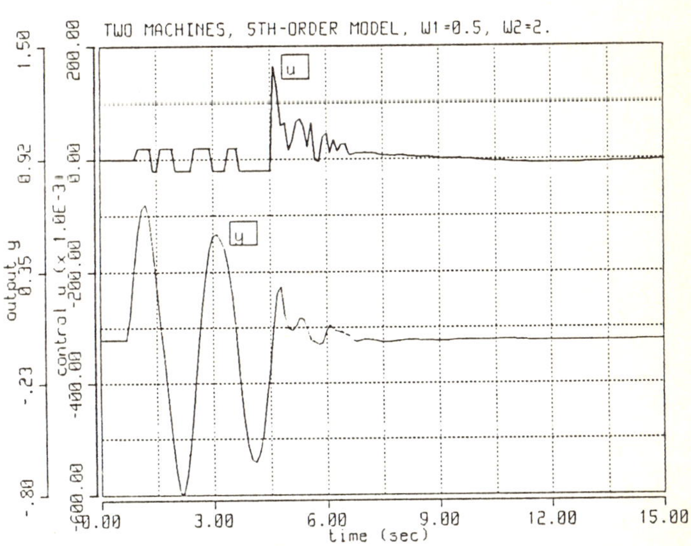

FIG. 3. Control of two-machine system with w1=0.5, w2=2, wf=50, wp=20.

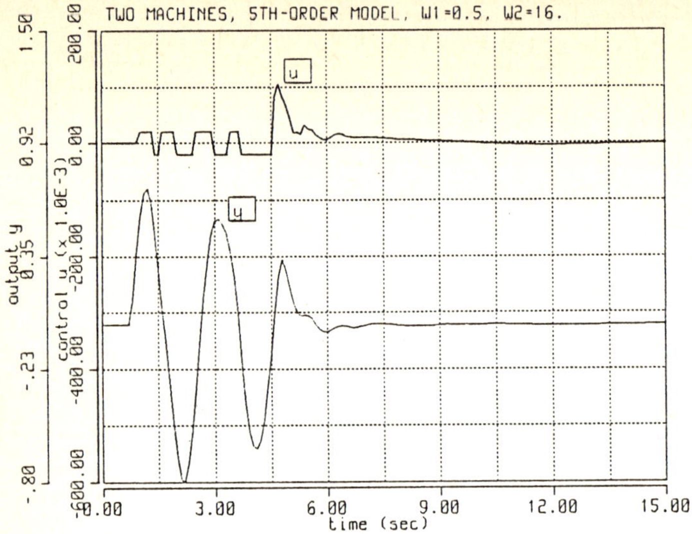

FIG. 4. Control of two-machine system with w1=0.5, w2=16, wf=50, wp=20.

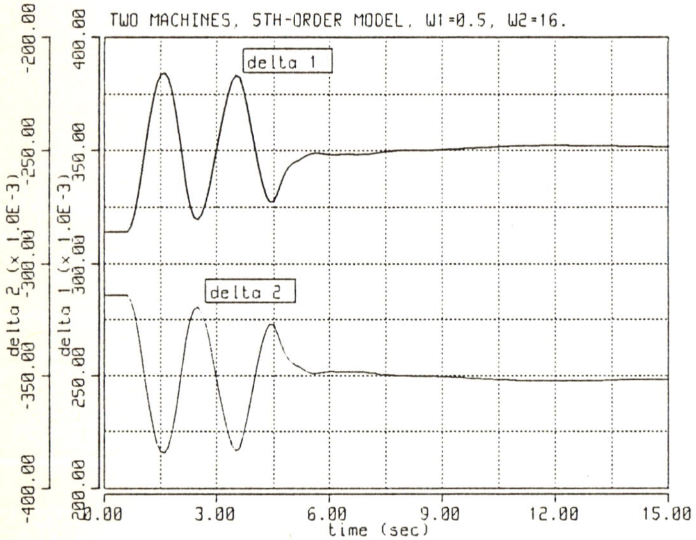

FIG. 5. Machine angles of the controlled two-machine system.

three machines is set to zero. Figs. 6 and 7 give system responses when no SVS control is applied. The responses stem from step changes in mechanical power delivered to the machines at t = 0.5 s. (Power to machines 1 and 3 is increased by 10% while mechanical power taken from machine 2 also is increased by 10%.)

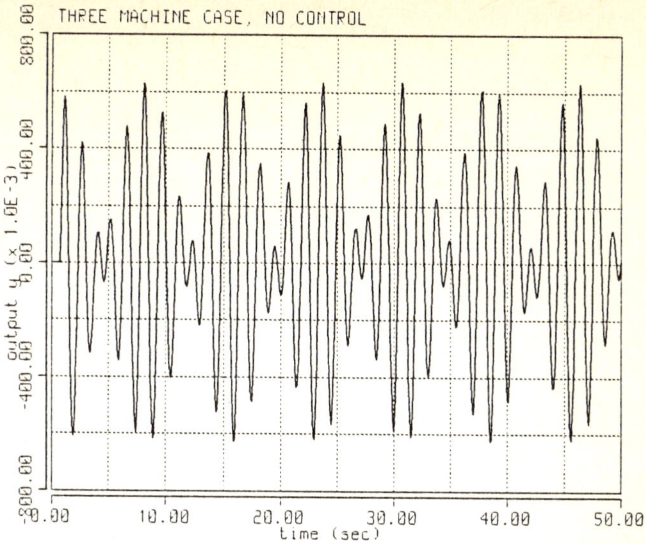

FIG. 6. Uncontrolled local output y from the three-machine system.

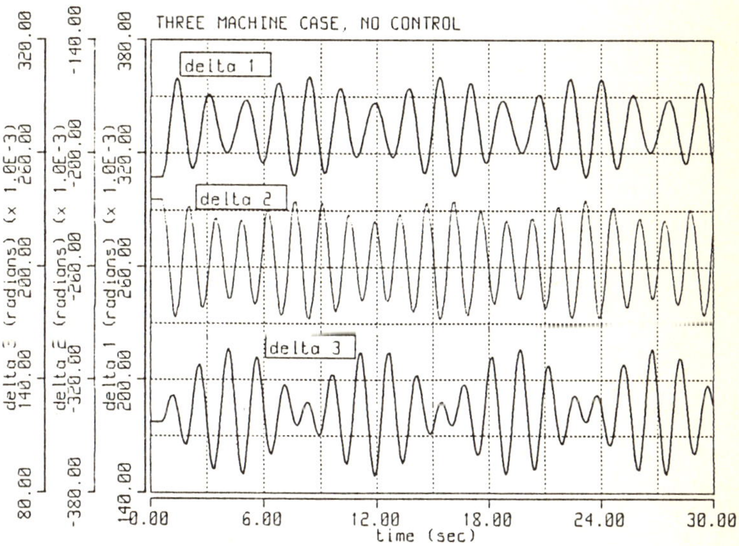

FIG. 7. Machine angles of the uncontrolled three-machine system.

Figs. 8 and 9 show the corresponding responses when adaptive SVS control is applied with a fifth-order model. Note that the output signal and the machine oscillations appear to be damping out, but that the rate of damping is small near the terminal time shown. Possible reasons for this decrease in damping are given in the next section.

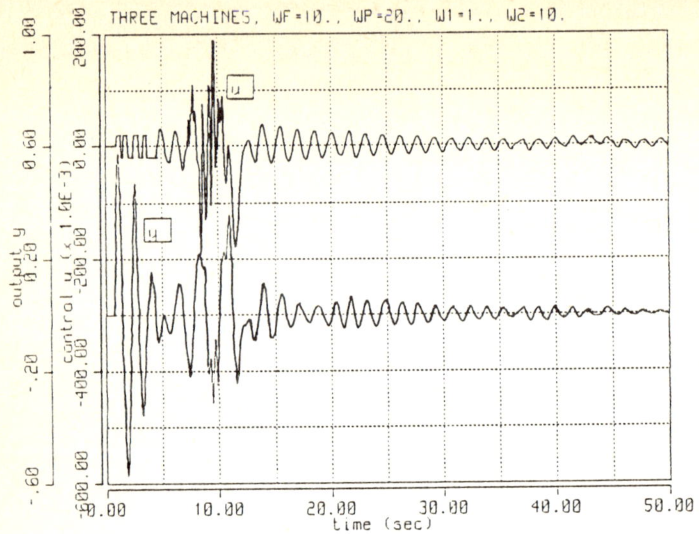

FIG. 8. Control u(t) and output y(t) from the three-machine system.

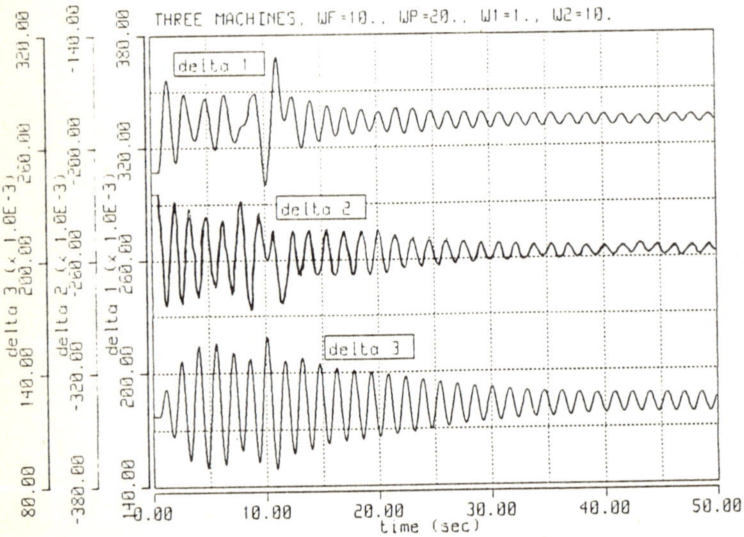

FIG. 9. Machine angles from the controlled three-machine system.

Additional three-machine cases were run with different weighting factors and different order models. Not all of these cases were satisfactory. In one third-order model case, damping over an initial time period was evident, but this was followed by a rapid increase (a bursting) of the control, followed again by damping. The bursting resulted when the trace of the covariance matrix was reduced to the unity level, in contrast to a starting value exceeding 10 to the 7th. This type of action can be avoided by including covariance reset features or certain constant-trace features (Sripada and Fisher, 1987) in the identification algorithm.

Five-Machine System

The five-machine system, including the output signal filtering and conditioning, is characterized by thirteen state variables. For the results displayed here, the damping coefficients on machines 2 through 5 are set at 0.1, while the natural damping on machine 1 is zero. Fig. 10 displays the system output with no control applied, which is to be compared with the adaptive control and output response of Fig. 11. The responses stem from step changes in mechanical power delivered to the machines at t = 0.5 s. (Power to machine 1 is increased by 5%, to machines 3, 4, and 5 by 10%, and mechanical power taken from machine 2 is increased by 8%.)

Note that slightly improved damping of the system output is evident in Fig. 11. When the machine angle responses are examined however, it is clear that the response of machine 4 is poorer with this particular adaptive control than without it (see Figs. 12 and 13). Part of the explanation for this is that only a 5th-order model is being used on a lightly damped 13th-order system. But even higher-order models failed to generate suitable damping in all cases that were run

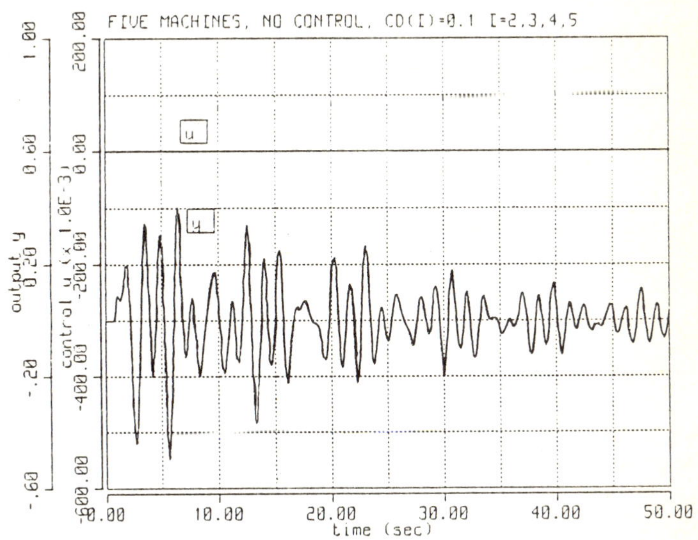

FIG. 10. Output y from the uncontrolled five-machine system.

with zero natural damping and with step changes in mechanical power inputs. In Section 7, we advance a more fundamental reason for the failure of the particular local control employed under the conditions of light natural damping and suggest a way to improve the control.

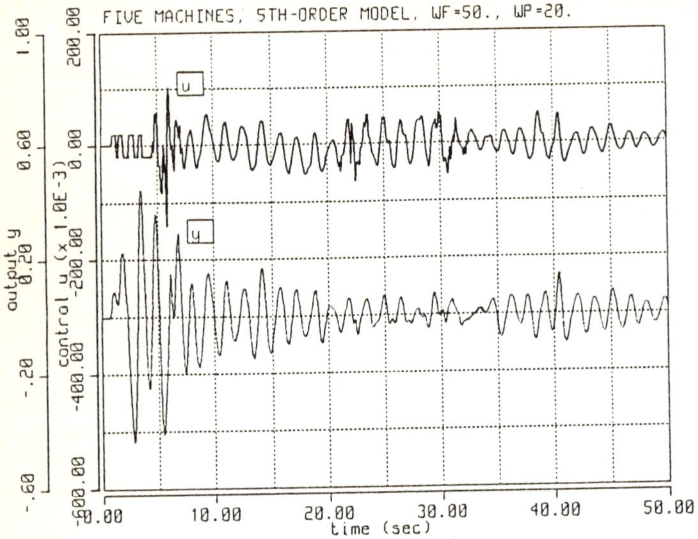

FIG. 11. Control u and output y from the five-machine system with w1=4, w2=4, wf=50, wp=20.

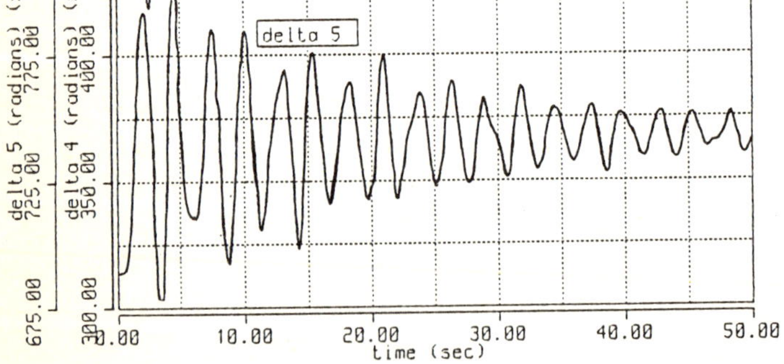

FIG. 12. Angles for machines 4 and 5, no-control case.

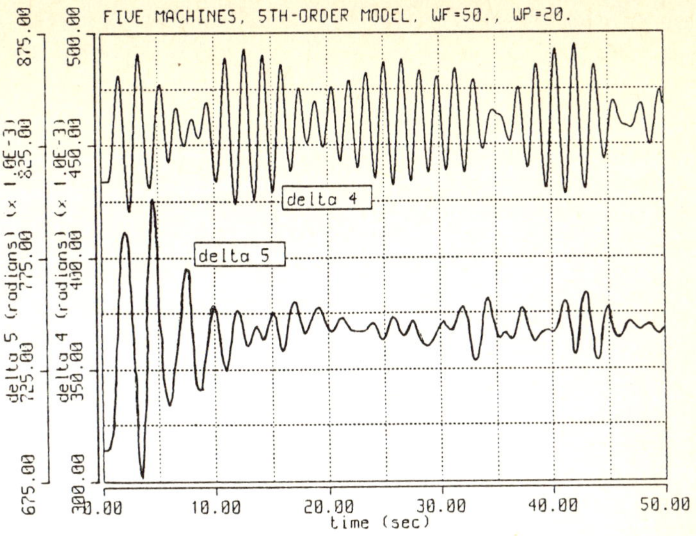

FIG. 13. Angles for machines 4 and 5, with local adaptive control applied on the line between machines 1 and 2.

An interesting observation on this five-machine system is as follows. When the system starts at equilibrium and no changes are made in the mechanical power inputs, and only probing is used to perturb the system away from equilibrium, the adaptive control described above brings the system back to equilibrium in rapid order (Pierre, 1987c). This is the case under conditions of zero natural damping and with $w_p = 0$. It appears that the transients that are initiated by the probing action of the control are more easily controlled from that point in the system than are transients that are initiated by other changes in the system.

7. An Explanation and an Improved Approach

In the three-machine responses of Fig. 9, damping of the oscillations appears to taper off as time progresses. In the five-machine case, the output response of Fig. 11 does not reflect the severity of the oscillation in machine number 4. The system output used in the previous section is a filtered linear combination of two local signals, a power-based signal v_p and a frequency-based signal v_f. Fig. 14 shows the two local signals over the final seconds in the five-machine case. Note that the signals are 180 degrees out of phase with one another, and that the weighted sum is relatively close to zero over much of the time. Thus, even though the adaptive control is working to dampen the output y, the signals v_p and v_f that are used to form y are not being damped out.

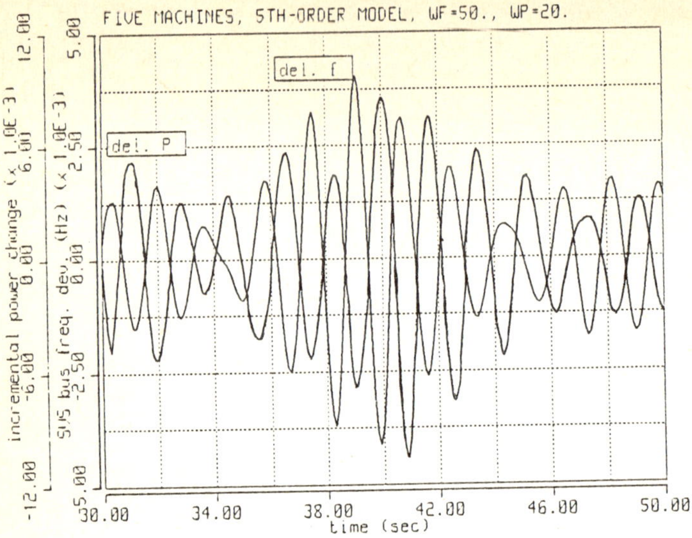

FIG. 14. The two local signals that are weighted, summed, and filtered to form the local output in the controlled five-machine case.

In the neighborhood of an equilibrium point, the signals v_p and v_f can be approximated by linear equations of the form

$$v_p \cong \sum_{i=1}^{M} \alpha_i \dot{\delta}_i \tag{27}$$

and

$$v_f \cong \sum_{i=1}^{M} \beta_i \dot{\delta}_i \tag{28}$$

in which the α_i's and β_i's are constants. Any weighted sum of v_p and v_f will be of the same form as (27) and (28). The performance measure (16) contains y squared, which in effect is the square of a filtered and weighted sum of $\dot{\delta}_i$'s. The minimization of y squared therefore does not guarantee that the individual $\dot{\delta}_i$'s will converge to zero.

The phenomenon just described is very evident in multimachine systems which contain no natural damping terms. In systems where other damping actions are included, the adaptive control based on y squared often adds to the damping, but not always.

A more effective damping control should result if filtered versions of v_p and v_f are treated as separate outputs y_1 and y_2, respectively, of the system, and the squares of both of these outputs are included in the performance measure. We base this statement on the following argument. The signals v_p and v_f are measured at a point (at a bus) on the transmission line that connects machines 1 and 2. Because of this, the first two terms in (27) and (28) will dominate under most conditions. If both $(\alpha_1\dot{\delta}_1 + \alpha_2\dot{\delta}_2)^2$ and $(\beta_1\dot{\delta}_1 + \beta_1\dot{\delta}_2)^2$ are forced to zero, then $\dot{\delta}_1$ and $\dot{\delta}_2$ will be forced to zero provided

$$\begin{vmatrix} \alpha_1 & \beta_1 \\ \alpha_2 & \beta_2 \end{vmatrix} \neq 0 \tag{29}$$

To treat y_1 and y_2 as separate outputs in the adaptive control, identifiers are run in parallel to obtain two system models, one for y_1 and the other for y_2. Observers for each are formed, and observer states are used to form the single control input u. The ELQ controller is readily extended to handle this case. Simulations using this multiple output approach are in progress.

8. Conclusion

In Section 2 the swing equation representation of power systems is described. A simulation program based on these equations is presented in Section 3; the simulation is structured for efficiency by partitioning the repetative algebraic network solution to take advantage of the fact that the only network susceptance that changes is the controlled susceptance. Local signals are derived (Section 4) at the control bus for use in adaptive control. An enhanced LQ self-tuning adaptive controller is described in Section 5. Simulation results (Section 6) demonstrate that the enhancement feature is effective in reducing erratic behavior in the control. The control used in Section 6 is based on one system output; it provides system damping in some cases, but its use on one five-machine system is not effective (see Figs. 12 and 13). A principal reason for this lack of effectiveness is described in Section 7 along with a proposed improved method of control.

The power system simulation used in this paper contains the nonlinear swing equation representation of synchronous machines but it does not include the dynamics of machine exciters and their controls. In recent work (Smith, et al., 1988, 1989), a more detailed nine-bus, three-machine system is simulated, and adaptive damping is examined in response to faults on the system. Additional practical issues are being addressed in ongoing work.

Acknowledgments

The developments in this paper were facilitated by frequent discussions with Dr. Don Rudberg, Dr. Roy Johnson, James Smith, Iraj Sadighi, and Anthony Johnson of Montana State University, and with Dr. John Hauer of the Bonneville Power Administration. Support for the work was provided by the Bonneville Power Administration and by the Engineering Experiment Station of Montana State University.

References

Anderson, P. M. and A. A. Fouad (1977). Power System Control and Stability. Iowa State University Press, Ames, Iowa.

Bierman, G. J. and C. L. Thornton (1978). Filtering and error analysis via the UDU covariance factorization. IEEE Trans. Aut. Control, vol. AC-23, pp. 901-907.

Clarke, D. W. (1980). Some implementation considerations of self-tuning controllers. pp. 81-110 of Numerical Techniques for Stochastic Systems, F. Archetti and M. Cugiani (editors), North Holland Pub. Co., New York.

Clarke, D. W., P. P. Kanjilal, and C. Mohtadi (1985a). A Generalized LQG approach to self-tuning control, Part I. Aspects of Design. Int. J. Control, vol. 41, no. 6, pp. 1509-1523.

Clarke, D. W., P. P. Kanjilal, and C. Mohtadi (1985b). A Generalized LQG approach to self-tuning control, Part II. Implementation and Simulation. Int. J. Control, vol. 41, no. 6, pp. 1525-1544.

Elgerd, O. I. (1982). Electric Energy Systems Theory (second edition). McGraw-Hill Book Co., New York.

Hauer, J.F. (1988). Robust damping controls for large power systems. IEEE Control Systems Magazine, Vol. 9, No. 1, pp. 12-18.

Johnson, C. D. (1968). Optimal control of the linear regulator with constant disturbances. IEEE Trans. Aut. Control, vol. AC-13, no. 4, pp. 415-421.

Mittelstadt, W. A. (1968). Four methods of power system damping. IEEE Trans. on Power Apparatus and Sys., vol. PAS-102, pp. 1323-1329.

Moore, J. B. and B. D. O. Anderson (1967). Optimal linear control systems with input derivative constraints. Proc. of the IEE, vol. 114, no. 12, pp. 1987-1990.

M'Saad, M., M. Duque, and I. D. Landau (1986). Practical implications of recent results in robustness of adaptive control schemes. Proceedings, IEEE Conf. on Decision and Control, pp. 477-481.

Pearson, J. B. and C. Y. Ding (1969). Compensator design for multivariable linear systems. IEEE Trans. Aut. Control, vol. AC-14, no. 2, pp. 130-135.

Pierre, D. A., J. R. Smith, J. P. Kuppinger, and T. F. Schiff (1987a). Research on adaptive var control in power systems. Electronics Research Laboratory Report no. 287, Montana State University, Bozeman, MT.

Pierre, D. A. (1987b). A perspective on adaptive control of power systems. IEEE Trans. Power Systems, vol. PWRS-2, no. 2, pp. 387-396.

Pierre, D. A. (1987c). Enhanced LQ controllers and adaptive control. Electronics Research Laboratory Report no. 687, Montana State University, Bozeman, MT.

Piwko, R. J. (chairman) (1987). Symposium on application of static var systems for system dynamic performance. IEEE Publication no. 87TH0187-5-PWR.

Samson, C. (1982). An adaptive LQ controller for non-minimum-phase systems. Int. J. Control, vol. 35, no. 1, pp. 1-28.

Smith, J. R., D. A. Pierre, D. A. Rudberg, and R. M. Johnson (1988a). Robust var unit control strategies to enhance damping of power system oscillations. Proceedings, IASTED Conference on High Technology in the Power Industry.

Smith, J. R., D. A. Pierre, D. A. Rudberg, I. Sadighi, A. P. Johnson, and J. F. Hauer (1988b). An enhanced LQ adaptive var unit controller for power system damping. Paper No. 88 SM 692-6, IEEE Power Engineering Society Summer Power Meeting.

Smith, J.R., D.A. Pierre, I. Sadighi, M.H. Nehrir, and J.F. Hauer (1989), A supplementary adaptive var unit controller for power system damping. Paper No. 89 WM 197-5, IEEE Power Engineering Society Winter Meeting.

Sripada, N. R. and D. G. Fisher (1987). Improved least squares identification for adaptive controllers. Proceedings of the American Control Conf., pp. 2027-2037.

Appendix. Per Unit System Values for the Examples

Machine data	#1	#2	#3	#4	#5
Machine impedance	0.2j	0.1j	0.2j	0.5j	0.2j
Voltage magnitude	2.0	2.0	2.0	2.0	2.0
Power coefficient (c)	4.0	4.0	4.0	4.0	4.0
Two-machine case					
Initial mechanical power	1.0	-1.0			
Initial equilibrium angle	18.0	18.0			
Damping coefficient (cd)	0.0	0.0			
Three-machine case					
Initial mechanical power	1.0	-1.5	0.5		
Initial equilibrium angle	17.6	-12.9	6.7		
Damping coefficient (cd)	0.0	0.0	0.0		
Five-machine case					
Initial mechanical power	1.0	-2.5	0.5	0.5	0.5
Initial equilibrium angle	29.1	-26.3	6.5	24.9	40.0
Damping Coefficent (cd)	0.0	0.1	0.1	0.1	0.1

SVS Unit Susceptance

Susceptance value when control $u=0.0$: $B=0.0$
Range of susceptance values under probing: $B = \pm 0.02$
Range of susceptance values under adaptive control: $B \in [-0.2, 0.2]$

Transmission line data

Two-machine case

Bus #'s	X	Bus #'s	X
1 - 6	0.8	2 - 6	1.25

Three-machine case

Bus #'s	X	Bus #'s	X
1 - 3	3.6	2 - 3	1.6
1 - 6	0.8	2 - 6	1.25

Five-machine case

Bus #'s	X	Bus #'s	X
1 - 4	1.6	2 - 3	1.6
1 - 5	1.6	2 - 6	1.25
1 - 6	0.8	3 - 4	2.0

ADAPTIVE CONTROL OF NONLINEAR MECHANICAL SYSTEMS

Koichi OSUKA

Department of Mechanical Engineering.
University of Osaka Prefecture, Sakai Osaka 591, JAPAN

ABSTRACT

In this paper, we show a design method of model reference adaptive control system for nonlinear mechanical systems using the special features of the mechanical system, without using any linearizations or approximations. The proposed MRACS assures the global asymptotic stability and the boundedness of internal signals of the whole system. Furthermore, to illustrate the effectiveness of the proposed MRACS, we show some experimental results using a direct drive arm.

INTRODUCTION

This paper is concerned with a model reference adaptive control system (MRACS) for nonlinear mechanical systems such as manipulators. And this paper is the English version of the Japanese paper published in 1986 by the author(Osuka,1986).

It is well known that the fundamental control method for the trajectory tracking problem of nonlinear mechanical systems is so-called computed torque method(Luh,1980). This method, however, requires the exact dynamical model of system to be controlled. This implies that the computed torque method does not ensure the robustness to modeling errors. Therefore, various kinds of control methods based on the adaptive control theory have been proposed.

Dubowsky and Forges(1979) showed an application of adaptive control scheme to robot manipulators. They regarded a manipulator, though which is highly nonlinear and coupled system, as a second order linear time-invariant decoupled model and constructed a MRACS for such linear system. Takegaki and Arimoto(1981) proposed a simple adaptive controller which has not needed any reference model. Koivo and Guo(1983) proposed an adaptive controller based on self-tuning regulator(STR) theory.

However, since their adaptive schemes were based on an adaptive control theory for linear time-invariant system, there is a problem that the global asymptotic stability was not assured. It seems that there are two reasons why such problem has been arisen;

In most adaptive control schemes for manipulators,
(1) the linearization or approximation has been used,
and
(2) the special features of mechanical systems have not been considered.
Here, the special features are stated as follows;
(a) Though the dynamical model of mechanical system is highly nonlinear system, it can be regarded as a linear system with respect to the constant parameters such as moment of inertia or mass of the links.
(b) Since the form of nonlinear terms in the dynamical model can be determined by the structure of the mechanical systems, we can know the form of nonlinear terms perfectly.
(c) The position, velocity and acceleration of the movement of each degree of freedom are detectable.

In this paper, we show the design method of MRACS for nonlinear mechanical systems using the special features (a), (b) and (c), without using any linearization or approximations. The proposed MRACS assures the global asymptotic stability and the boundedness of internal signals of the whole system. Though the similar adaptive control scheme has been proposed by Craig in 1987(Craig,1987), we had already published the results shown in this paper in 1986 (Osuka,1986 in Japanese)

Then, to illustrate the effectiveness of the MRACS proposed in this paper, we show some experimental results using DARM-1(Direct Drive Arm with two degrees of freedom).

2. PROBLEM STATEMENT

We consider a nonlinear mechanical system of the form,

$$\begin{pmatrix} I & 0 \\ 0 & J(x) \end{pmatrix} \begin{pmatrix} \dot{x}_1 \\ \dot{x}_2 \end{pmatrix} = \begin{pmatrix} x_2 \\ F(x)+u \end{pmatrix}, \qquad [1]$$

where $x=[x_1^T\ x_2^T]^T$, $x_1 \in R^n$, $x_2 \in R^n$, $u \in R^n$, $J(x) \in R^{nxn}$, and $F(x) \in R^n$ are nonlinear matrices of the variable x. The matrix $I \in R^{nxn}$ is an unit matrix and $0 \in R^{nxn}$ is a zero matrix. We also consider the following linear time invariant system as a reference model;

$$\dot{x}_M = Ax_M + Br \qquad [2a]$$

where $x_M \in R^n$, $r \in R^n$,

$$A=\begin{pmatrix} 0 & I \\ k_1 I & k_2 I \end{pmatrix}, \qquad B=\begin{pmatrix} 0 \\ I \end{pmatrix} \ (k_1, k_2 > 0) \qquad [2b]$$

The objective of this paper is to construct an adaptive controller which ensures the asymptotic stability of the error

$$e = x_M - x \qquad [3]$$

with the bounded internal signals.

Now, because the system [1] is a mechanical system, we can assume the following assumptions.

[A1] The nonlinear matrices J(x) and F(x) are linear with respect to physical parameters. Namely,

$$J(x) = \sum_{k=1}^{m} a_k J_k(x) \qquad [4a]$$

$$F(x) = \sum_{k=1}^{q} a_k F_k(x) \qquad (m \leqq q) \qquad [4b]$$

where $a_k (k=1,\cdots,q)$ represent the uncertain physical parameters. The matrices $J_k(x)$ and $F_k(x)$ are nonlinear but do not contain any uncertain parameters.

[A2] The structure of the matrices $J_k(x)$ and $F_k(x)$ are known and they have the following properties;

(a) The all elements of each matrix are piecewise continuous nonlinear functions.

(b) The all elements of $J_k(x)$ are bounded for arbitrary x, and their upper and lower bounds can be calculated in advance, namely;

$$| J_k^{i,j}(x) | \leqq M_J \text{ for } \forall x \ (k=1,\cdots,m) \qquad [5]$$

(c) The all elements of $F_k(x)$ are bounded for arbitrary bounded x.

[A3] The upper and lower bounds of a_k can be estimated in advance, namely;

$$| a_k | \leqq a_{max} \quad (k=1,\cdots,q) \qquad [6]$$

[A4] The state variable x, $\dot{x}$, and x_M can be detectable.

[A5] The input r is bounded.

For example, it can be shown that the dynamical equations of manipulator satisfies the above assumptions (see Chapter 4).

3. ADAPTIVE CONTROLLER FOR NONLINEAR MECHANICAL SYSTEMS

In this chapter, we construct a MRACS for the system [1]-[3]. The result of this paper is summarized as the following proposition.

《PROPOSITION》 Consider the nonlinear mechanical system [1] and the input,

$$u=\lambda\,[-\dot{x}_2+k_1x_1+k_2x_2+r]$$
$$+\sum_{k=1}^{m} \hat{a}_k[J_k(x)\dot{x}_2-F_k(x)]+\sum_{k=m+1}^{q} \hat{a}_k[-F_k(x)] \qquad [7a]$$

$$\hat{a}_k = -\Gamma_k \int_0^t \{\sum_{j=1}^{n} Z_{jk} y_j\} d\tau + \hat{a}_k(0) \qquad [7b]$$

$$\Gamma_k > 0 \quad |\hat{a}_k(0)| \leqq a_{max} \quad (k=1,\cdots,q) \qquad [7c]$$

$$Z = -\frac{1}{\lambda}[J_1(x)\dot{x}_2 - F_1(x) \,\vdots\, \cdots \,\vdots\, J_m(x)\dot{x}_2 - F_m(x) \,\vdots\, -F_{m+1}(x) \,\vdots\, -F_q(x)] \qquad [7d]$$

$$y = De. \qquad [7e]$$

If the matrix D of Eq.[7e] is chosen to make the transfer function $W(s)=D(sI-A)^{-1}B$ to be strictly positive real, then we can prove the existence of a constant value λ which assures the boundedness of the all internal signals in the whole system and

$$\lim_{t \to \infty} e = 0. \qquad [8]$$

proof)Substituting Eqs.[4] and [7] into Eq.[1], we get

$$\begin{cases} \dot{x}_1 = x_2 & [9a] \\ \dot{x}_2 = k_1x_1 + k_2x_2 + r + \sum_{k=1}^{m} (\hat{a}_k - a_k)[J_k(x)\dot{x}_2 - F_k(x)]/\lambda \\ \qquad + \sum_{k=m+1}^{q} (\hat{a}_k - a_k)[-F_k(x)]/\lambda. & [9b] \end{cases}$$

Further, subtracting Eq.[9] from Eq.[2] and using Eq.[3], we have the following error equation,

$$\begin{cases} \dot{e} = Ae + B[Z\phi] & [10a] \\ y = De. & [10b] \end{cases}$$

where $\phi = [\phi_1, \cdots, \phi_k]^T = [\hat{a}_1 - a_1, \cdots, \hat{a}_q - a_q]^T$. Since the transfer function W(s) is chosen to be strictly positive real, using the Kalman-Yakubovich-Popov Lemma, there exists a positive definite matrix P which satisfies the equations,

$$PA+A^TP=-Q \quad (\forall Q>0) \qquad [11a]$$

$$B^TP=D. \qquad [11b]$$

Next, consider the positive definite function

$$V(t)=e^TPe+\phi^T\Gamma^{-1}\phi \geq 0 \qquad [12]$$

where $\Gamma = \mathrm{diag}[\Gamma_1,\cdots,\Gamma_q]$. After differentiation of V(t) along the solution of Eq.[10], substitution of Eq.[11] yields

$$\dot{V}(t)=-e^TQe \leqq 0. \qquad [13]$$

From Eqs.[12] and [13], the following properties can be obtained;

(I) e is bounded. [14a]

(II) ϕ is bounded and $\lim_{t\to\infty} \phi = \phi^*$ (constant) [14b]

(III) $V(t) \leqq V(0)$ [14c]

To assure $\lim_{t\to\infty} e=0$, we must prove that the signal e is continuous. On the other hand, the continuity of e can be proved by assuring the boundedness of the internal signals in the whole system.

(Boundedness of the internal signals)

Let rewrite the system [9] as follows:

$$\begin{cases} \dot{x}_1=x_2 & [15a] \\ [\lambda I-\sum_{k=1}^{m} \phi_k J_k(x)]\dot{x}_2=\lambda[k_1x_1+k_2x_2+r] & \\ \qquad + \sum_{k=m+1}^{p} \phi_k[-F_k(x)]. & [15b] \end{cases}$$

Since r, k_1 and k_2 are bounded, it is clear that x_{M1} and x_{M2} are bounded. And so, from Eq.[3] and the property (I), x_1 and x_2 are bounded. Therefore, from [A3], the all elements of $J_k(x)$ and $F_k(x)$ are bounded. Moreover, form the property (II), ϕ_k are bounded.

In this way, we can prove that the both sides of Eq.[15a], the all elements of the right hand side of Eq.[15b] and the matrix defined as

$$R(x)=\lambda I-\sum_{k=1}^{m} \phi_k J_k(x) \qquad [16]$$

are bounded. Therefore, to assure the boundedness of the internal signals, we show that the existence of a constant value λ which assures the non-singularity of $R(x)$, namely;

$$\det|R(x)| \neq 0. \qquad [17]$$

Actually, we can show the claim as follows. From Eqs.[6] and [7c], we get

$$\phi_k(0)^2=[\hat{a}(0)-a_k]^2 \leqq 4a_{max}^2 \quad (k=1,\cdots,q). \qquad [18]$$

Therefore, the following inequality can be derived from Eqs.[12], [18] and the proposition (III);

$$\begin{aligned}\phi^T\Gamma^{-1}\phi &= \sum_{k=1}^{q} \phi_k^2/\Gamma_k \\ &\leqq e(0)^T Pe(0)+\sum_{k=1}^{q} \phi_k(0)^2/\Gamma_k \\ &\leqq e(0)^T Pe(0)+\sum_{k=1}^{q} 4a_{max}^2/\Gamma_k \\ &\equiv M_\phi . \end{aligned} \qquad [19]$$

Define Γ_{max} as

$$\Gamma_{max}= \max\{\Gamma_1,\cdots,\Gamma_q\}, \qquad [20]$$

then from Eqs.[19] and [20], we have

$$|\phi_k| \leqq \sqrt{\Gamma_{max} M_\phi} . \qquad [21]$$

Now from Eqs.[5],[16] and [21], we get

$$R_{ii} \geqq \lambda - mM_J\sqrt{\Gamma_{max}M_\phi} \quad (i=1,\cdots,n) \qquad [22a]$$

$$|R_{ij}| \leqq mM_J\sqrt{\Gamma_{max}M_\phi} \quad (i,j=1,\cdots,n \;\; i\neq j) \qquad [22b]$$

where R_{ij} denotes the i-j element of $R(x)$.

Therefore, if we chose λ as

$$\lambda > (n+1) m M_J \sqrt{\Gamma_{max} M_\phi} \qquad [23]$$

so that the matrix R(x) may be a row sum dominant matrix, then Eq.[17] can be satisfied.

In this way, the existence of λ which assures the boundedness of the internal signals in the whole system can be shown. Moreover, the value of λ can be calculated from Eq.[23].

(Asymptotic stability of e)

Since the boundedness of the internal signals has been proved, it is obvious from Eq.[10] that the error signal e is continuous. Therefore, from Eqs.[12] and [13], $\lim_{t\to\infty} e=0$ can be derived. ∎

The structure of the proposed adaptive control system is shown in Fig.1.

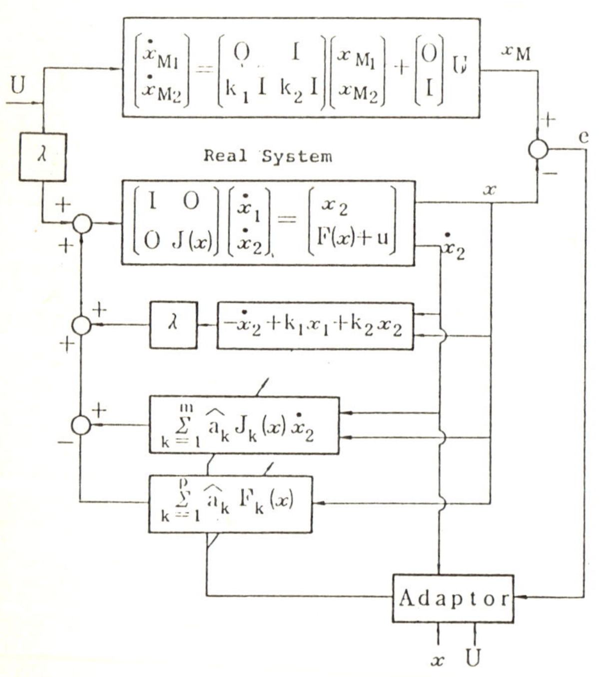

Fig.1 Adaptive Controller for Nonlinear Mechanical Systems

4.APPLICATION TO THE CONTROL OF A DIRECT DRIVE ARM

In order to show the effectiveness of the proposed control system, we applied the controller to solve the control problem of a direct drive arm.

4.1 Direct Drive Arm

Fig.2 shows the photograph of the direct drive arm named DARM-1. This arm is a SCARA-Type manipulator and it has two joints which rotate about the vertical axes and both joints are of direct drive type. The rare earth D.C torque motors are used for joint actuators and links are made of aluminum. Each joint has a photo-encoder for measuring the joint angle and a tacho-genetator for measuring the angular velocity.

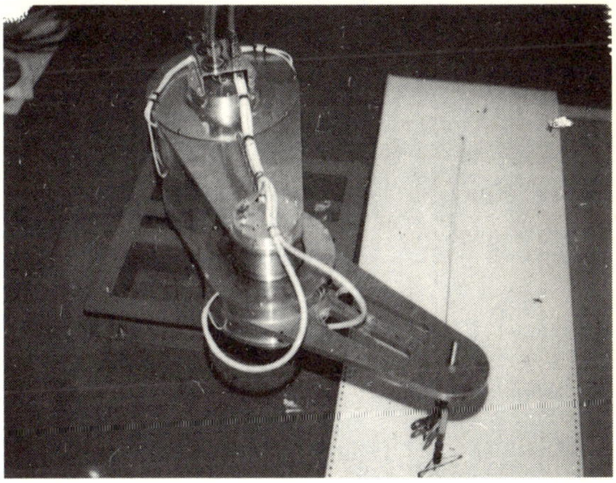

Fig.2 Direct Drive Arm (DARM-1)

Moreover, an accelerometer is mounted on the edge of each link as shown in Fig.3. We can obtain the joint angular acceleration using the signals from these accelerometers as follows:

$$\ddot{\theta}_1=\alpha_1/d_1 \qquad [24a]$$

$$\ddot{\theta}_2=[\alpha_2-\alpha_1(\cos\theta_2+d_2/d_1)-d_1\dot{\theta}_1^2\sin\theta_2]/d_2 \qquad [24b]$$

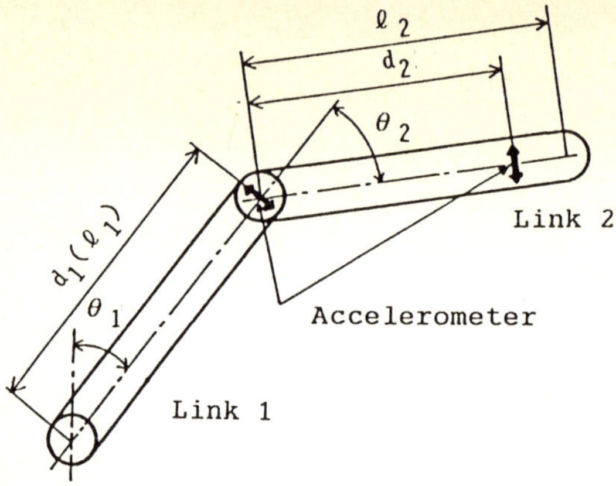

Fig.3 Model of the D.D.Arm

where α_i denote the acceleration signals from the accelerometers.

Now, the dynamical model of this arm can be derived as follows:

$$[J_1+2R_{21}\cos\theta_2]\ddot{\theta}_1+[J_{21}+R_{21}\cos\theta_2]\ddot{\theta}_2 -2R_{21}\dot{\theta}_1\dot{\theta}_2-R_{21}\dot{\theta}_2^2\sin\theta_2+B_1\dot{\theta}_1+f_{c1}\mathrm{sgn}\,\dot{\theta}_1=\tau_1 \quad [25a]$$

$$[J_{22}+R_{22}\cos\theta_2]\ddot{\theta}_1+[J_2]\ddot{\theta}_2+R_{22}\dot{\theta}_1^2\sin\theta_2 +B_2\dot{\theta}_2+f_{c2}\mathrm{sng}\,\dot{\theta}_2=\tau_2. \quad [25b]$$

where $J_{21}=J_{22}=I_2+m_2r_2^2$, $J_1=J_{21}+I_1+m_1r_1^2+m_2\ell_1^2$, $R_{21}=R_{22}=m_2r_2$, and here I_i, m_i, B_i and f_{ci} denote the moment of inertia of link i about the center of mass, the mass of link i, the viscous friction and the Coulomb friction, respectively. And r_i denotes the distance from joint i to the center of mass of link i. Then consider the reference model;

$$\ddot{\theta}_{M1}-k_1\theta_{M1}-k_2\dot{\theta}_1=r_1 \quad [26a]$$

$$\ddot{\theta}_{M2}-k_1\theta_{M2}-k_2\dot{\theta}_2=r_2 \quad [26b]$$

where $k_1<0$, $k_2<0$.

Here, if we set $x_{11}=\theta_1$, $x_{12}=\theta_2$, $x_{21}=\dot{\theta}_1$, $x_{22}=\dot{\theta}_2$,

$x_1=[x_{11}\ x_{12}]^T, x_2=[x_{21}\ x_{22}]^T, x=[x_1\ x_2]^T, x_{M11}=\theta_{M1}, x_{M12}=\theta_{M2}, x_{M21}=\dot\theta_1, x_{M22}=\dot\theta_2, x_M=[x_{M11}\ x_{M12}\ x_{M21}\ x_{M22}]^T$ and

$$[a_1,\cdots,a_{10}]=[J_1,J_{21},J_{22},J_2,R_{21},R_{22},B_1,B_2,f_{c1},f_{c2}] \quad [27a]$$

$$[J_1(x)\vdots\cdots\vdots J_6(x)]=\left[\begin{array}{cc|cc|cc|cc|cc|cc} 1 & 0 & 0 & 1 & 0 & 0 & 0 & 0 & 2\cos x_{21} & \cos x_{21} & 0 & 0 \\ 0 & 0 & 0 & 0 & 1 & 0 & 0 & 1 & 0 & 0 & \cos x_{21} & 0 \end{array}\right] \quad [27b]$$

$$[F_1(x)\vdots\cdots\vdots F_{10}(x)]=\left[\begin{array}{ccccc|ccccc} 0 & 0 & 0 & 0 & 2x_{21}x_{22}\sin x_{12}+x_{22}^2\sin x_{12} & 0 & x_{21} & 0 & \mathrm{sgn}\,x_{21} & 0 \\ 0 & 0 & 0 & 0 & 0 & -x_{21}^2\sin x_{12} & 0 & x_{22} & 0 & \mathrm{sgn}\,x_{22} \end{array}\right], \quad [27c]$$

we can get Eqs.[1] and [4]. This implies that the assumption [A1] can be satisfied. Moreover, from Eqs.[27b] and [27c], we can understand easily that the assumption [A2] can be satisfied.

Thus, we can see that the D.D.Arm satisfies the assumptions [A1] to [A5]. Therefore, we can construct the following adaptive controller for the D.D.Arm;

$$\tau_1=\lambda[-\ddot\theta_1+k_1\theta_1+k_2\dot\theta_1+r_1]+\bar J_1\ddot\theta_1+\bar J_{21}\ddot\theta_2 +\bar R_{21}[2\ddot\theta_1 C_2+\ddot\theta_2 C_2-2\dot\theta_1\dot\theta_2 S_2-\dot\theta_2^2 S_2] +B_1\dot\theta_1+f_{c1}\mathrm{sgn}\,\dot\theta_1 \quad [26a]$$

$$\tau_2=\lambda[-\ddot\theta_2+k_1\theta_2+k_2\dot\theta_2+r_2]+\bar J_{22}\ddot\theta_1+\bar J_2\ddot\theta_2 +\bar R_{22}[\ddot\theta_1 C_2+\dot\theta_1^2 S_2]+B_2\dot\theta_2+f_{c2}\mathrm{sgn}\,\dot\theta_2 \quad [26b]$$

$$y_1=d_1[\theta_{M1}-\theta_1]+d_2[\dot\theta_{M1}-\dot\theta_1] \quad [27a]$$

$$y_2=d_1[\theta_{M2}-\theta_2]+d_2[\dot\theta_{M2}-\dot\theta_2] \quad [27b]$$

$$\bar J_1=-\Gamma_1\int_0^t y_1\ddot\theta_1 ds+\bar J_1(0) \quad [28a]$$

$$\bar{J}_{21}=-\Gamma_2\int_0^t y_1\ddot{\theta}_2 ds+\bar{J}_{21}(0) \qquad [28b]$$

$$\bar{J}_{22}=-\Gamma_3\int_0^t y_2\ddot{\theta}_1 ds+\bar{J}_{22}(0) \qquad [28c]$$

$$\bar{J}_{2}=-\Gamma_4\int_0^t y_2\ddot{\theta}_2 ds+\bar{J}_{2}(0) \qquad [28d]$$

$$\bar{R}_{21}=-\Gamma_5\int_0^t y_1[2\ddot{\theta}_1 C_2+\ddot{\theta}_2 C_2-2\dot{\theta}_1\dot{\theta}_2 S_2 -\dot{\theta}_2{}^2 S_2]ds+\bar{R}_{21}(0) \qquad [28e]$$

$$\bar{R}_{22}=-\Gamma_6\int_0^t y_2[\ddot{\theta}_1 C_2+\dot{\theta}_1{}^2 S_2]ds+\bar{R}_{22}(0) \qquad [28f]$$

$$B_1=-\Gamma_7\int_0^t y_1\dot{\theta}_1 ds+B_1(0) \qquad [28g]$$

$$B_2=-\Gamma_8\int_0^t y_2\dot{\theta}_2 ds+B_2(0) \qquad [28h]$$

$$f_{c1}=-\Gamma_9\int_0^t y_1 \mathrm{sgn}\,\dot{\theta}_1 ds+f_{c1}(0) \qquad [28i]$$

$$f_{c2}=-\Gamma_{10}\int_0^t y_2 \mathrm{sgn}\,\dot{\theta}_2 ds+f_{c2}(0) \qquad [28j]$$

where $S_2=\sin\theta_2$, $C_2=\cos\theta_2$.

4.2 Experimental results

To examine the performance of the proposed controller, we made the D.D.Arm write a straight line by the edge of the arm as shown in Fig.4. The construction of the whole experimental system is shown in Fig.5.

The experiments are executed under the following four cases;

Case 1) The desired maximum speed of the edge of the arm is 0.5(m/s), and no payload is mounted.

Case 2) The desired maximum speed of the edge of arm is 0.5(m/s), and the payload mounted on the edge of arm is 2 kg.

Case 3) The desired maximum speed of the edge of arm is

0.5(m/s), and the payload mounted on the edge of arm is 4 kg

Case 4) The desired maximum speed of the edge of the arm is 1.0(m/s), and no payload is mounted

Case 5) The desired maximum speed of the edge of arm is 1.0(m/s), and the payload mounted on the edge of arm is 2 kg.

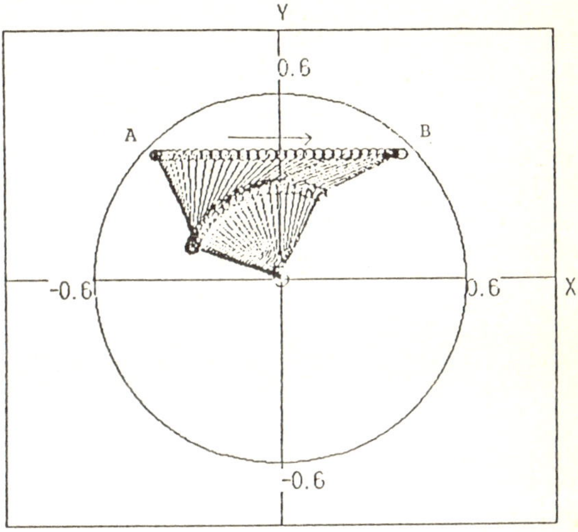

Fig.4 Arm Work Space

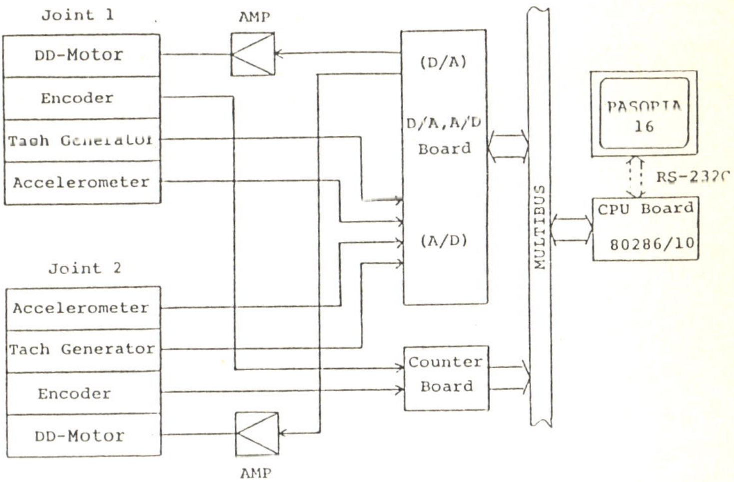

Fig.5 System Configuration

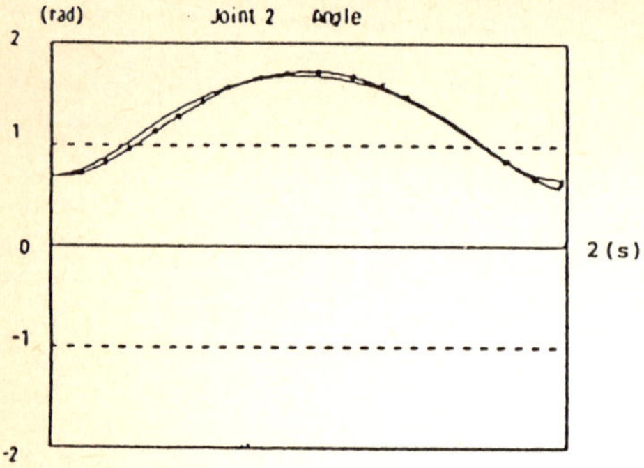

(a)-1 Case 1. Non-adaptive

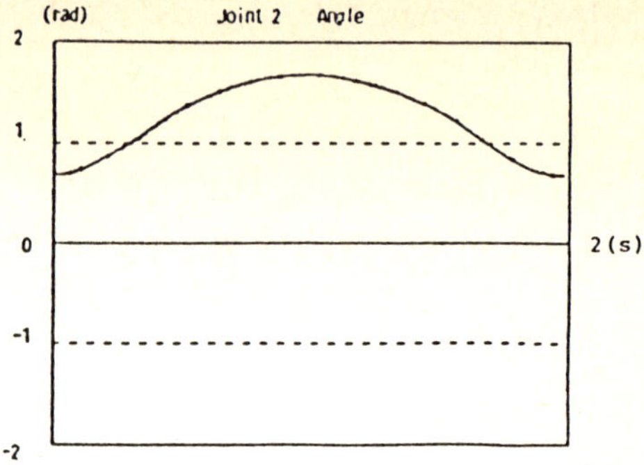

(b)-1 Case 1. Adaptive

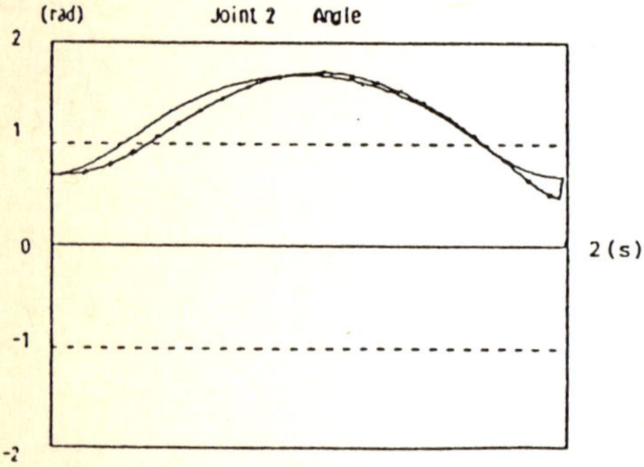

(a)-2 Case 2. Non-adaptive

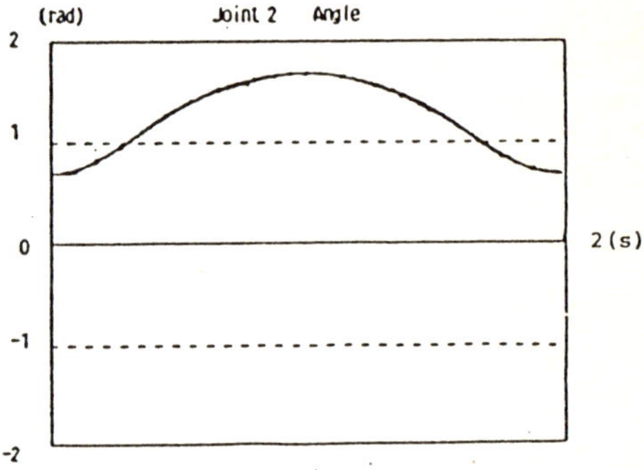

(b)-2 Case 2. Adaptive

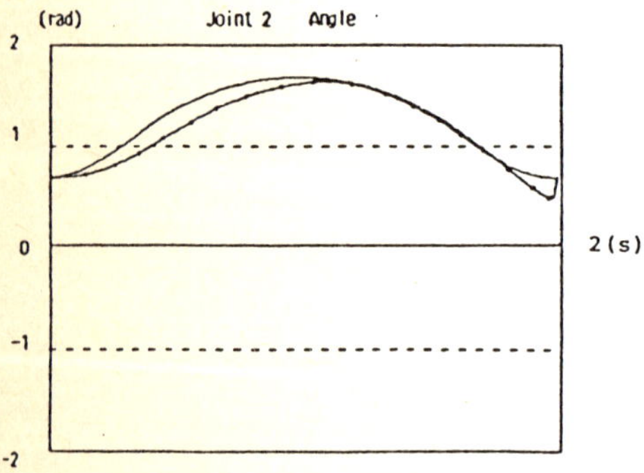

(a)-3 Case 3. Non-adaptive

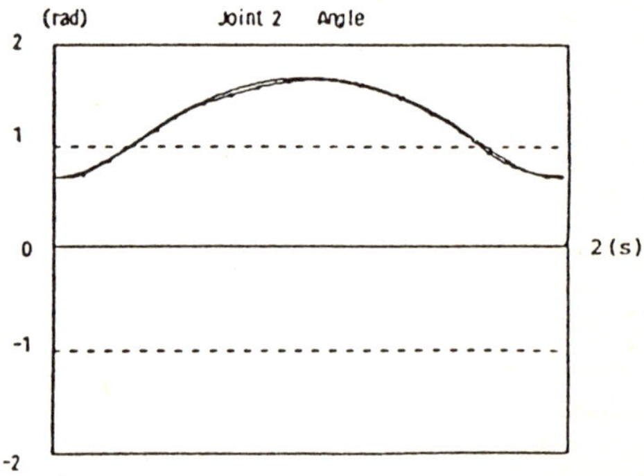

(b)-3 Case 3. Adaptive

Fig.6 Experimental Results (case 1 to case 3)

—— model, —•—•— actual

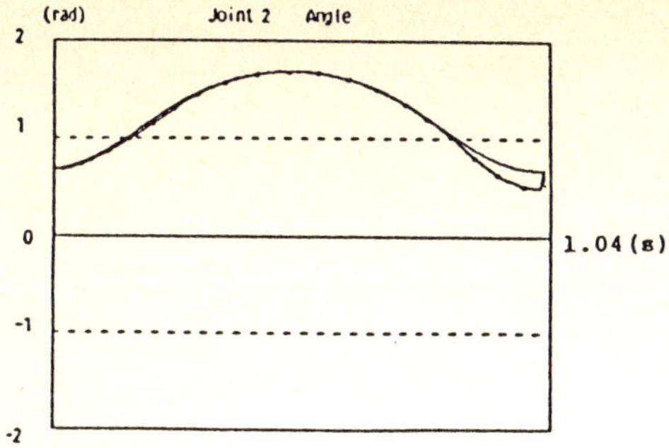

(a)-4 Case 4. Non-adaptive

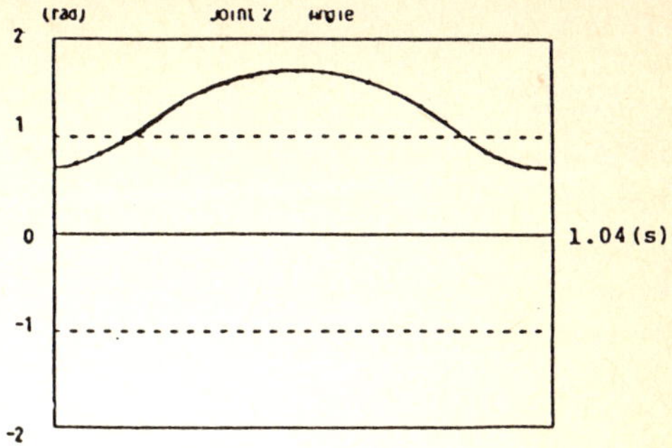

(b)-4 Case 4. Adaptive

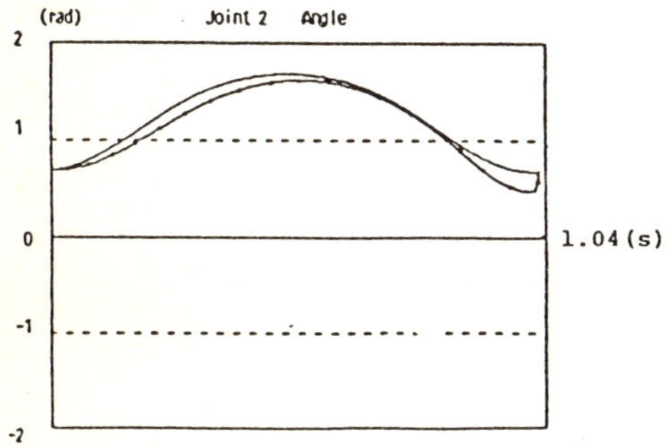

(a)-5 Case 5. Non-adaptive

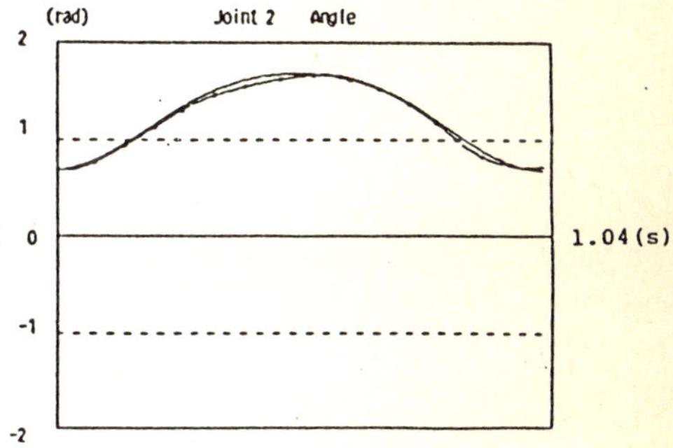

(b)-5 Case 5. Adaptive

Fig.7 Experimental Results (case 4 and case 5)

——— model, —•—•— actual

For each case, we compared with the experimental results due to the proposed control scheme and the results due to the conventional control scheme, i.e., nonlinear compensation method or computed torque method. In Figs.6 and 7, we show the experimental results for the cases 1 to 5, respectively. In each figure, ——— shows the desired trajectory and —•—•— show the actual trajectory. Moreover, Fig.8 shows the photograph of one of the result (case 5)that the D.D.Arm wrote a straight line on the paper with a pen. In the figure, (a) shows the result using the non-adaptive control scheme and (b) shows the result using the adaptive control scheme.

These experimental results indicates the effectiveness of the proposed control scheme.

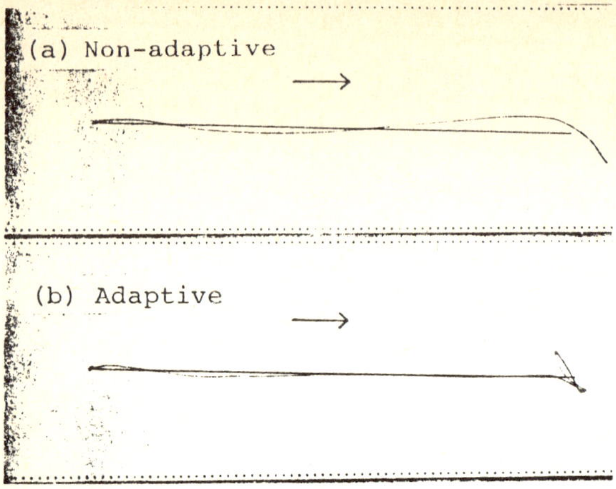

Fig.8 Experimental Result (case 4)

5. CONCLUSION

We have proposed a design method of an adaptive control system for nonlinear mechanical systems and shown the effectiveness of the proposed method through several experiments using the D.D.Arm.

Our method has the following desirable features;

(a) No linealizations or approximations are used for designing the controller.

(b) Therefore, the global asymptotic stability of the error between the states of reference model and the actual states has been assured.

(c) The boundedness of the internal signals in the whole system has been ensured.

ACKNOWLEDGMENT

The author would like to express his appreciation to Professor Toshiro Ono, Univ. of Osaka Prefecture, for instructive comments in preparing the English version of the paper.

REFERENCES

Craig,J.J., P.Hsu and S.S.Sastry(1987). Adaptive Control of Mechanical Manipulators, I. J. Robotics, Vol.2, No.2, 16/28.

Dubowsky,S and D.T.Desforges(1979). The Application of Model Referenced Adaptive Control to Robotic Manipulators. Trans. ASME, J. Dynamical Systems, Measurement, and Control, 101-9, 193/200.

Koivo,A.J. and T.H.Guo(1983). Adaptive Linear Controller for Robotic Manipulators, IEEE Trans. on Automatic Control, AC-28-2, 162/171

Luh,J.Y.S, Walker,M.W and PAul,R.P.C(1980). Resolved Acceleration Control of Mechanical Manipulators. IEEE Trans. Automatic Control, AC-25-3, 468/474.

Osuka,K(1986). Adaptive Control of Nonlinear Mechanical Systems. Trans. of The Society of Instrument and Control Engineers, 22-7, 756/762(in Japanese).

Takegaki,M and Arimoto,S(1981). An Adaptive Trajectory Control of Manipulators, Int. J. Control, 34-7, 219/230.

FLEXIBLE MECHANICAL STRUCTURES ADAPTIVE CONTROL

M. M'SAAD, M. DUQUE and SH. HAMMAD

Laboratoire d'Automatique de Grenoble
CNRS - GRECO (69) "SARTA"
E.N.S.I.E.G. - I.N.P.G.
B.P.46
38402 - SAINT-MARTIN-D'HERES

ABSTRACT .

This paper presents an applicability study of a partial state model reference adaptive controller involving a flexible arm and a flexible transmission system. The partial state model reference concept is particularly used to incorporate a suitable tracking capability into the generalized predictive control design. The latter has shown to be a potential approach for adaptive control.

1.INTRODUCTION.

A general purpose adaptive controller has been developped in [1] bearing in mind the available robust adaptive control theory as well as those fundamental design features as large applicability , good control design stability margin, offset-free performance, performance specification simplicity and robust parameter estimation. The core of such an approach consists of: 1) Using the partial state model reference concept to incorporate an appropriate tracking capability in those receding horizon linear quadratic criteria minimization. 2) Invoking the internal model principle to incorporate an offset-free performance capability in the control design. 3) Using a robust parameter estimator with respect to those ubiquitous non-idealities, as bounded disturbances, time varying plant model parameters and reduced order modelling errors. Some applicability studies have been already reported in [2]-[5] to demonstrate the feasibility of the involved adaptive control approach.

In this paper, the authors aim at presenting two real-time applications to emphasize the proposed adaptive control performance. These applications involve a flexible arm and a flexible transmission system, which represent a challenging opportunity to evaluate the adaptive control applicability in realistic situations.

The paper is organised as follows. The considered implementation approach is briefly discussed in section 2. The involved experimental evaluation is described in section 3.

2. THE ADAPTIVE CONTROL APPROACH

In this section we describe the main components of the considered adaptive control approach. The class of plants that can be considered is presented with particular emphasis on the design assumptions. The control objective is given together with suitable interpretation of its underlying control laws. The adaptive control algorithm is briefly described.

2.1 The plant model.

We consider those plants that can be described by the following discrete-time model:

$$A(\theta,q^{-1})\ y(t) = B(\ \theta,q^{-1})\ u(t-1-d) + v(t) + w(t)$$
$$D(q^{-1})\ v(t) = e(t)$$

with

$$\theta^T = [a_1 \ldots\ldots\ldots a_{na}\ b_0 \ldots\ldots\ldots b_{nb}]$$
$$A(\theta,q^{-1}) = 1 \ + a_1\ q^{-1} + \ldots\ldots + a_{na}\ q^{-na}$$
$$B(\theta,q^{-1}) = b_0 + b_1\ q^{-1} + \ldots\ldots + b_{nb}\ q^{-nb}$$
$$D(q^{-1}) \ = 1 \ + d_1\ q^{-1} + \ldots\ldots + d_{nd}\ q^{-nd}$$

where u(t) and y(t) denote the plant input and output, respectively, d is the minimal expected value of the plant dead time in sampling periods, q^{-1} is the backward shift operator, v(t) denotes the external disturbances, {e(t)} is a uniformly bounded zero mean sequence, w(t) represents the unmodeled plant response. The polynomials $A(q^{-1})$ and $B(q^{-1})$ define the nominal pulse transfer function which is at least stabilizable. The polynomial $D(q^{-1})$ is introduced, according to the internal model principle to provide a load disturbances compensation capability into the design. The underlying control problem is well posed if the following assumptions are at least satisfied:

A1. The plant model $B(z^{-1})$ / $A(z^{-1})D(z^{-1})$ is admissible.

A2. {w(t)/m(t)} is sufficiently small in the mean.

where m(t) is any input-output norm. The assumption A1 is standard one in any control design, it stipulates that the control law, to be considered, stabilizes the plant model. The assumption A2 characterizes a relatively important class of plant-model mismatch, including those due to parameter variations and reduced order modeling. Such a class has been first defined in [6] and subsequently used in recent robust adaptive control investigations.

2.2 The control objective.

The control objective consistsof minimizing, in a receding horizon sens, the linear quadratic cost function

$$J(t,ph,ch,sh) = \sum_{j=sh}^{ph} (s(t+j))^2 + \lambda(t)\ (e(t+j-hi))^2$$

assuming that $e(t+i) = o$ for $ch \leq i \leq ph$

where ph and ch denote the so- called prediction and contol horizons respectively , sh is the starting prediction horizon which should be at least greater than the time delay d, $\{\lambda(t)\}$ is a positive sequence, $\{s(t)\}$ and $\{e(t)\}$ represent input and output sequences of suitable plant model reparametrization depending on the considered control design, namely

- $s(t) = P(q^{-1})(y(t) - y^*(t))$ and $e(t) = D(q^{-1})P(q^{-1})u(t)$ for the generalized predictive control (GPC) design proposed in [7].

- $s(t) = P(q^{-1})(y(t) - B(\theta,q^{-1})\ \beta(\theta)y^*(t))$ and $e(t) = D(q^{-1})P(q^{-1})(u(t) - A(\theta,q^{-1})\ \beta(\theta)y^*(t+d+1))$ for the partial state model reference control (PSMRC) design propoded in [1] .

where $P(q^{-1})$ and $\{y^*(t)\}$ are arbitrary chosen Hurwitz polynomial and uniformly bounded reference sequence, the scalar $\beta(\theta)$ is introduced for unitary static gain purpose; i.e $\beta(\theta) = [B(\theta,1)]^{-1}$. Notice that the variables $P(q^{-1})(y(t)-B(\theta,q^{-1})\beta(\theta)y^*(t))$ and $P(q^{-1})(u(t)-A(\theta,q^{-1})\beta(\theta)y^*(t+d+1))$ represent the output and input tracking errors when a pole placement or partial state model reference control objective is considered (See [4] and [8] for more details).

2.3. Control law interpretation.

The above control objectif can be carried out using the generelised predictive control approach [7]. Whatever is the considered control design, the underlying control law is linear and may be written as follows:

$$S(\theta,q^{-1})\ D(q^{-1})u(t) + R(\theta,q^{-1})\ y(t) = T(\theta,q^{-1})\ y^*(t+d+1)$$

The resulting control system is stable if the following property is satisfied

$$P^*(\theta,q^{-1})=A(\theta,q^{-1})D(q^{-1})S(\theta,q^{-1})+q^{-1-d}B(\theta,q^{-1})R(\theta,q^{-1}) = 0 \Rightarrow |q| < 1$$

Such a property defines clearly the model admissibility assumption A1, the stabilizability of the transfer function $B(z^{-1}) / A(z^{-1})D(z^{-1})$ is necessary. When the partial state model reference control is considered, the polynomial $T(\theta,q^{-1})$ is given by

$$T(\theta,q^{-1}) = \beta(\theta)(A(\theta,q^{-1})D(q^{-1})S(\theta,q^{-1})+q^{-1-d}B(\theta,q^{-1})R(\theta,q^{-1}))$$

and can be factorized as follows

$$T(\theta,q^{-1}) = \beta(\theta)A^*(\theta,q^{-1})P(q^{-1})$$

where the polynomial $A^*(\theta,q^{-1})$ depends on the design parameters ph,ch, sh and $\{\lambda(t)\}$. The partial state model reference control law allows to separate the tracking behavior (e.g. the reference sequence choice) and the regulation response (e.g. the input output tracking errors'shaping choice). The tracking dynamics is specified via the reference sequence $\{y^*(t)\}$ while the regulation dynamics is given in terms of the polynomial $T(\theta,q^{-1})$. The latter depends on the design parameters involved in the performance criteria but can ge adjusted by the polynomial $P(q^{-1})$.

2.3. The adaptive control algorithm.

The adaptive control algotithm is obtained by invoking the certainty equivalence principle, wich consists of simply replacing the plant model parametrers θ by their estimates. The considered parameter estimation is performed by least squares parameter adaptation algorithm, involving input-output signal filtering, data normalization, adaptation gain regularization and coditionnal parameter estimates contration (See [1] and [4] for further details).In all these experiments, the adaptive control system is implemented as follows:

1. Sample the plant output.
2. Update the plant model parameters.
3. Evaluate the control law using the GPC or PSMRC design.
4. Wait for the clock pulse.
5. Implement the control signal and go to step 1.

3. EXPERIMENTAL EVALUATION.

The involved adaptive control approach has been implemented on a Hp-1000/900 computer providing a software package called CATRES. Several real time experiments, involving two flexible mechanical structures have been carried out to evaluate the applicability of the proposed adaptive control approach ([2]-[5]). In the following, we will describe each of the considered flexible mechanical system and present the corresponding adaptive control performances.

3.1 The flexible arm.

The underlying experimental device is depicted in figure 1. This consists of one link flexible arm moving in the horizental plan and driven by a dc motor. The motor axle is locally controlled with a standard position and velocity analog control system. An IR light emiter diode is mounted at the free tip for position measurement purpose; this measure is obtained by a silicium photodiode. The plant to be controlled consists in the flexible arm together with its actuator; i.e the dc motor and its corresponding local control system. The input and output of this plant are the actuator set point sequence and the free tip position,respectively.

This experimental flexible arm provides an interisting adaptive control problem. In particular, it exibits an osillatory behavior with a very low damping rate, non-minimum phase behavior and significant dynamics'variations under inertia changes. Moreover it is disturbed by the frictions occuring in the bearings.

3.2 The flexible transmission system.

The involved flexible transmission system represents the first two vibrations modes of flexible belt transmission system. It consists, as depicted in figure 2, of three pulleys which are linked via flexible system; the flexibility being provided by the springs. The angular position of the first pulley is driven by a dc motor, whose axle position and velocity are locally controlled by a standard analog feedback control system. It is possible to add friction on any pulley axle and change the inertia of the last pulley, providing hence an interesting adaptive control problem [3].

3.3 Experimental results.

Several experiments involving the above presented pilot plants have been carried out to emphasize the applicability of the proposed adaptive control approach. We will particularly present those concerning the tracking capability of the control design and the adaptation alertness of the parameter estimator. The effects of the input-output signal filtering, data normalization and adaptive control robustness with respect reduced order modeling are reported in [4] and [5].

In all these experiments, the adaptive control is started with an open loop identification involving a suitable pseudo random binary sequence, and the input -output signal have been filtered and normalized. The filter and the normalized factor have been chosen as follows:

$$D(z^{-1})G(z^{-1}) / F(z^{-1}) = (1-z^{-1})(1+0.98z^{-1}) / (1-0.98z^{-1})(1-0.5z^{-1}).$$

$$\eta(t) = 0.95\,\eta(t-1) + .05 \max(\|\phi_f(t)\|, 0.01) \text{ with } \eta(0) = 1.$$

For the tracking capability, a comparative study between the partial state model reference adaptive control (PSMRAC) and the generalized predictive adaptive control (GPAC) has been made. Figures 3, 4 and 5 show the performances of the PSMRAC and GPAC for the flexible arm when changing, in real time, the weighting sequence $\{\lambda(t)\}$, the prediction horizon ph and the control horizon ch, respectively. In each experiment the same design parameters are used for the PSMRAC and GPAC. Figure 6 show the performances of the PSMRAC and GPAC for the flexible transmission system, using the same design parameters.

The ability of the parameter estimator to deal with time varying dynamics under inertia changes are shown in figure 7 for the flexible arm. Similar results have been obtained for the flexible transmission system [5]. Figure 8 shows the parameter adaptation alertness with respect to friction changes for the flexible transmission system. The friction changes are more difficult to deal with than the inertia changes. In all these figures, the linear control means that the parameters are frozen before changing the inertia or the friction.

REFERENCES

[1]. M'Saad, M. (1987). "Sur l'applicabilité de la commande adaptative". Thèse d'Etat de l'Institut National Polytechnique de Grenoble.

[2]. Dugard, L., M. M'Saad and M. Duque (1987). "Adaptive LQ control of a flexible transmission system". 3rd ICAR, Versailles, France, October 1987.

[3]. M'Saad, M., Duque and E. Irving (1987)."Thermal process robust adaptive control : an experimental evaluation". Proc of the 10th IFAC World Congress , Munich, July 1987.

[4]. M'Saad, M., E. Van den Bossche, A.J. Montano and L. Dugard (1988)."A long-rangepredictive adaptive control : performance enhancements and experimental evaluation". 1988 IFAC Workshop on Robust Adaptive Control, Newcastle, Australia 1988.

[5] M'Saad, M., M.Duque, A.J. Montano and Sh. Hammad (1988). "A flexible mechanical structures adaptive control". Workshop on Adaptive Control Strategies for Industrial Use, Alberta, Canada, Juin 1988.

[6]. Praly, L. (1982). "MIMO stochastic adaptive control : stability and robustness" ; CAI Report, Ecole des Mines de Fontainebleau, France.

[7]. Clarke, D.W., P.S. Tuffs and C. Mohtadi (1987). "Generalized predictive control", Automatica, vol. 23, n° 2, pp. 137-160.

[8]. Irving, E., C.M. Falinower and C. Fonte (1986). "Adaptive generalized predictive control with multiple reference model". 2nd IFAC workshop on Adaptive systems in control and signal processing, Lund, Sweden 1986.

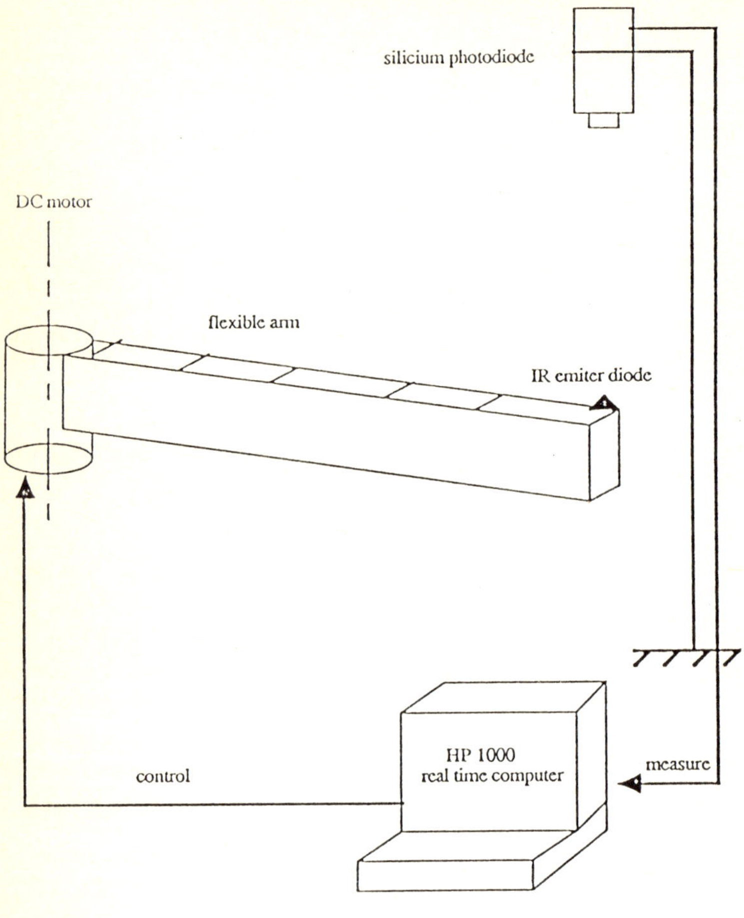

Figure 1. flexible arm experimental device

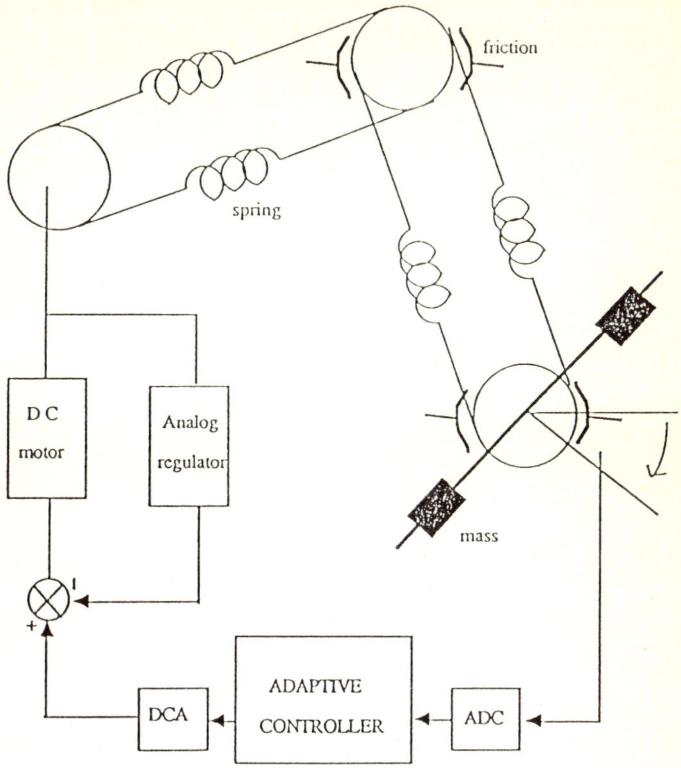

Figure 2. Flexible transmission system experimental device

GPAC

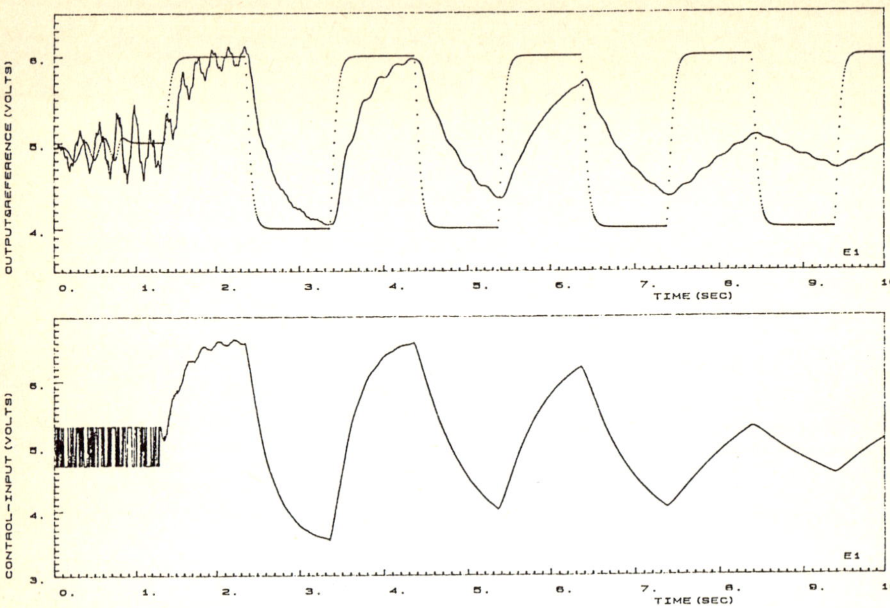

PSMRAC

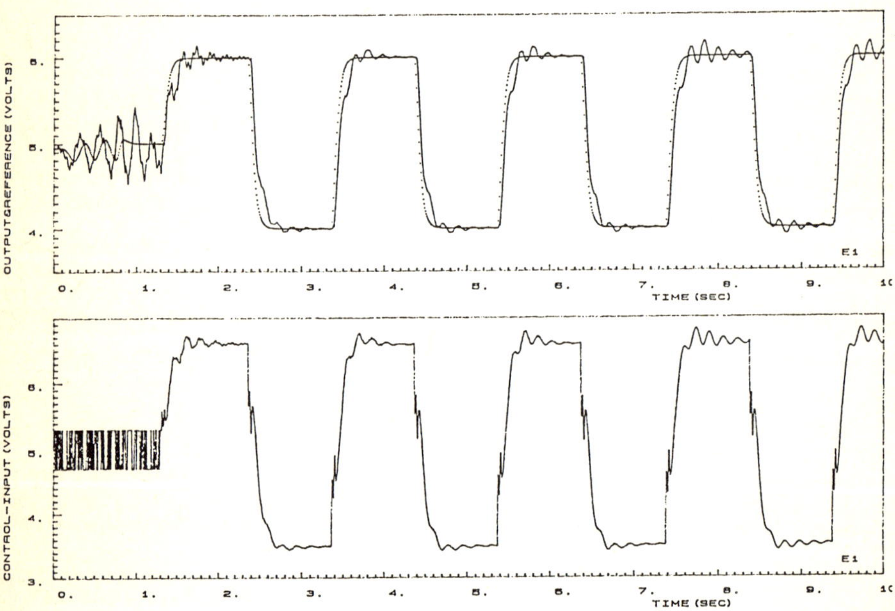

Figure 3. Performances of the PSMRAC and GPAC when changing λ

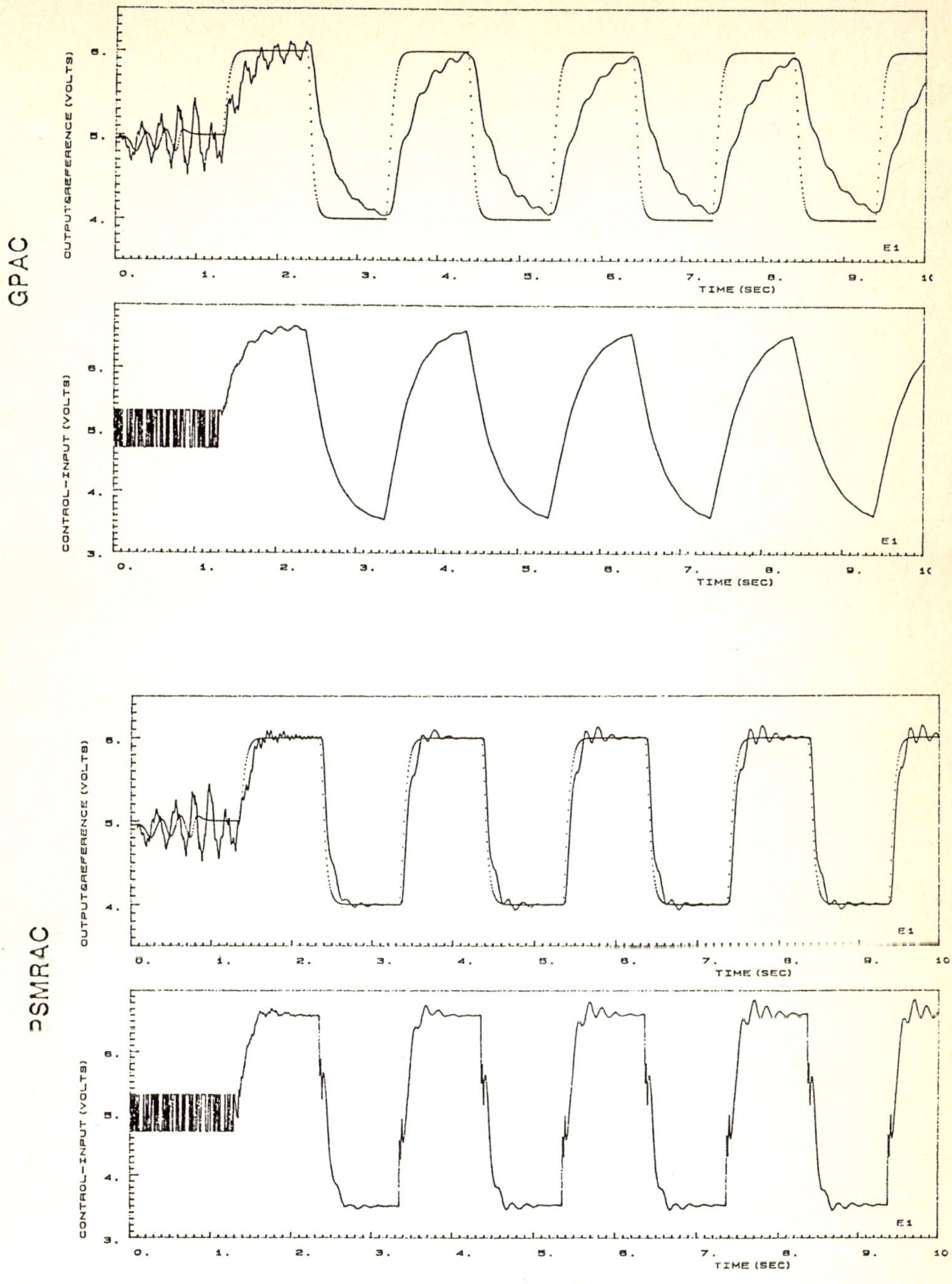

Figure 4. Performances of the PSMRAC and GPAC when changing ph

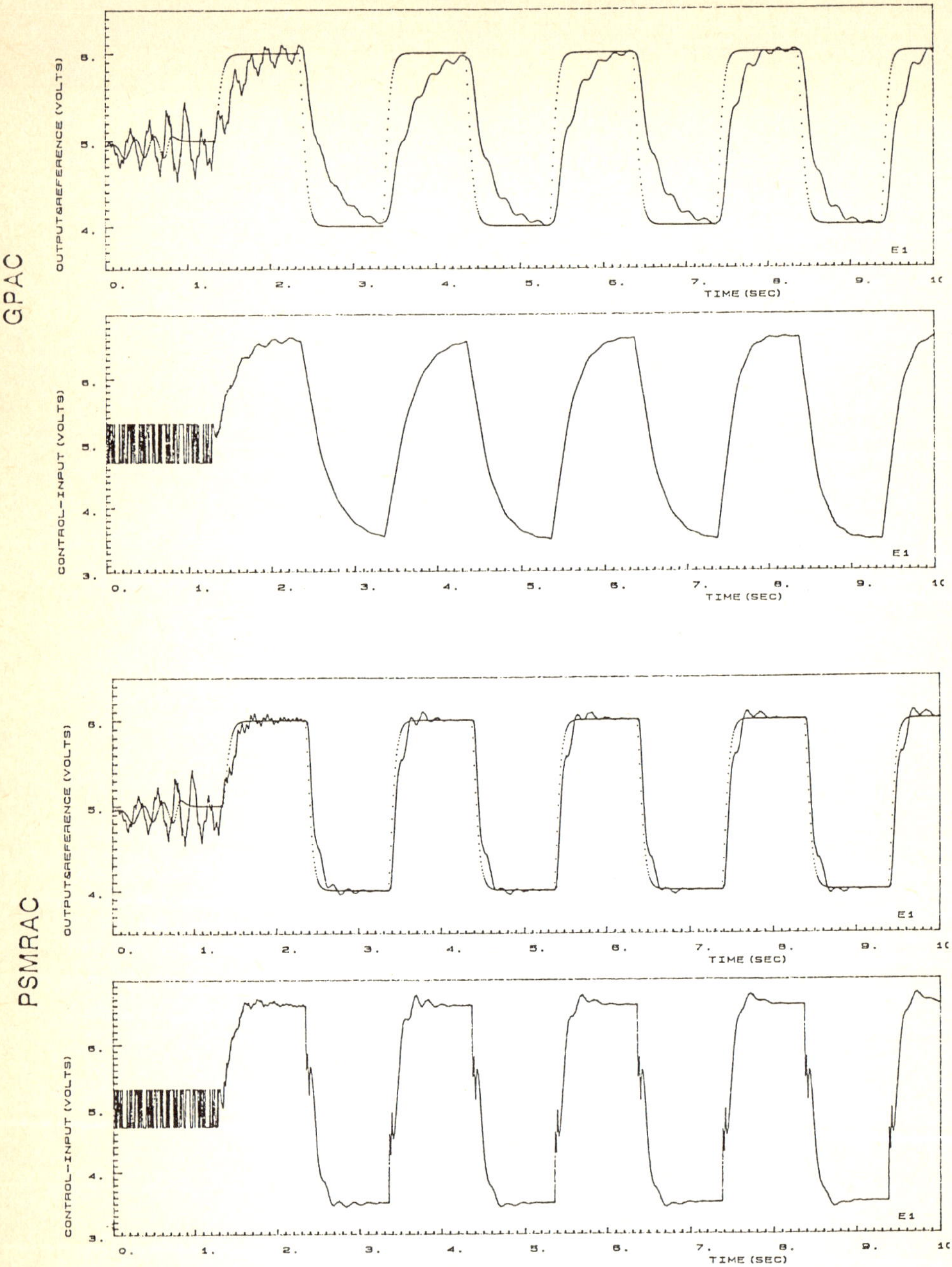

Figure 5. Performances of the PSMRAC and GPAC when changing ch

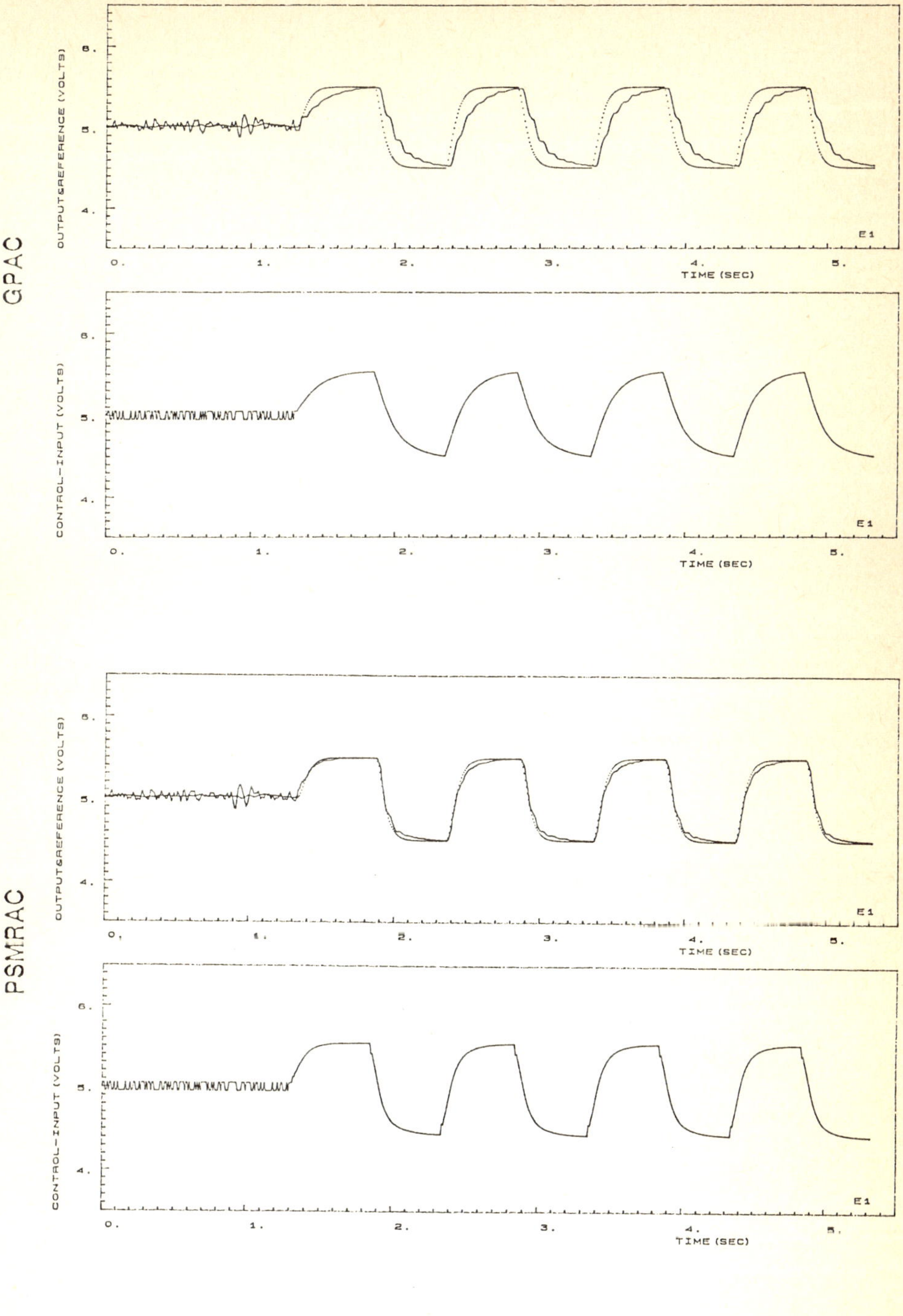

Figure 6. Performances of the PSMRAC and GPAC for the flexible transmission system

ADAPTIVE CONTROL

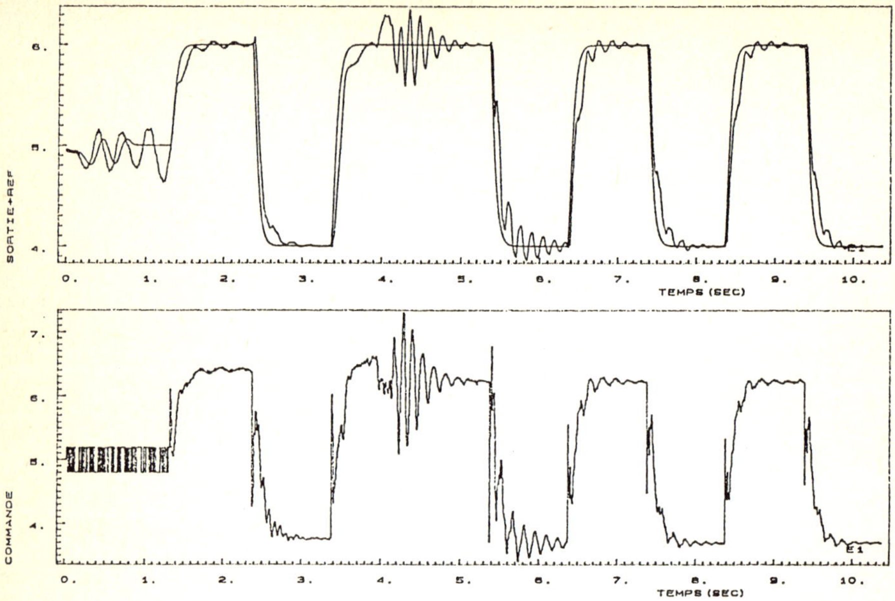

LINEAR CONTROL

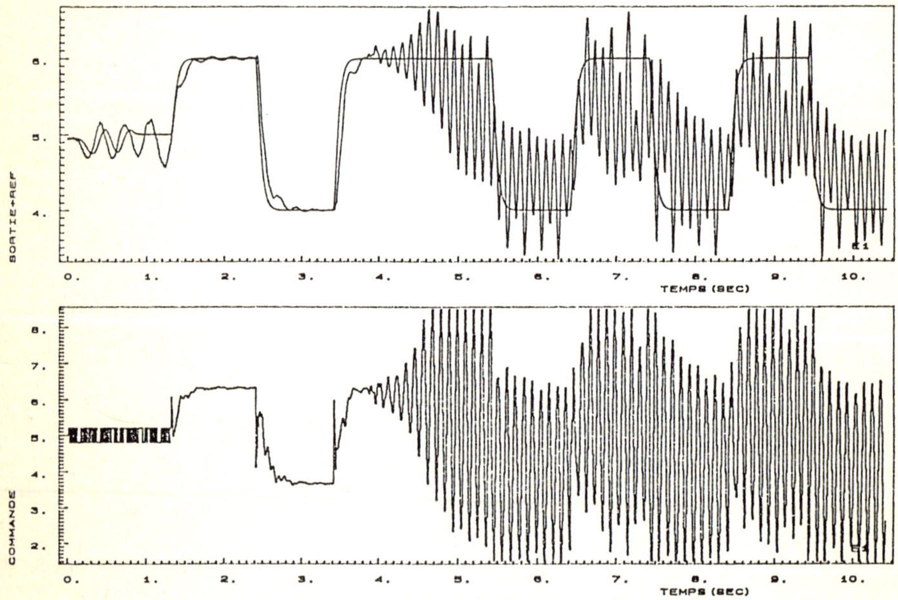

Figure 7. Adaptive control performances under inertia changes

ADAPTIVE CONTROL

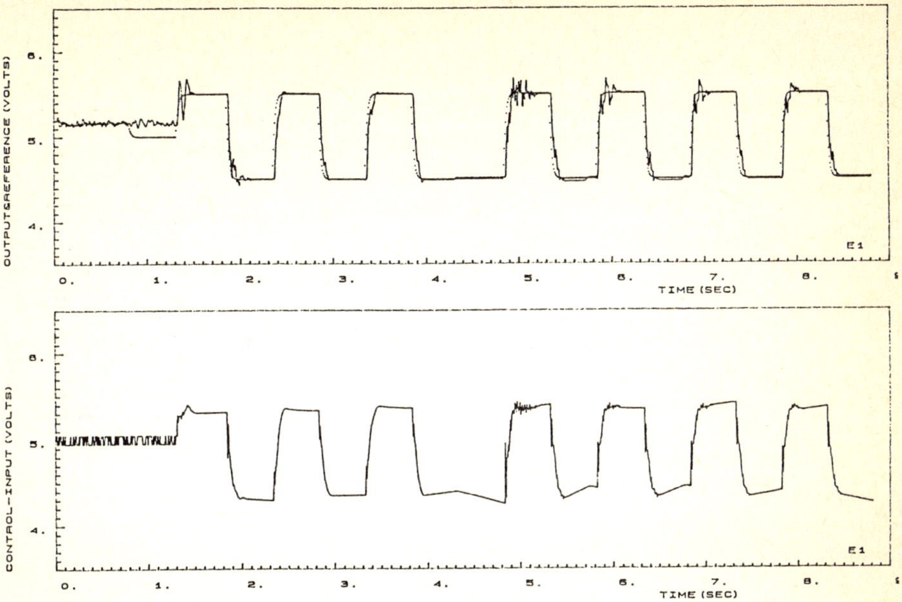

LINEAR CONTROL

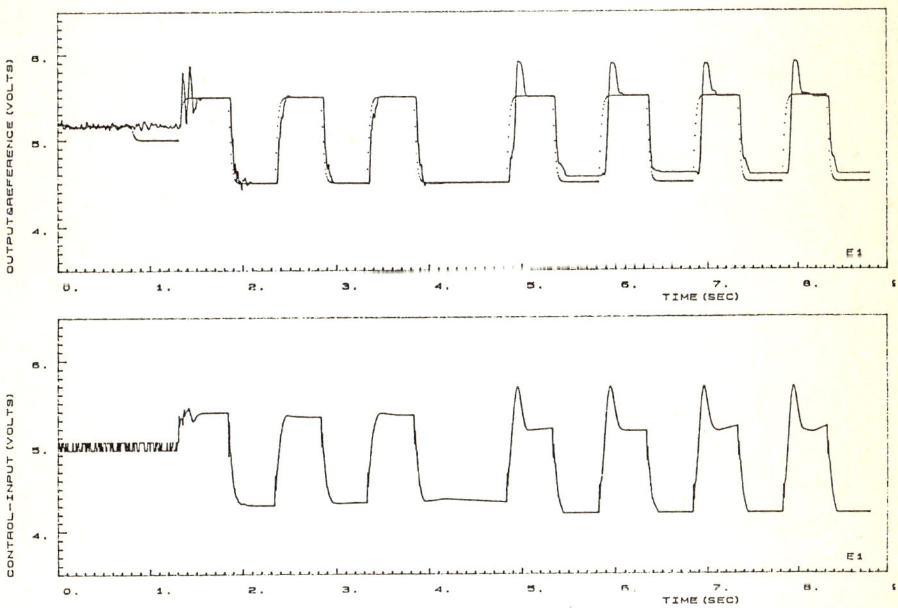

Figure 8. Adaptive control performances under friction changes

AI and Industrial Adaptive Controllers

EVALUATION OF AN INDUSTRIAL PID AUTOTUNER

E. Goberdhansingh and W.R. Cluett
Department of Chemical Engineering
University of Toronto
Toronto, Canada M5S-1A4

Abstract

The automatic tuning method proposed by Åström and Hägglund (1984) for PID controllers has been incorporated into an industrial autotuner. The tuning method is based on the precondition that a process connected in feedback with a relay will exhibit a stable limit cycle. The theoretical basis for the design of the autotuner is derived in this paper from the results of the describing function analysis coupled with phase and gain margin criteria. The design basis is then verified by applying the autotuner to an analog simulator where the process model is well known. Finally, the autotuner is tested on a pilot plant heat exchanger.

1 Introduction

The rapid advancement of microprocessors and computer technology has enabled the application of advanced control techniques, such as self-tuning or adaptive control, in the chemical process industries (see Seborg et al (1986) for a survey). Despite recent theoretical and practical developments of adaptive control systems, the design of these algorithms still requires the selection of several types of parameters associated with the process model structure, the control law and the parameter estimation algorithm. These requirements make it difficult for the nonexpert to consider adaptive control as a viable control option. As pointed out by Seborg et al (1986), the user should only have to specify a minimum amount of information. Recently progress has been made towards a more general purpose adaptive algorithm for satisfactory control of a wide range of processes (e.g. Clarke et al (1987)).

The motivation for using adaptive control in industry has evolved from the fact that many chemical processes are non linear (their behaviour varies over the range of operation) and time- varying (their behaviour varies with time). Adaptive strategies provide industry with a control technique which attempts to compensate for these changes in the process via self-adjustment of its controller settings. Nevertheless, in industry the PID controller still dominates. These non-adaptive controllers, when properly tuned, often give satisfactory control performance in the presence of process uncertainty and changes. In fact, processes with uncertain dynamics do not necessarily require adaptive control. Classical design specifications such as gain and phase margins coupled with the fundamental principle of feedback control, often provide the PID controller with sufficient robustness. The issue of comparative robustness of non-adaptive versus adaptive controllers is discussed by Gawthrop (1985).

Motivated by the premise that a well tuned PID controller is robust, the main objective of this paper is to describe and evaluate an automatic method for tuning these controllers, developed by Åström and Hägglund (1984), which has been incorporated into a commercially available digital controller. The autotuning technique examined in this paper is not self- tuning in that the tuning function must be initiated by the operator. However, it is worth noting that the idea of autotuning has evolved from self-tuning and adaptive ideas where one of the earliest motivations for considering adaptive control was to find strategies which could tune themselves (see Åström and Wittenmark (1973)). No comparison is done in this paper with other commercially available controllers which have autotuning or adaptive features. Therefore, our choice of this particular device does not imply that this unit is better or worse in any way than similar units. This controller was selected because the PID controller is still the most common controller structure used in industry and we feel that the analysis of an automatic method for tuning its parameters deserves attention.

2 Basic Idea of Autotuning

The autotuning method of Åström and Hägglund (1984) has evolved from the combination of two well-known concepts; the Ziegler- Nichols method for tuning PID controllers, and relay feedback control. Originally proposed by Ziegler and Nichols (1942), their method involves connecting a proportional controller in feedback with the process (see Figure 1 for a schematic). The gain of the controller is then adjusted until a stable oscillation in the process output is achieved. The gain at which this stable oscillation occurs, referred to as the critical gain, and the period of oscillation provide the necessary information for choosing the PID parameters. As pointed out by Åström and Hägglund (1984b) this method is difficult to do automatically and can be very time consuming. Another drawback of this method is that the amplitude of the oscillation cannot be easily controlled.

Early work on relay feedback was done by Hamel (1949) and Tsypkin (1958). Consider the system in Figure 2 where the proportional controller in Figure 1 has been replaced with a relay. One of the important results derived for a process under relay control is the describing function approximation. This technique provides a means for estimating the critical gain in the case of a pure relay, or a specific point on the process Nyquist diagram in the case of a relay with hysteresis.

The new method for tuning PID controllers proposed by Åström and Hägglund (1984) is based on the fact that for a process with a phase lag of at least 180° at some frequency may oscillate under relay control. If such an oscillation exists, and it is stable, the describing function approximation may be used to predict a point on the Nyquist diagram. From this single point, the PID parameters can be designed for specific gain and phase margins. The major advantage of this technique over the

Ziegler-Nichols method is that the oscillations are automatically generated by the relay. Furthermore, by controlling the relay amplitude, it is possible to control the amplitude of the oscillations. Since this method depends on the existence of a limit cycle, it is imperative that the conditions under which a stable limit cycle exists be known. To this end Åström and Hägglund (1984a) provide a proof of the necessary conditions.

The commercial product, the SattControl Autotuner, is a digital controller sampling at a rate of 0.2s in the control mode. Among the built-in functions is the ability to gain schedule in three regions based on the reference or process output value. The controller is placed in the tune mode by pressing the tune button from the manual or automatic mode. Tuning is initiated only if the process value is sufficiently close to the reference. It is interesting to note that during tuning the sampling period is approximately three times that in the control mode.

3 Theory- The Describing Function Analysis

In the describing function method, the relay, which is a non- linear component, is represented by the Fourier series expansion of its action on an input sinusoid. One can view the method as a representation of the relay as a transfer function. The describing function approximation (DFA) is obtained by using the first term of the Fourier series. Therefore, if this approximation is to be accurate, the first term of the series must dominate the expression. The DFA for a pure relay and with hysteresis is presented below and a complete derivation is given by Goberdhansingh (1988).

(a) For a pure relay

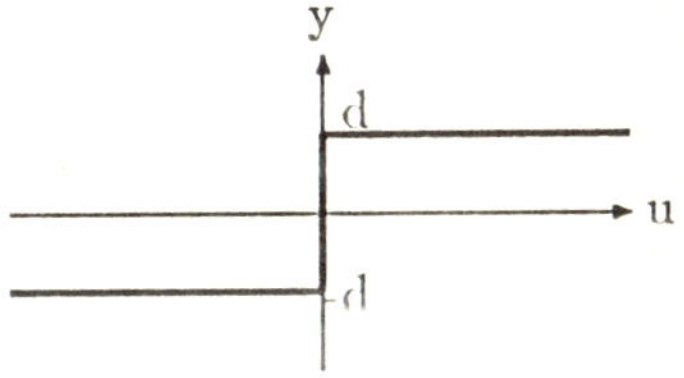

$$Na = \frac{4d}{\pi a}$$

(b) For a relay with hysteresis

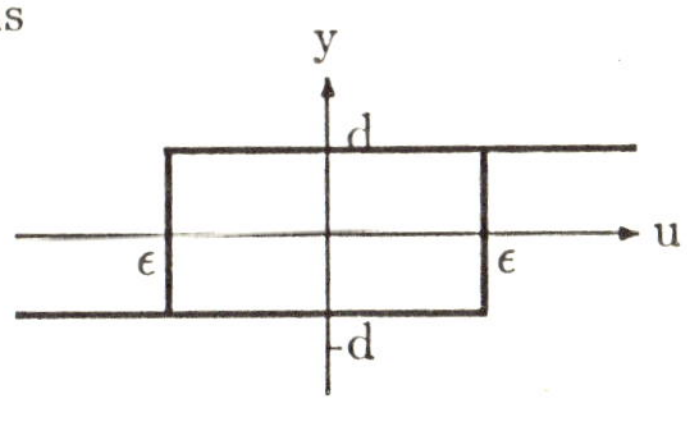

$$Na = \frac{4d}{\pi a^2}\sqrt{a^2 - \epsilon^2} - i\frac{4d\epsilon}{\pi a^2}$$

4 Locating a point on the Nyquist Diagram

Consider the case of a relay connected in feedback with the process. Assuming the system exhibits a stable limit cycle, sinusoid in nature with frequency ω, the non-linear relay component can be approximated using the describing function analysis (Figure 3). In terms of the frequency domain, the closed loop response of the system in Figure 3 is as follows

$$y(i\omega) = -y(i\omega) \cdot Na \cdot G(i\omega) \tag{1}$$

$$y(i\omega) \cdot [1 + Na \cdot G(i\omega)] = 0 \tag{2}$$

Since $y(i\omega) \neq 0$, (2) implies that

$$G(i\omega) = -\frac{1}{Na} \tag{3}$$

Therefore a point on the Nyquist diagram can be determined from the negative reciprocal of the describing function approximation. For a pure relay :

$$G(i\omega) = -\frac{\pi a}{4d} \tag{4}$$

For a relay with hysteresis

$$G(i\omega) = -\frac{\pi}{4d}\sqrt{a^2 - \epsilon^2} - i\frac{\pi\epsilon}{4d} \tag{5}$$

The above is the standard result used in the literature (Åström and Hägglund (1984b)).

5 Controller Design

Given that a process under relay feedback control exhibits a stable limit cycle, the describing function estimates a point on the Nyquist diagram. In conjunction with the frequency response analysis of the PID controller, the autotuner selects the parameters based on a phase margin design. This technique will be explained in this section (see Åström and Hägglund (1983) for this approach and other design techniques based on the frequency response).

The open loop transfer function (G_{ol}) of a process in series with a standard PID controller is given by

$$G_{ol} = K(1 + \frac{1}{T_i i\omega} + T_d i\omega)G(i\omega). \tag{6}$$

The argument of the process frequency response, $G(i\omega)$, may be represented as

$$\arg G(i\omega) = -\pi + \theta. \tag{7}$$

where θ is the angle relative to the negative real axis determined from the point on the Nyquist diagram. For a specified phase margin, for example $\frac{\pi}{4}$, the argument of the open loop transfer function is

$$\arg(G_{ol}) = -\pi + \frac{\pi}{4} = -\frac{3}{4}\pi \tag{8}$$

and the controller argument must be $\frac{\pi}{4} - \theta$. Therefore, from the frequency response of the PID controller

$$\tan(\frac{\pi}{4} - \theta) = T_d\omega - \frac{1}{T_i\omega} \tag{9}$$

One possibility (see Åström and Hägglund (1984)) is to choose T_i and T_d such that

$$T_d = \frac{T_i}{\alpha} \qquad \alpha > 0 \tag{10}$$

Substituting (10) in (9) gives

$$\tan(\frac{\pi}{4} - \theta) = T_d\omega - \frac{1}{\alpha T_d\omega} \tag{11}$$

$$\Rightarrow \tan(\frac{\pi}{4} - \theta) = \frac{T_d^2\omega^2\alpha - 1}{\alpha T_d\omega} \tag{12}$$

and solving for T_d gives

$$T_d = \frac{\tan(\frac{\pi}{4} - \theta) + \sqrt{\frac{4}{\alpha} + \tan^2(\frac{\pi}{4} - \theta)}}{2\omega} \tag{13}$$

A suitable choice for the value of α is 4 which corresponds to that suggested by Ziegler and Nichols (1942). Equations (9) through (13) enable the T_i and T_d parameters to be selected for a specific phase margin.

Based on the desired phase margin, the controller gain is now selected such that the modulus of $G_{ol}(i\omega)$ is equal to unity. Defining $\phi = \frac{\pi}{4} - \theta$

$$|G_{ol}| = K|1 + i\tan(\phi)||G(i\omega)| = 1 \tag{14}$$

or

$$K = \frac{\cos(\phi)}{|G(i\omega)|} \tag{15}$$

A more conservative controller design would be to select a controller gain which is less than the K value in (15), i.e.

$$K = \beta\frac{cos(\phi)}{|G(i\omega)|} \qquad \beta < 1 \tag{16}$$

6 Analog Simulation Results

The autotuner was connected to an analog simulator in which three transfer functions of orders 2, 3 and 4 were used for an evaluation of its design basis. The transfer functions are listed below. The use of the analog simulator enabled us to directly evaluate the accuracy of the DFA (as shown in Example 1) and to calculate the overall phase margin for the design. The results are presented in Table 1. The tuning cycle was repeated three times for each transfer function.

Example 1: Accuracy of the DFA

$$G(s) = \frac{1}{(10s+1)^2}$$

Results of the autotuner:

$$\begin{array}{ll} a = 1.275\% & K = 4.12 \\ T = 21.1s & Ti = 11s \\ \epsilon = 0.82\% & T_d = 2.7s \\ d = 10.25\% & \\ w = 0.298\frac{rad}{s} & \end{array}$$

DFA : From combining (5) and the results of the autotuner

$$-\frac{1}{Na} = 0.0748 - 0.0628i$$

$$|-\frac{1}{Na}| = 0.098 \quad (cf. \arg|G(i\omega)| = 0.101)$$

$$\arg(-\frac{1}{Na}) = -140^\circ \quad (cf. \arg G(i\omega) = -143^\circ)$$

$$(a) \quad G(s) = \frac{1}{(10s+1)^2} \qquad (b) \quad G(s) = \frac{1}{(10s+1)^3} \qquad (c) \quad G(s) = \frac{1}{(10s+1)^4}$$

7 Experimental Results for Heat Exchanger

Figure 4 illustrates the pilot plant heat exchanger on which the autotuner was evaluated. The unit has two passes on the water (tube) side and only one on the steam (shell) side. A computer connected in parallel to the autotuner logged all relevant signals independent of the autotuner. The computer logging was accomplished using standard interface equipment sampling at a rate of 0.1 sec. Figure 5 shows the tuning cycle (0 - 300s) followed by several set point changes.

8 Discussion

8.1 Autotuner Design Basis

The analysis of the data from the analog simulators provides insight into the design criteria used in this controller. In order to examine the design, it is first necessary to calculate the point on the Nyquist diagram (as shown in Example 1). This requires computation of the period and amplitude of the oscillations which was accomplished using a least squares fit. The additional phase shift provided by the PID controller may be calculated from (11) using the T_i and T_d values selected by the autotuner. The controller phase shift together with the phase angle at the particular point on the Nyquist diagram gives the overall phase margin. The gain ratio β is determined by comparing the value of K selected by the controller and the value calculated by (15).

The results in Table 1 indicate that the autotuner selects a phase margin in the range 50° to 65°. The autotuner has three different autotune modes (slow, medium and fast) which allow the operator to effect the speed of response. The results in Table 1 were based on the medium mode, which is the default mode, and it is seen that the gain ratio was approximately 0.5. The gain ratio for the slow and fast modes was 0.25 and 1.0 respectively.

A phase margin of 60° with a gain ratio of 0.5 is somewhat of a conservative design basis. Support for this remark is illustrated by the heat exchanger results where it is observed that the close loop response to a set point change is overdamped.

8.2 Selection of PI vs PID

In most runs performed with the autotuner a PID controller was selected. However in some cases the autotuner opted for a PI structure. The overall phase shift contributed by a PI controller is in the range −90° to 0°. Therefore if the point on the Nyquist diagram has a phase shift between −120° and −180°, then a phase margin design of 60° cannot be achieved with a PI controller. On the other hand, the overall phase shift due to a PID controller is between −90° and 90°. Therefore the PID controller has the flexibility to add either negative or positive phase shift and a phase margin design of 60° can be achieved from any Nyquist diagram point with a phase shift between −90° and −180°.

8.3 Limitations of the Autotuner

The autotuner is of course still basically a PID controller and will no doubt inherit the limitations of this control algorithm. Apart from this, one possible limitation of the autotuner comes from the fact that there is no interface between the operator and controller after tuning is completed. As such, there exists the possibility of some erroneous PID parameters being implemented automatically.

9 Conclusion

The new technique of Åström and Hägglund (1984) has proven to be successful for automatically tuning the PID controller. However the design basis appears to be a bit on the conservative side. While the background theory of this method is by no means new, the application is indeed innovative. Its importance can be illustrated by the fact that the PID class of controllers works well for many cases in process control. In fact many control loops do not require the sophistication of adaptive control but would certainly benefit from an autotuning feature.

Acknowledgement : The authors wish to acknowledge the financial support of the Natural Science and Engineering Research Council of Canada and of Imperial Oil Limited.

10 References

Åström, K. J. and T. Hägglund(1983). Automatic Tuning of Simple Regulators for Phase and Amplitude Margins Specifications. Proceedings IFAC Adaptive Systems in Control and Signal Processing, San Francisco, 271-276.

Åström, K. J. and T. Hägglund(1984a). Automatic Tuning of Simple Regulators. Proceedings IFAC 9th World Congress, Budapest, 1867-1872.

Åström, K. J. and T. Hägglund(1984b). Automatic Tuning of Simple Regulators with Specifications on Phase and Amplitude Margins. Automatica, 20, 645-651.

Åström, K. J. and B.Wittenmark(1973). On Self-tuning Regulators. Automatica, 9, 185-199.

Clarke, D. W., C. Mohtadi and P. S. Tuffs(1987). Generalized Predictive Control-Part 1: The Basic Algorithm Automatica, 23, 137-148.

Gawthrop, P. J.(1985). Comparative Robustness of Non-adaptive and Adaptive Control. Proceedings of Control '85, Cambridge.

Goberdhansingh, E.(1988). Autotuning for PID and GMV controllers. MASc thesis, Department of Chemical Engineering, University of Toronto, Toronto, Canada.

Hamel, B.(1949). Contribution a l'etude mathematique des systems de reglage par tout-ou-rien. C.E.M.V. Service Technique Aeronautique 17.

Seborg, D. E., T. F. Edgar and S. L. Shah(1986). Adaptive Control Strategies for Process Control: A Survey. AIChE Journal, 32, 881-913.

Tsypkin, J. A.(1958). Theorie der relais systeme der automatischen regelung. R. Oldenburg, Munich.

Ziegler J. G. and N. B. Nichols(1942). Optimum Settings for Automatic Controllers. Trans A.S.M.E., 759-768.

Table 1: Experimental Results for Analog Simulator

	T(sec)	a(%)	ϵ(%)	d(%)	K	T_i	T_d
	21.100	1.3214	0.8046	10.272	4.0600	11.00	2.70
a	21.100	1.2756	0.8199	10.254	4.1200	11.00	2.70
	21.200	1.2970	0.7955	10.254	4.0600	11.00	2.70
	45.700	2.5513	0.4800	10.242	1.8900	31.00	7.70
b	44.800	2.4536	0.6999	10.242	2.0000	30.00	7.40
	45.900	2.5696	0.7813	10.248	1.8600	31.00	7.70
	73.050	2.7405	0.7365	6.726	1.1100	49.00	12.00
c	73.050	2.7283	0.7660	6.717	1.130	49.00	12.00
	73.050	2.7618	0.7792	6.714	1.100	49.00	12.00

T(sec)	ω	Hyst. lag.	Control. shift	Phase margin	ϵ/d	β
21.10	0.2979	37.512	27.608	65.1208	0.0783	0.4629
21.10	0.2979	39.996	26.044	66.0410	0.0799	0.4481
21.10	0.2979	37.830	27.630	65.4614	0.0775	0.4553
45.70	0.1375	17.448	45.044	62.4967	0.0468	0.5233
44.80	0.1372	16.573	39.372	55.8453	0.0683	0.4861
45.90	0.1369	17.700	39.478	57.1788	0.0762	0.4746
73.05	0.0860	15.589	35.734	51.3236	0.1095	0.4376
73.05	0.0860	16.305	38.284	54.5897	0.1140	0.4592
73.05	0.0860	19.387	36.254	55.6424	0.1160	0.4407

Nomenclature

a = amplitude of the process oscillation (%)
d = relay amplitude (%)
G = process transfer function
K = proportional gain (%)
Na = the describing function approximation
s = Laplace operator
T = period of process oscillation (sec)
T_i = integral time (sec)
T_d = derivative time (sec)
ω = frequency of the oscillation (rad/sec)
β = gain ratio
ϵ = the hysteresis width (%)

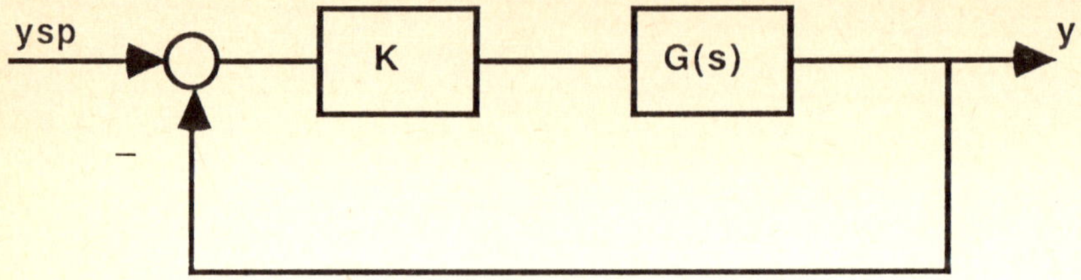

Figure 1: Proportional Feedback Control

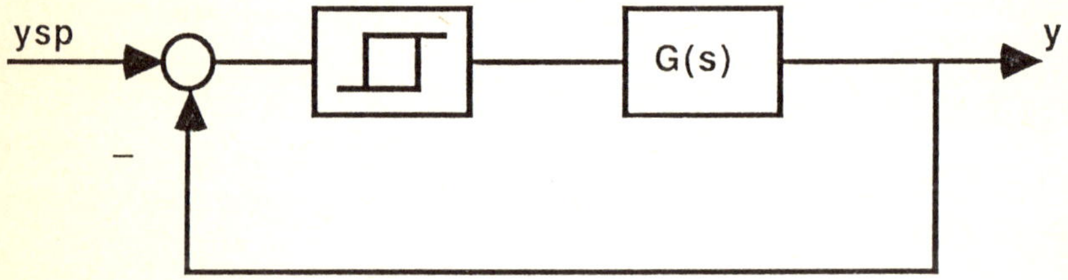

Figure 2: Relay Feedback Control

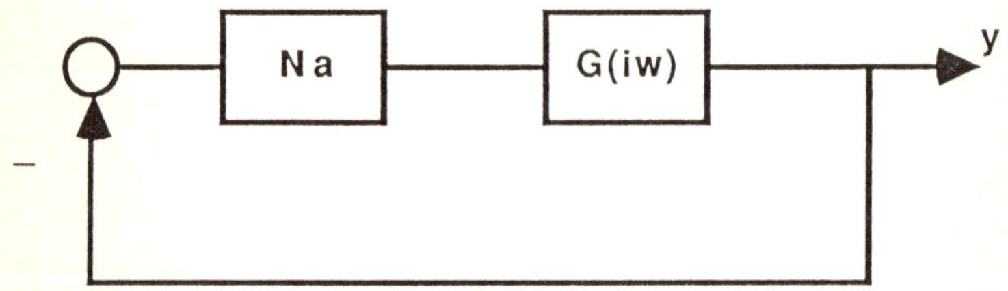

Figure 3: Relay Feedback with DFA

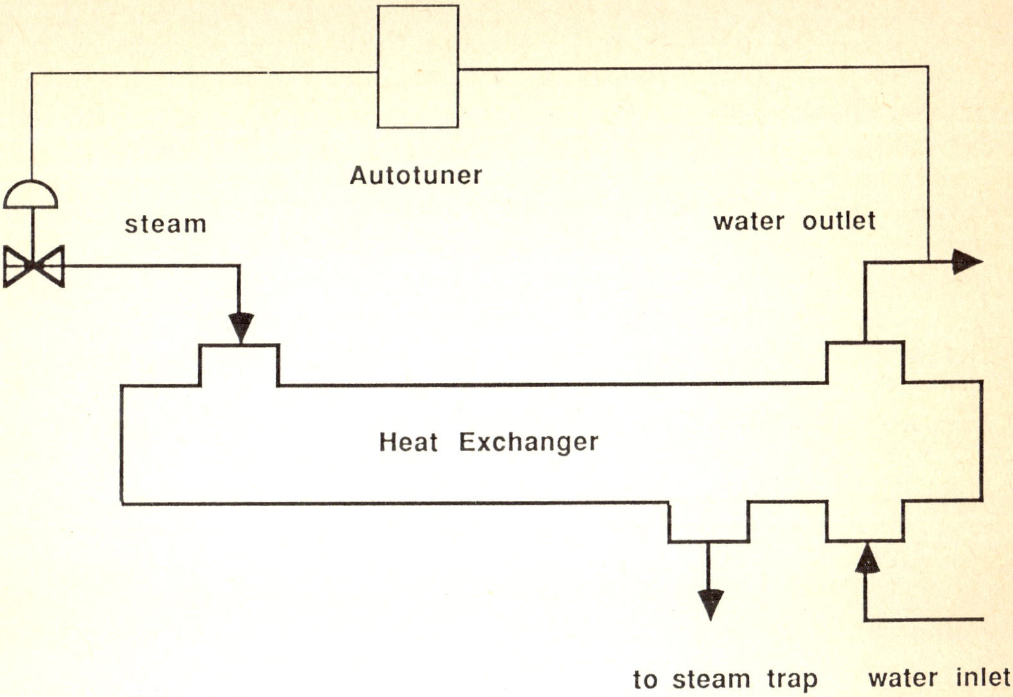

Figure 4: Heat Exchanger Schematic

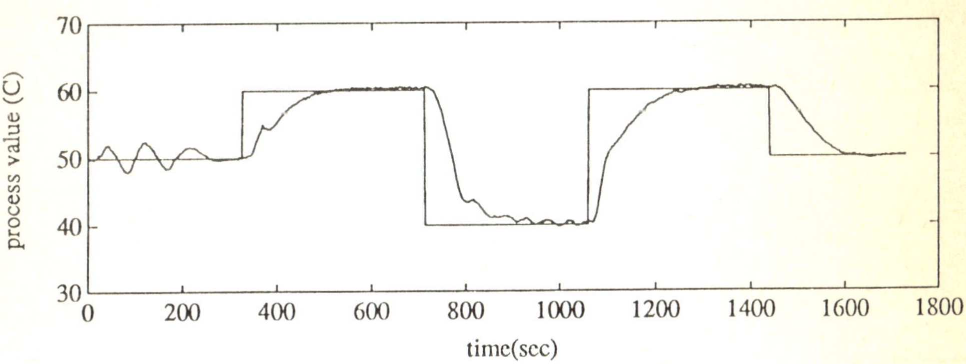

Figure 5: Tuning Cycle and Setpoint Response

SELF-TUNING VERSUS ADAPTIVE-PREDICTIVE CONTROLLERS*

D. Grant Fisher+ and Brendan John Minter#
Department of Chemical Engineering, University of Alberta
Edmonton, Alberta, Canada, T6G2G6

ABSTRACT

The performance of Foxboro's Expert Adaptive Controller Tuning (EXACT) controller and Turnbull Control System's (TCS) Auto-Tuning Controller were compared versus the Generalized Minimum Variance (GMV) controller of Clarke and Gawthrop(1979). The EXACT controller's pattern recognition approach to automatic tuning of PID parameters gave good control for a wide range of operating conditions on two real processes. However, it was sometimes slow to adapt because it adjusts PID parameters only once per set point or disturbance transient. The TCS controller did not automatically update the PID controller parameters and hence acted more as an advisor than a self-tuning controller. The GMV controller is adaptive-predictive, but <u>not</u> self-tuning in the sense that parameters in the 1/Q control law (P=R=1) are user specified and not adapted. It is shown that a self-tuning strategy analogous to that used by the EXACT's tuner can be combined within the GMV algorithm so that the P+I controller parameters in 1/Q can be tuned based on measured closed loop performance.

(#)Currently with Syncrude Canada Ltd. (Advanced Applications Group)
(+)Author to whom correspondence should be directed.

(*) A version of this paper was presented at the American Control Conference, Atlanta, June 1988.

INTRODUCTION

The industrial demand for robust adaptive and/or self-tuning controllers is growing as plant engineers tackle the problem of controlling time-varying or unknown processes. Several instrumentation companies now market single loop adaptive PID controllers, some of which are gaining widespread acceptance due in part to their improved performance as well as industry's familiarity with the three term controller.

At the same time, both the industrial and academic communities are continuing research into the development of more general adaptive control schemes. These advanced control schemes can give better performance than conventional PID controllers and offer greater design flexibility. Most, however, have not been widely accepted for industrial applications due to their relative complexity and unproven performance. Also, many of the implementations have suffered from a lack of practical design features to accommodate real process problems and their effects on controller performance.

The main objectives for an adaptive controller are guaranteed closed loop stability, rejection of set point error and user-specified performance. A practical adaptive controller must be able to deliver this user specified performance and adjust its dynamic behavior in response to changes in process conditions. Ideally, the performance criteria should ensure that the actual closed-loop performance meets the user's specifications. Time-domain criteria such as percent overshoot are particularly easy to interpret and to verify.

In this work, two industrial adaptive PID controller's, The Foxboro Company's "Exact" and Turnbull Control System's "Auto-Tuning" controller, were experimentally evaluated and compared with an "academic" adaptive controller. The purpose of these experiments was to identify the controllers' self-tuning mechanism, evaluate practical design features and define their useful range of application.

COMPARISON OF CONTROL STRATEGIES

Predictive Control Historically, the most common form of single loop feedback control has been the three term PID control law, in its various forms, tuned using conventional methods and heuristic rules. The major drawback of the PID controller is its inability to compensate for processes with time delay, plus the fact that it waits until there is an error in the actual process output before taking corrective action.

Without time delay compensation, controller gains must be decreased to maintain closed loop stability at the expense of sluggish control performance. To compensate for processes with time delays, the model-based Smith Predictor scheme (Smith, 1957) was introduced. It uses model based predictions of plant outputs to remove the process delay from the feedback loop, thus permitting tighter tuning. Prediction, like derivative action introduces phase lead into the closed loop which offsets (to some degree) the destabilizing effect of time delays. The reader should note that with a predictive controller, even when a perfect model is available (and hence perfect prediction), parameters in the PID control law still require retuning when process conditions change.

Adaptive Predictive Control The Smith predictor is an example of non-adaptive **predictive** control which gives improved performance for plants with known process models. However, for nonlinear, time-varying plants a fixed parameter process model is not adequate. The Generalized Minimum Variance (GMV) controller used in this work, and shown in Figure 1, is one example of an **adaptive predictive** controller. The control law is given by:

$$u(k)=\frac{1}{Q}[e^{*}(k+d)]$$

where:

$$e^{*}(k+d)=[Rw(k)-Py^{*}(k+d)|_{k}] \qquad (1)$$

where w=desired value and $Py^{*}(k+d)$ is a weighted estimate of the predicted plant output. This controller minimizes the cost function:

$$J=E\{[Py^{*}(k+d)-Rw(k)]^{2}+\frac{e_0}{q_0}[Qu(k)|_{k}]^{2}]\}+\sigma^{2} \qquad (2)$$

The reader is referred to (Minter, 1987) for complete details of the above derivations. An estimate of the predicted output term, $Py^{*}(k+d)$ is obtained from an implicit formulation given by:

$$Py^{*}(k+d)|_{k}=Eu(k)+Fy'(k)+HPy^{*}(k+d)|_{k}+\delta \qquad (3)$$

using recursively estimated values for E, F and H. This formulation for $Py^{*}(k+d)|_{k}$ is, in turn, based on an assumed process model:

$$Ay(k)=Bz^{-d}u(k)+Lz^{-q}v(k)+C\xi(k)+\bar{d} \tag{4}$$

where A and C are monic, $b_0 \neq 0$ and $\bar{d}$ is the steady state process output when $u(k)=\xi(k)=0$. The following Diophantine identity:

$$PC=GA+z^{-d}\ \frac{F}{P_d} \tag{5}$$

is assumed to hold, where G is monic and of order "d-1", F is of order "ni-1", ni=MAX(na+npd,nc+npn-d+1) with $f_0 \neq 0$. Note that P, Q and R in equation (2) are user specified discrete transfer functions and C is assumed to be bounded.

In the practical implementation of equation (1), there are two important issues: robust long-term estimation of $Py^*(k+d)$ and proper selection of P, Q and R. If estimated values of $Py^*(k+d)$ do not asymptotically approach the true values, this prediction error will result in steady state control error or offset. Even with perfect predictions it can be shown that $Q|_{z=1}$ must equal zero to guarantee zero steady state controller offset. The simplest choice of Q=0 gives minimum variance control for set point following when P=1 and R=1. The controller attempts to drive the process measurement to the desired set point in no less than 'd' sample intervals. The characteristic equation is given by CB=0 which indicates that closed loop stability will depend on the roots of B. This demonstrates that minimum variance controllers can not stabilize plants that are non-minimum phase either in the continuous time domain or the discrete time domain due to fast sampling rates (Clarke,1984). As shown in Figure 2, adjustment of K_i when $Q=K_i(1-z^{-1})$ results in a direct tradeoff between speed of response and degree of damping in controlled process outputs, analogous to the effect of proportional gain in conventional PID controllers.

A comparison of (1) with conventional PD control (derivative on measurement only) written as:

$$u(k)=K_c\cdot[w(k)-y(k)]+K_d/T_s\cdot[y(k)-y(k-1)] \tag{6}$$

shows the similarity in controller structure. If $K_d=[-dT_s]$ in (9) and P=R=1 in (1) we can write:

$$\text{GMV:}\quad u(k)=\frac{1}{Q}[w(k)-y^*(k+d)|_k] \tag{7}$$

$$\text{PD:}\quad u(k)=K_c[w(k)-[y(k)+d(y(k)-y(k-1))]]=K_c[w(k)-y_p(k+d)] \tag{8}$$

where $y_p(k+d)$ is a straight line approximation of the predicted process output. From (7) and (8) it is clear that PD control is equivalent to proportional control acting on a predicted error. Likewise, GMV control with P=R=1 and $Q=1/K_c$ is equivalent to a fixed gain proportional controller acting on a predicted process output $y(k+d)|_k$. However, in GMV, the prediction is based on an adaptive model of the process rather than simple linear extrapolation of the output y(k). (Note that y^* in (7) is a function of u(k) which is not the case in (8).) Figure 1 is drawn to emphasize that GMV is a **fixed-gain** controller with an *adaptive-predictive* mechanism to produce $y^*(k+d)$.

This interpretation suggests that for direct comparison with conventional controllers 1/Q should have the structure of a PI controller so that the user has sufficient tuning flexibility to increase the speed of response and/or decrease oscillations in the manipulated variable. The desired form of Q is therefore:

$$Q=\frac{\lambda(1-z^{-1})}{q_{d0}-q_{d1}z^{-1}} \tag{9}$$

The meaning of the adjustable parameters q_{d0} and q_{d1} can easily be determined by comparing equation (6) with the velocity form of a PI controller corresponding to:

$$u(k)=u(k-1)+K_c(1+\frac{T_s}{K_i})e(k)-K_c e(k-1) \tag{10}$$

Given user specified controller gain, K_c and reset time, K_i, q_{d0} and q_{d1} are given by:

$$K_c=\frac{q_{d1}}{\lambda}$$

$$K_i=[\frac{q_{d1}}{(q_{d0}-q_{d1})}]\cdot T_s \tag{11}$$

The above Q design results in a very practical controller. It meets the necessary conditions for removal of controller offset, contains an adjustable parameter that can be used to stabilize non-minimum phase systems and has advantages over a purely integrating form in that larger λ values (higher penalty on control) can be used with less overshoot for transient responses.

Figure 3 demonstrates how adjustment of the effective controller gain and integral time affects transient responses to set point changes. As in conventional PI control, higher gains give faster responses and increasing overshoot. Likewise, increasing integral action gives more oscillatory behavior.

GMV's Predicted Error vs Conventional Error Driven PI Control Figure 4 demonstrates the effect of process nonlinearities on closed loop performance for both conventional PI (time 0-485s) and adaptive predictive GMV (PI) control (time 495-960s). The process was very nonlinear over the normal range of operating conditions and hence it was necessary to retune the controller constants for each set point change. Using conventional PI control, manual detuning was required prior to each negative change in set point to avoid unacceptable oscillations. The subsequent positive set point change (at t=420s), made without adjusting the controller constants, gave very sluggish control. At t=485s control was switched to GMV (PI) using the same original PI constants. Much less manual detuning (of q_{d0} and q_{d1} in (11)) was required and better overall control was achieved. (The performance of GMV without manual resetting of q_{d0} and q_{d1} was unsatisfactory [3])

This result follows from the discussion in the preceding section since the "derivative" effect of prediction provides a stabilizing influence on closed loop performance.

Model Based Self-Tuning PID Control This type of controller uses a model based estimation scheme to predict future process outputs. Calculated **model** parameters are then incorporated into a self-tuning design procedure which generates PID controller parameters such that the model has the desired performance, eg critically damped response, pole placement, etc. Figure 5 shows a block diagram structure for this type of controller. Note that the error input to the controller is based on the actual process measurements, i.e. is not predictive. The Turnbull Control Systems' model 6355 controller features continual calculation of recommended PID parameters based on parameter estimates, θ^* obtained from a statistically based process model estimation scheme. The most important limitation of the above type of self-tuning scheme is that PID constants are adjusted on the basis of changes in **estimated model** parameters instead of on the basis of **actual** closed loop performance. The accuracy of θ^* is dependent on the performance of the identification algorithm, the assumed model structure (likely second order) and the current process conditions. Nonlinear, high order processes are much harder to accurately model than low order, linear plants. Hence, the recommended PID constants may not give good control.

To illustrate the above points, the closed loop performance of the model 6355 is shown in Figure 6 for the first order, linear plant and in Figure

7 for the high order, nonlinear temperature process. With the easier to model linear plant, the controller gave smooth, highly damped responses to step changes in set point. With the nonlinear temperature process, control was also well damped with small oscillations at steady state when the controller was well tuned. (In Figure 7 the set point changes from 0 to 750s were generated by the TCS controller as part of an operator requested tuning cycle. In general, control was not as satisfactory at steady state following larger set point changes away from the base conditions.)

One important advantage of this type of self-tuning approach is is that PID parameters can be calculated as often as the estimated model parameters, θ^{*}. With perfect modelling the 6355 controller would always have "optimum" closed loop performance, unlike the "once per transient" approach used by the Exact controller. However, current technology can not give perfect modelling of real processes. Therefore, self-tuning schemes that are sensitive to model parameter estimates or modelling performance are not practical for complex processes. For practical applications, there must be feedback of closed loop performance to the self-tuning block so that PID parameters are based on process I/O data and performance measures rather than solely on model based predictions.

<u>Performance Based Self-Tuning PID Control</u> The most important feature of this type of self-tuning controller is that the PID parameters are adjusted based on actual <u>measured</u> closed loop performance. Estimated PID parameters are not a function of model estimates and therefore do not suffer from the negative effects of poor modelling. Foxboro's Exact controller falls into this self-tuning category, since it uses values of overshoot and damping of closed loop responses to disturbance or set point changes, as performance criteria. Figure 8 illustrates the Exact controller's block diagram structure. Note that it is a conventional feedback system with an added mechanism for tuning the PID constants. There is no parameter estimation or output prediction. Performance evaluation and controller parameter generation are based directly on actual process measurements and heuristic tuning rules (cf. Expert System).

Unlike model based self-tuners, PID parameters are adjusted only after "closed loop performance" has been measured. In the Exact's case PID constants are tuned <u>once</u> per underdamped transient response, since overshoot and damping are "once-per-transient" measures of performance. Figure 9 illustrates the EXACT controller's closed loop performance with the nonlinear temperature process. The specified values for overshoot and damping were 0.25 and 0.20, respectively. For this particular transient, the controller delivered an overshoot of 0.23 and a damping ratio of 0.17. The adapted controller parameters had remained unchanged over five previous set point changes.

The ability of the EXACT controller to deliver a user-specified response pattern is shown in Figure 10. The controller was commissioned with conservative band and reset time of 100 and 0.5 minutes, subsequently resulting in sluggish set point tracking. As PID parameters were adjusted following each transient, performance improved until the specified overshoot (0.25) and damping (0.20) were achieved after fourteen transients. The trajectories for controller band and reset time are shown in Figure 11, indicating a better than linear rate of convergence (near quadratic) of PI parameters. The P+I values obtained using an initial open-loop PRETUNE procedure (shown as horizontal lines) are reasonably close to the final values and would have required only five to six tuning transients before converging. These results indicate the utility of an initial tuning option in delivering good initial values for parameters.

The once per transient approach to tuning has its advantages and disadvantages. Figure 12 illustrates closed loop performance for a large set point change into a high process gain region (nonlinear process). Control about the new operating point was initially very oscillatory until a series of three peak responses were recognized and the controller detuned. This Figure simply illustrates that a user must carefully consider the implications a particular self-tuning design will have for his application. Another important aspect of this self-tuning design is the implicit assumption that a process can be forced to overshoot and oscillate in a feedback control scheme, given an appropriate set of PID parameters. Because controller output, u(k), is not considered in the performance index, overshoot specifications may not be an appropriate performance index for highly damped processes. The EXACT's tuner will continue to increase controller gains in an effort to force the process output to overshoot following subsequent set point or disturbance transients. If the process is so damped that it can not be forced to overshoot, eventually the tuner will give very high controller gains. This will result in large, high frequency oscillations in u(k) and excessive wear on actuator devices. This phenomena, described as gain wind-up, is shown in Figure 13. In selecting a self-tuning controller, it is very important to consider what criteria are used by the tuner and their appropriateness for a given application. (The EXACT has an option to detune the controller if there are persistent, large oscillations in the controller output. It works well, in general, but can "fight" the controller tuning mechanism in applications like this.)

Self-Tuning Adaptive Predictive Control Evaluation work with the Exact showed that a tuning mechanism that adjusts controller parameters based on measured closed loop performance results in a practical self-tuning control system. At the same time, results with the GMV controller demonstrated the advantage of dead time compensation and adaptive prediction over conventional error driven feedback control schemes. Analysis of the GMV algorithm (e.g. the PI form in 8) shows that there is no adaptation of controller parameters based on closed loop performance specifications. Although the control law is based on the minimization of equation 2, once Q is non zero, GMV is no longer minimum variance and the effect of Q on closed loop performance can not be predicted for unknown processes. The proper choice of Q is process dependent and for nonlinear processes also dependent on the operating point. Regions with a high static process gain require a lower controller gain to maintain overall closed loop gain.

On-line adjustment of Q obviously affects the characteristics of closed loop transient responses to set point changes. Through the proper specification of gain and integral time, Q can be used to maintain stability of the closed loop, provided that predicted outputs are asymptotically correct. These statements clearly suggest a need for an adaptation loop that will adjust something in the GMV's control law of (1) so that user specified closed loop performance specifications are realized. The present "adaptive" mechanism attempts to give accurate predictions, $y^*(k+d)$ by adjusting $\theta^*(k)$ but does not provide any kind of closed loop adaptation of the controller constants.

The block diagram in Figure 14 is for a self-tuning adaptive predictive controller which combines a closed loop self-tuner (cf. Foxboro Exact in Figure 8) and an adaptive predictive feedback loop (cf. GMV in Figure 1). This combination gives better overall performance at the cost of a more complicated implementation.

For demonstration purposes, a closed loop adaptive scheme was implemented to automatically adjust Q to maintain a user specified overshoot for set point transients. Using a fixed reset time, K_c was adjusted in response to observed changes in process dynamics or user specified overshoot. For a series of +5% set point changes, the tuner demonstrated its ability to maintain a user specified overshoot of 0.30. Figures 15 and 16 show that the desired closed loop performance was maintained (overshoot=0.3) over five set point changes on the nonlinear process by "self-tuning" of K_c.

This demonstrates the combination of "self-tuning" and "adaptive-predictive" control functions recommended by the authors. The adaptive-predictive controller is designed so that the model-based estimates of future process values satisfy the specified performance criterion, $J(y^*)$. The self-tuning is accomplished by a slower "outer-loop" which adjusts constants (eg. $Q(z)$) of the "inner" adaptive-predictive loop so that the performance of the actual measured plant output, $J(y)$, approaches the specified process performance in spite of model-process mismatch or non-idealities.

PRACTICAL CONSIDERATIONS

The generality and improved performance of the self-tuning, adaptive-predictive controllers is achieved by adding additional design flexibility. Some people have argued that this leads to an excessive number of parameters that the user must specify even to obtain the equivalent of a PID controller. To counter this argument, software was written which reduced the number of parameters that <u>must</u> be specified by the user to one - the maximum allowable change in the manipulated input, $u(k)$. The remaining parameters are set to default values or determined by on-line process tests carried out by the controller. One possible procedure is illustrated by the data in Figure 17. After a brief period of open-loop steady-state operation a process step response is obtained. This data is used to estimate the process gain, time-constant, time-delay, and noise band. These values are then used to estimate: Q (eg. from PID constants obtained using Cohen and Coon procedures in conjunction with equation 6); R by using a first-order filter with a time-constant equal to 75% of the process time-constant; filter constants for $y(k)$ and $u(k)$; and initial parameter estimates, $\theta^*(0)$. The parameter estimates can be improved, if necessary, by introducing further PRBS excitation into $u(k)$ as illustrated in Figure 17. The controller can then be put into "normal operation" and the self-tuning feature should ensure that the desired performance is obtained. (Note that this is only one of many start-up procedures that could be used.)

In addition to being easy to commission and use a practical controller for industrial use should include options: to constrain the estimated process parameters; for signal conditioning to filter measurement noise, adjust the frequency content of the process data to be consistent with the assumed model and handle abnormal data (eg. random zeros); to switch to an independent set of backup parameters; and to monitor its own operation.

Some of the results of current research that should improve future versions of commercial controllers are: improved parameter estimation algorithms; multi-step predictive controllers that do a better job of handling unknown and/or time-varying time-delays; the ability to handle constraints on u(k) and y(k); improved algorithms for performance tuning; plus more complete theoretical and analytical studies which define stability and performance limits.

The best of the current generation of adaptive controllers can be profitably applied in industry. Future products will be even better and justify a broader range of applications.

CONCLUSIONS

Most "academic" adaptive controllers are not self-tuning. They are adaptive predictive controllers, e.g. $u(k)=1/Q[e^*(k+d)]$ where the predicted error is based on estimated process model parameters (explicit case). The control law structure and its parameters are determined by the user specified, constant parameters in discrete filters such as P, Q and R. Even with adaptive prediction, these controllers can not give satisfactory performance on highly nonlinear processes such as used in the experimental runs shown in this paper. An additional tuning mechanism based on measured closed loop performance is required to adjust parameters such as Q to maintain user specified performance. As shown, model based adaptive predictive control (such as GMV) can be effectively combined with closed loop performance based self-tuning control (e.g. Foxboro EXACT) to give significantly better control over a much broader operating region.

LITERATURE CITED

Clarke, D.W.; Gawthrop, P.J.(1979) "Self-Tuning Control", Proc. IEE, IEE, 126, No. 6, June.

Clarke, D.W.(1984). "Self-Tuning Control of Non-Minimum Phase Systems", Automatica, 20, No. 5, 501-517.

Minter, B.J.(1987). "A Comparison of Adaptive Controllers: Academic vs Industrial", Dept. of Chem. Eng., University of Alberta.

Smith, O.J.M.(1957). "Close Control of Loops With Dead-Time", Chem. Eng. Prog., 53, No. 5, p. 217.

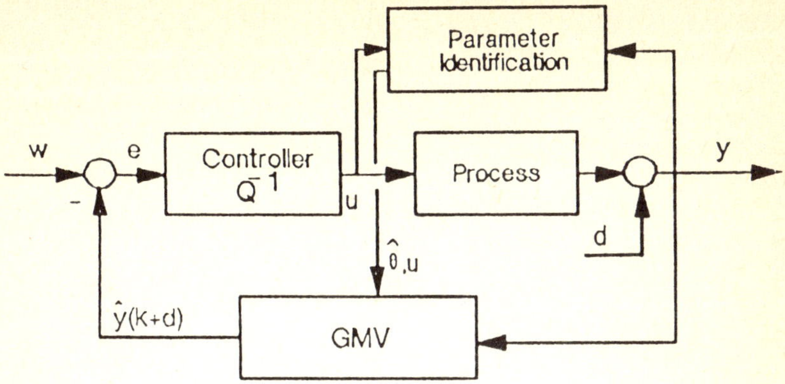

Figure 1.Block Diagram of the GMV Controller Emphasizing Fixed-Gain Predictive Structure

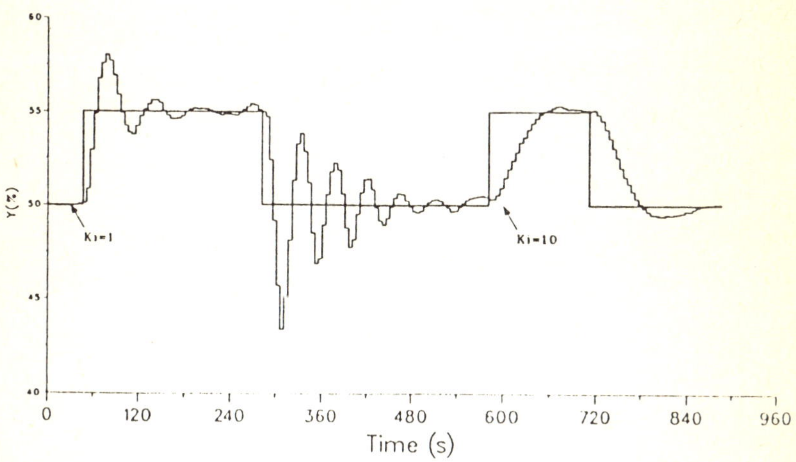

Figure 2.Adaptive Predictive, GMV (integral) Control of a Nonlinear Temperature Process

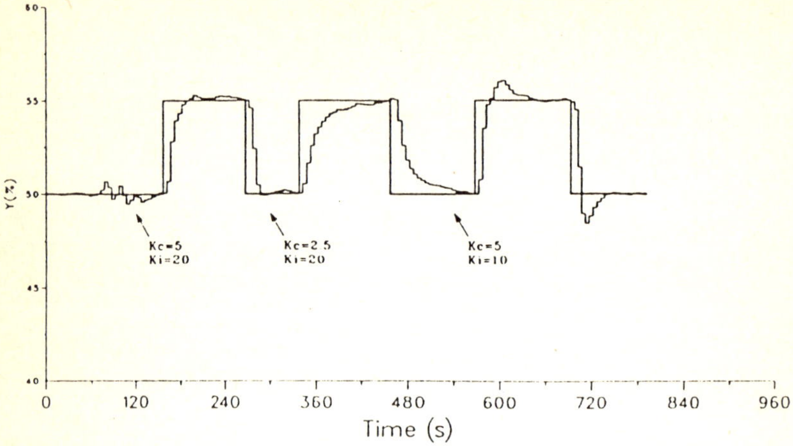

Figure 3.Adaptive Predictive, GMV (P+I) Control of a Nonlinear Temperature Process

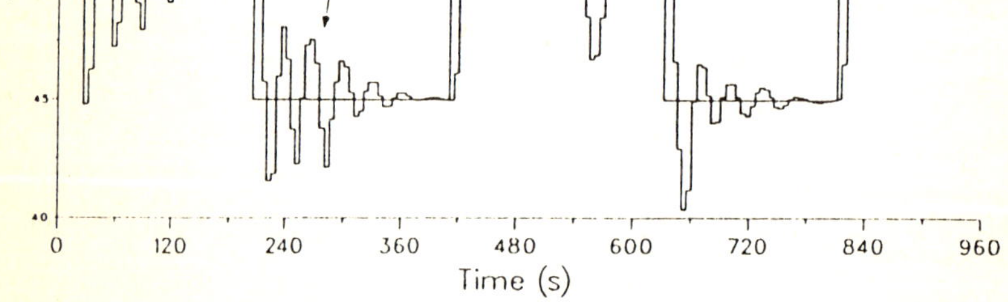

Figure 4.Conventional versus GMV P+I Control - Effect of Process Nonlinearities

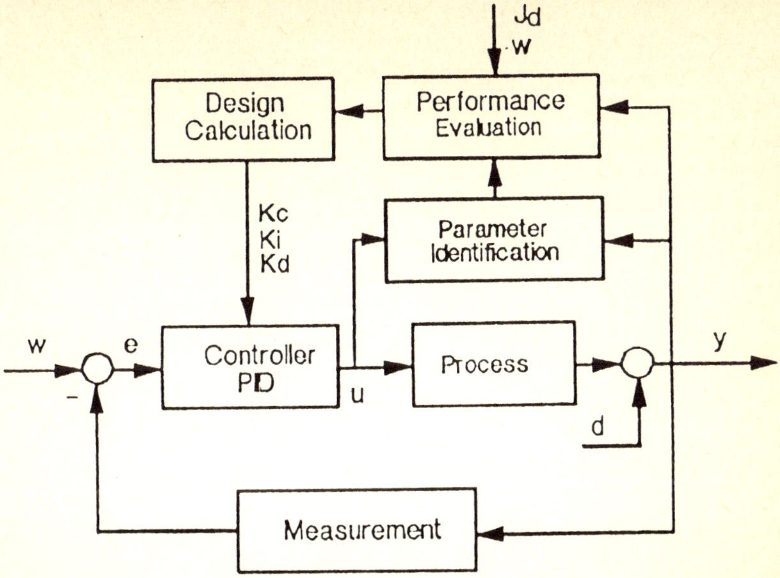

Figure 5.Model-based, Self-Tuning Control to Meet Desired Performance, J_D

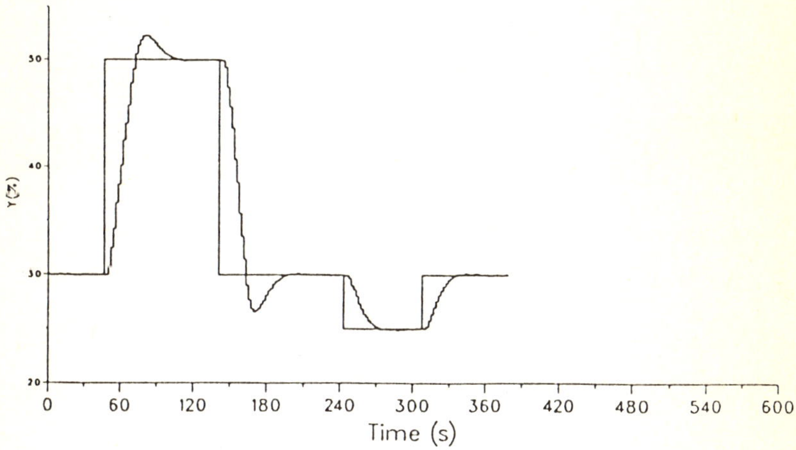

Figure 6.Self-Tuning PID Control Using the TCS Controller on a Linear, First Order Process

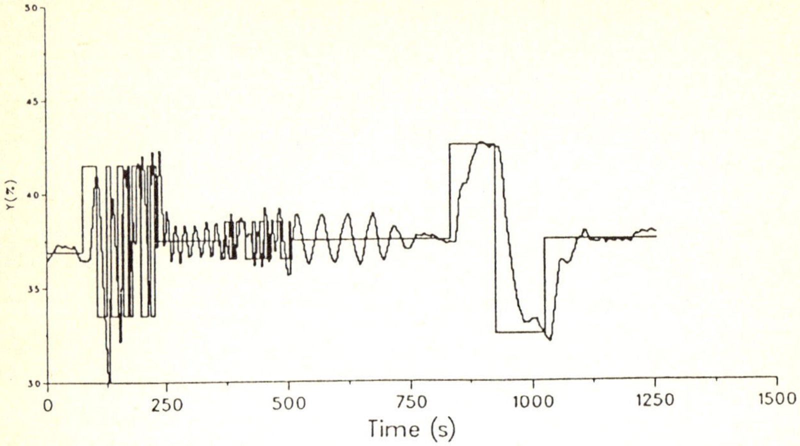

Figure 7. Self-Tuning PID Control Using the TCS Controller on a Nonlinear, High Order Process

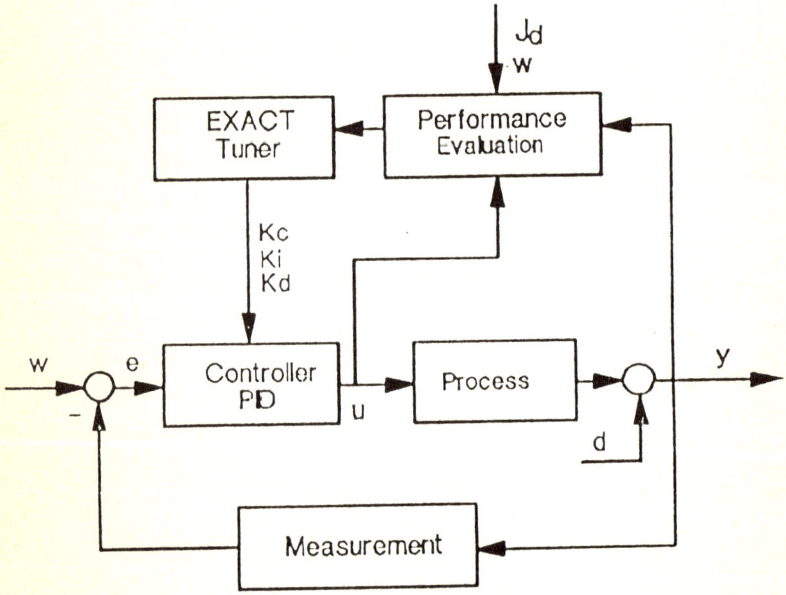

Figure 8. Structure of a Performance-based, Self-tuning PID Controller (cf. Foxboro EXACT)

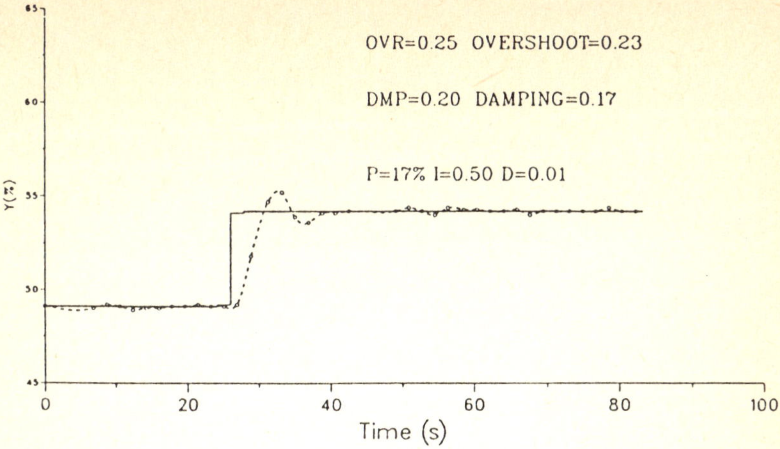

Figure 9. Self-tuning PID Control Using the EXACT Controller

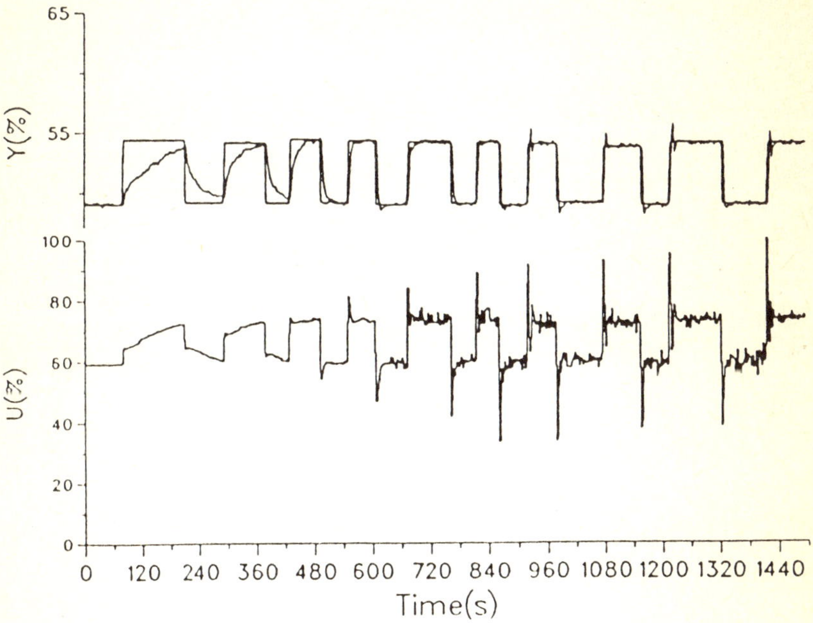

Figure 10. Performance-Based Self-Tuning (EXACT) PID Control: Convergence to User-Specified Overshoot (0.25) and Damping (0.20).

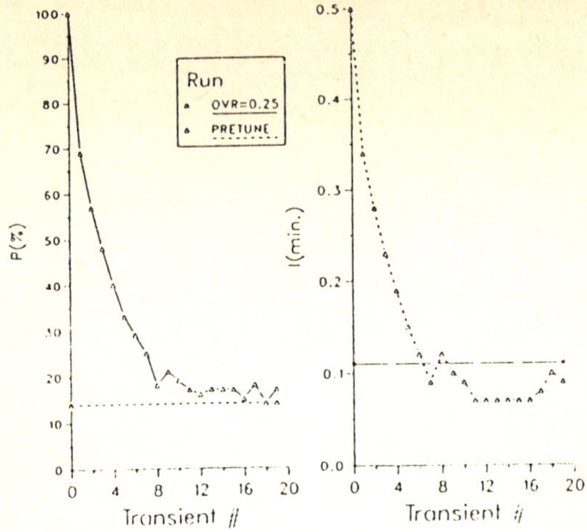

Figure 11.Convergence of P+I Parameters for a Series of ±5% Setpoint Changes (EXACT).

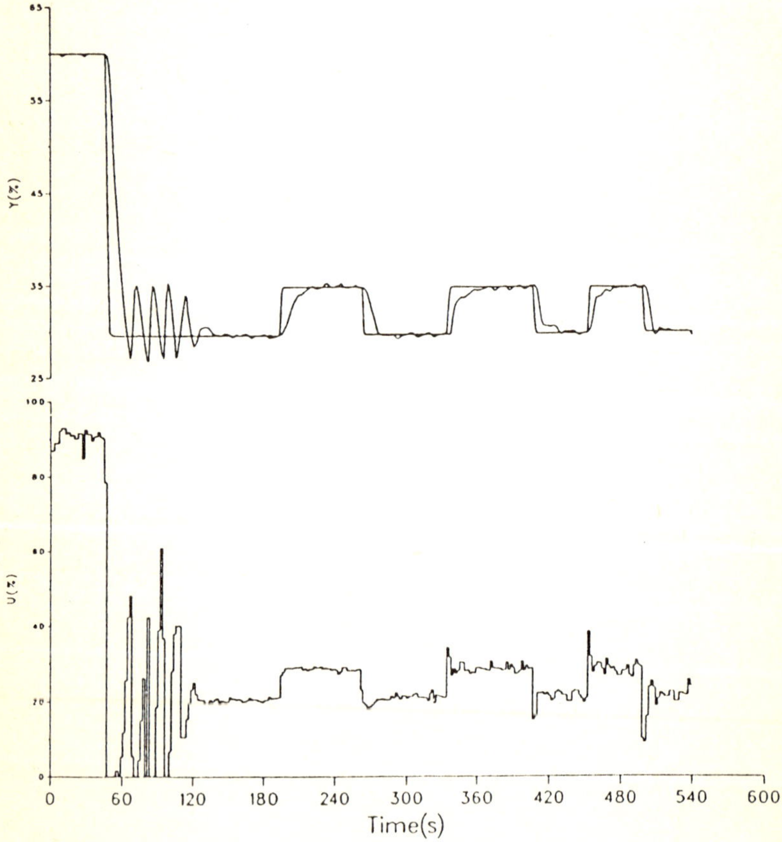

Figure 12.Foxboro EXACT Controller - Effect of Nonlinear Process Gain on Closed Loop Performance

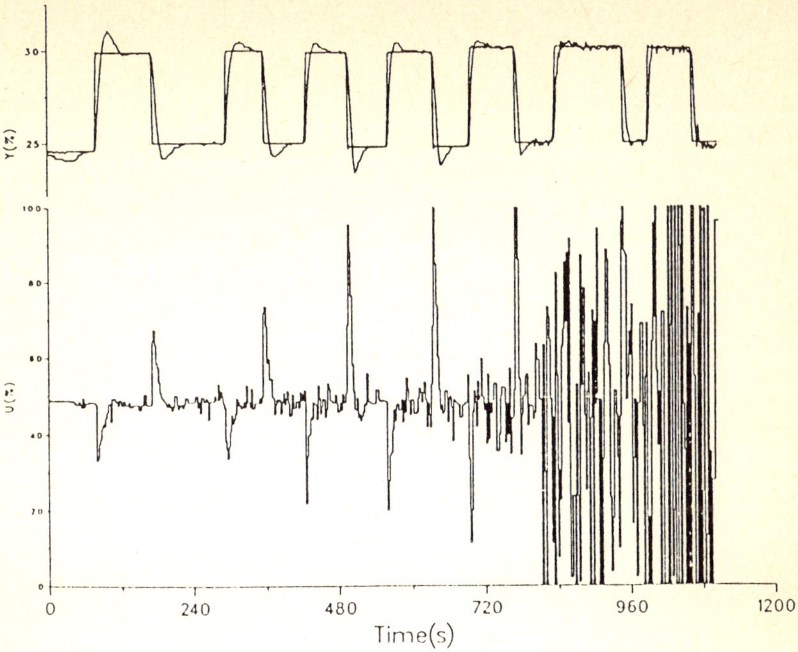

Figure 13.Self-Tuning Feedback Control of the First-Order Level Process Using the Foxboro Exact (with default specification of 50% overshoot)

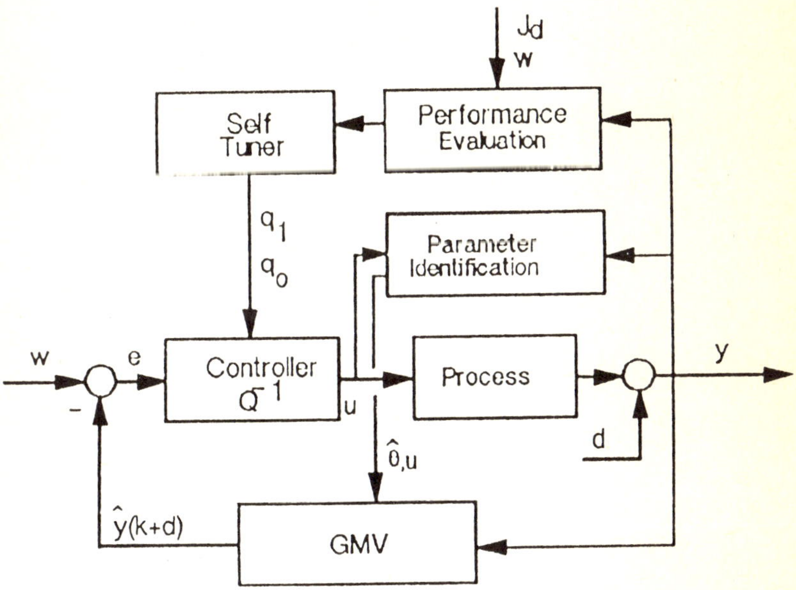

Figure 14.Block Diagram Structure of an Adaptive Predictive (GMV), Self-Tuning (Exact) Controller

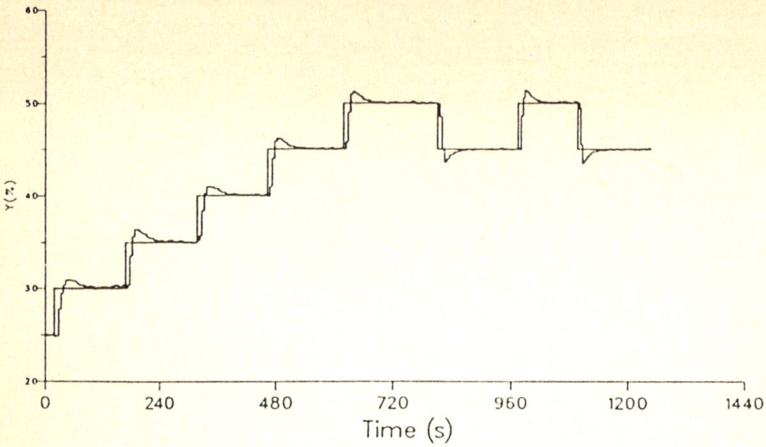

Figure 15. Self-Tuning, Adaptive Predictive GMV Control of a First Order Process

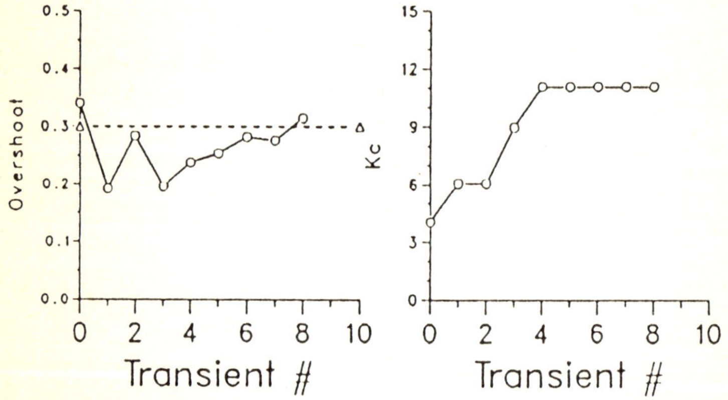

Figure 16. Trajectories for Overshoot and K_c for the Series of +5% Set Point Changes in Figure 15.

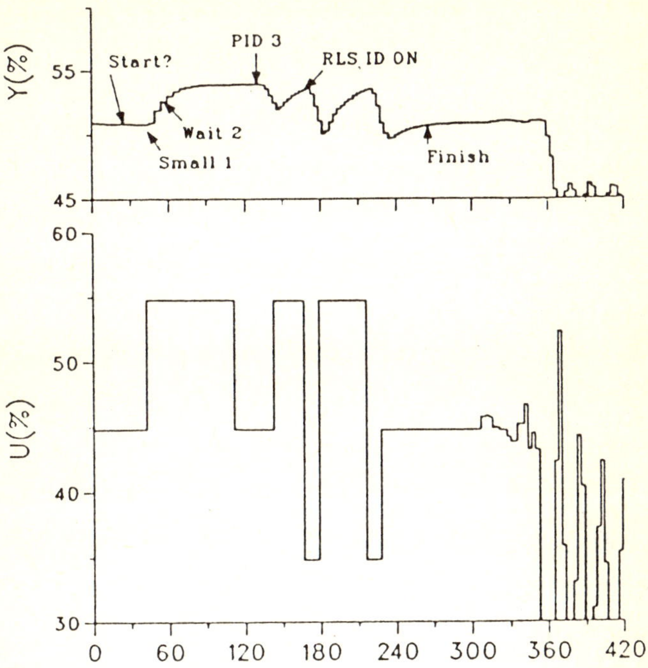

Figure 17. Initial Open-Loop Identification Using the GMV (P+I) Controller and User-Specified Limits on Allowable Changes in Process Output.

SYMBOLICALLY ENHANCED PARAMETER ESTIMATION FOR RELIABLE ADAPTIVE CONTROL

D. J. Cooper, A. M. Lalonde and R. Pae
Department of Chemical Engineering
The University of Connecticut
Storrs, CT 06269-3139, USA

ABSTRACT

Adaptive controllers with linear estimators are often implemented on processes that exhibit nonlinear, nonstationary behavior. As a result, the sensitivity of these estimators must be continually adjusted to remain responsive to current process behavior. This work takes a novel approach to the practice of adjusting sensitivity. Here, the recent behavior of the process is used as the primary indicator for making decisions about when and how to make such adjustments. The diagnosis of process behavior is made at a qualitative level to place current operation into one of several categories. An inferencing system directs the diagnosis and uses the result to implement appropriate adjustment actions. The result is a preemptive action that indicates improved controller reliability. A preliminary validation is presented in the adaptive control of a nonlinear reactor simulation.

INTRODUCTION

In adaptive process control, linear estimators are used even though many chemical processes exhibit nonlinear, nonstationary behavior. The popular methods that enable linear estimators to reasonably describe the current behavior of such processes are based on modifications of recursive least-squares (RLS). These modifications maintain sensitivity to the changing character of chemical processes through algorithmic techniques such as forgetting factors and covariance resetting. Traditionally, the RLS modifications have used indicators such as the error of estimation and the trace of the covariance matrix to make decisions about when and how much to adjust estimator sensitivity (e.g. Shah and Cluett, 1987; Seborg *et al.,* 1986).

In this work, it is the recent behavior of the process that is used as the primary indicator for making decisions about when and how much to adjust sensitivity. The approach is based on the view that process behaviors such as large and rapid changes in process operating level are the leading indicators that signal potential problems associated with RLS estimation. Estimation error, although a certain problem, is caused by, and thus lags behind, the change in process behavior. A continual diagnosis of process behavior therefore offers a means by which preemptive adjustments to estimator sensitivity can be made before potential problems become certain problems.

The diagnosis of process behavior is a qualitative evaluation designed to place current operation into one of several categories. The category which results from a diagnosis then indicates appropriate actions for adjusting estimator sensitivity. To illustrate, consider a process whose level of operation is slowly drifting with time, such as when a reactor temperature changes as a result of a decay in catalyst activity or a vessel that has a changing pressure drop because internals foul. Although there may be a band of operating conditions within which process dynamics can be well described using all data and a linear estimator, processes such as those mentioned will eventually move out of this band.

When a process is experiencing drift in the level of operation, it is desirable to judiciously adjust estimator sensitivity so that the estimator will remain alert to the changing process character. This can be achieved through the use of a forgetting factor (Fortesque *et al.,* 1981; Ydstie *et al.,* 1985), or with covariance resetting (Vogel, 1982; Goodwin *et al.,* 1983). In the case of covariance resetting, the diagonal elements of the covariance matrix are augmented by modest amounts so that new data is weighed more heavily, yet the influence of older data is still reflected in the parameter estimates.

When a change in setpoint or a disturbance imparts a large and sudden change to the level of operation, it is desirable to significantly increase estimator sensitivity. This can be achieved by resetting the covariance matrix elements to large values immediately after such an occurrence. The dramatically increased sensitivity is analogous to re-starting the estimator. Convergence of the estimator to the process character at the new operating level is rapid, as the influence of older data is substantially disregarded.

The ability to take preemptive corrective actions in adjusting estimator sensitivity requires the ability to diagnose process behavior on a timely basis. Presented are details of a method for performing such a diagnosis and then selecting and implementing estimator sensitivity adjustments based on the diagnosis. A preliminary validation of the system is also presented in application to the adaptive control of a nonlinear reactor simulation.

RLS ESTIMATION

RLS estimation theory and its modifications for adaptive control are well documented in the literature (e.g. Seborg *et al.*, 1986; Goodwin and Sin, 1984; Ljung and Soderstrom, 1983). Following the development of these authors, a linear, discrete-time difference equation used to describe a general process is the ARMAX (autoregressive, moving average with auxiliary inputs) model, expressed:

$$\begin{aligned} y(t) = {} & a_1 y(t-1) + a_2 y(t-2) + \ldots + a_n y(t-n) \\ & + b_1 u(t-k-1) + b_2 u(t-k-2) + \ldots + b_m u(t-k-m) + d(t) \end{aligned} \qquad (1)$$

where y(t) is the system output, u(t) is the system input, d(t) is the unmeasured load disturbance, t is an integer representing the sampling instant, and k is the delay time expressed as a multiple integer of the sampling period, Δt.

Fitting the model of Eqn (1) to current process data requires the determination of the a_i and b_i model parameters. For the real-time computation of these parameters, the most popular successive regression technique is recursive least-squares. By defining a system variable vector:

$$\psi^T(t-1) = [y(t-1) \ldots y(t-n)\ u(t-k-1) \ldots u(t-k-m)\ 1] \qquad (2)$$

and a parameter vector:

$$\theta^T(t-1) = [a_1 \ldots a_n\ b_1 \ldots b_m\ d] \qquad (3)$$

then Eqn (1) can be recast in the form:

$$y(t) = \psi^T(t-1)\theta(t-1) + \epsilon(t) \tag{4}$$

where $\epsilon(t)$, the modeling error, is assumed to be statistically independent of y(t) and u(t).

The estimation problem is now to determine $\theta'(t)$, the estimates of the unknown parameters $\theta(t)$, which minimize the cost function:

$$J(t) = \sum_{i=1}^{t} [y(i) - y'(i)]^2 \tag{5}$$

where y'(t) is the estimated value of output y(t) and is computed from the parameter vector estimates, $\theta'(t)$. Minimization of Eqn (5) enables computation of $\theta'(t)$ via the equation:

$$\theta'(t) = \theta'(t-1) + P(t)\psi(t-1)[y(t) - \psi^T(t-1)\theta'(t-1)] \tag{6}$$

where P(t), the covariance matrix of the estimation error, is:

$$P(t) = P(t-1)-P(t-1)\psi(t-1)[\psi^T(t-1)P(t-1)\psi(t-1)+1]^{-1}\psi^T(t-1)P(t-1) \tag{7}$$

The estimator of Eqns (6) and (7) is designed for linear, stationary processes. RLS estimators can be made to describe the local behavior of nonlinear, nonstationary processes through the use of a forgetting factor, λ, which weighs new data more heavily than older data. Estimator robustness can be increased by varying λ (Fortesque *et al.*, 1981; Ydstie *et al.*, 1985) such that smaller values for λ are used as the prediction error increases. When using forgetting factors, the covariance matrix, P(t), can grow large, dramatically increasing estimator sensitivity to the point that random disturbances cause excessive fluctuation in parameter estimates. Thus, an upper bound on P(t) is imposed.

Since large values of the elements in P(t) result in large adjustments to the parameter estimate elements in $\theta'(t)$, an alternate mechanism for regulating algorithm sensitivity involves direct adjustments to the elements of P(t). One approach is covariance resetting (Seborg *et al.*, 1986; Goodwin *et al.*, 1983), in which the covariance matrix is augmented by a constant matrix D, whenever the trace of P(t) falls below some minimum value. More recently, Ortega *et al.*, (1985) have proposed a matrix regularization scheme that combines a covariance resetting feature with a forgetting factor and enforced lower and upper bounds on P(t). Goodwin *et al.*, (1985) have also suggested adjusting λ such that P(t) maintains a constant trace and thus remains bounded.

ADAPTIVE CONTROLLER SPECIFICATION

The self-tuning regulator (Astrom and Wittenmark, 1973) is designed to minimize the variance of the output variable, y(t). The self-tuning controller (Clarke and Gawthrop, 1979; Clarke, 1981) expands upon the self-tuning regulator by employing a more general performance objective. Two additional objectives that can be included in the self tuning controller are a penalty on control effort and a filter for setpoint changes.

This work employs a self tuning controller with the performance objective:

$$J = [y'(t+1) - y^f_{sp}(t)]^2 + Q(t)[u(t) - u(t-1)]^2 \tag{8}$$

where $y'(t+1)$ is the output estimate computed from the RLS estimator, $y^f_{sp}(t)$ is the filtered setpoint, and Q(t) is a time varying penalty or weight on the incremental input change. The setpoint, $y_{sp}(t)$, is filtered externally to the controller using the first order form:

$$y^f_{sp}(t) = \alpha y^f_{sp}(t-1) + [1-\alpha]y_{sp}(t) \tag{9}$$

where α is a number between zero and one that governs the filter rate. In the simulation examples to follow, α has been set to 0.8.

One heuristic followed in this work is that RLS estimators for chemical processes should be second order in input and output. Such a specification has proven satisfactory both in our experience and the experience of others (e.g. Seborg *et al.*, 1986). The simulation examples presented in this manuscript have negligible dead time. Hence, minimizing the performance objective expressed in

Eqn (8) with respect to the controller action at current sample time t, and using a second order linear difference model of the form of Eqn (1), results in the adaptive controller:

$$u(t) = \frac{b_1(t)}{Q(t)+[b_1(t)]^2}\,[y^f_{sp}(t)-a_1(t)y(t)-a_2(t)y(t-1)] + \frac{Q(t)-b_1(t)b_2(t)}{Q(t)+[b_1(t)]^2}\,[u(t-1)] \tag{10}$$

The specification is complete with a description of Q(t), the time varying penalty on incremental input change. Inspection of Eqn (8) reveals that one logical choice for Q(t) is as a function of steady state process gain, K_p. This would cast the two terms of Eqn (8) into consistent units and work toward keeping them proportionate in size. Hence:

$$Q(t) = [K_p(t)]^2 \tag{11}$$

Because K_p is a varying parameter in general for nonstationary processes, and is time varying for nonlinear processes that experience changes in level of operation, K_p must be continually updated in Eqn (11). The appropriate gain is obtained in this work through the use of a gain schedule which is automatically generated and maintained during process operation.

K_p is computed for placement on the schedule whenever, based on a qualitative diagnosis of process behavior, the recent behavior of the process is determined to be sufficiently exciting. Sufficient excitation is required because K_p is computed from the nonlinear parameter fit of a second order plus dead time (SOPDT) model to current input/output data. The SOPDT model used in this work assumes the overdamped form:

$$\tau_1\tau_2 \frac{d^2y(t)}{dt^2} + (\tau_1+\tau_2)\frac{dy(t)}{dt} + y(t) = K_p u(t-t_d) \tag{12}$$

where τ_1 and τ_2 are process time constants and t_d is process dead time. When a diagnosis indicates that sufficient excitation exists for a determination of K_p (or

τ_1, τ_2 or t_d), a parameter estimation is initiated. To determine the parameters, Eqn (12) is integrated using the process input record from the previous 200 samples of u(t). The integration scheme is a 4th order Runge-Kutta with a step size equal to sample time, Δt. The integration produces SOPDT model output values which are then subtracted from the actual process output record in a sum of squared errors computation. The parameter search, which is accelerated using a multidimensional secant method (Press *et al.,* 1986), determines values for the four parameters which produces the smallest sum of squared errors residual.

Whenever K_p is computed, it is stored on a u(t) vs K_p gain schedule (e.g. Stephanopoulos, 1984). Over time, the process passes through a range of operating levels and the schedule grows toward completeness. There is reasonable uncertainty associated with these nonlinear fits, however. Thus, when the process re-enters a level of operation where a gain has been previously computed and stored, the new data provides the opportunity to refine the gain schedule.

AUTOMATED CONTROLLER DESIGN

Given the general algorithms for estimation and control developed in the previous sections, what remains for the design of the adaptive controller is the specification of algorithm parameters. In the examples to follow, some assumptions have been made. These include that the process is continuous, open-loop stable, available for perturbation during startup, that the signal to noise ratio is large, and that dead time is negligible.

A final assumption is that an operator is available to supply some information, including the true minimum and maximum values for input, u(t), an estimate of the overall process time constant, and limits on how much u(t) may be perturbed during identification. The system must know the minimum and maximum values for u(t) so that controller computations can be clipped. An estimate of the process time constant is important as it enables the system to collect process identification samples at a reasonable rate. The system must also know how much it may perturb u(t) from its present value during automated process identification without creating safety or operational problems.

When the operator supplies this information and gives permission to proceed, the system begins an automated process identification study. The system first perturbs the process in one direction with a step increase in u(t) to the limit imposed by the operator. The system maintains the elevated input signal for 20 time constants as estimated by the operator, and then returns to the original value, where u(t) is held constant for an additional 20 time constants.

A four parameter fit is made using the input/output record and the SOPDT model of Eqn (12) with the multidimensional secant search as described in the previous section. The result is estimates for τ_1, τ_2, t_d and K_p which best describe the process in the region of the perturbation. τ_1 is defined here as the smallest time constant. After the fit, the values for τ_1 and τ_2 are considered so that sample rate, Δt, can be updated. The heuristic employed is; if $\tau_1 < 0.05\tau_2$, then $\Delta t=0.10\tau_2$, otherwise, $\Delta t=0.10\tau_1$. The basis for this is that if the parameter fit produces one time constant that is considerably smaller than the other, the small time constant is assumed a mathematical fabrication of the fit rather than a true characteristic of the process. Thus, sampling is to be at a rate of 10 samples per smallest believable time constant.

With an improved value for process time constant and associated sample rate, the system initiates a the procedure with an input perturbation step in the direction opposite from the first. The result is a set of parameters for each of perturbation direction. The system averages the τ_1, τ_2 and t_d's from the two perturbations, and the sample rate heuristic is applied to the averaged time constants to obtain a Δt for estimator/controller implementation. Also, the two values for K_p become the first two entries on the gain schedule.

Parameters must also be specified for RLS estimator design. As mentioned, a linear difference model that is second order in input and output is employed. Because t_d is assumed negligible in the examples, n=m=2 and k=0 in Eqn (1). The other specifications for RLS estimator design include initial estimates for the parameter vector, $\theta(0)$, and the covariance matrix, P(0). These values are easily obtained because an input/output record of process dynamics exists from the identification study. After the study, sample time is known. Hence, $\theta(0)$ is initialized to a vector of zeros, and P(0) is initialized to 100*I where I is the identity matrix. The complete input/output record is then rerun through the estimator. By the end of this procedure, $\theta(t)$ is well converged and P(t) has diminished to a level reflecting an appropriate degree of estimator sensitivity. The adaptive controller can now be safely implemented.

REACTOR SIMULATION

Continuous stirred tank reactors (CSTR's) have been simulated for the examples. The reaction is the 2nd-order reaction 2A → B, which produces a very nonlinear process behavior. As illustrated in Figure 1, three CSTR's are cascaded in series to produce a third order process. A balance on component A for reactor 1 yields simultaneous ordinary differential equations, one equation describing reactor volume and one equation describing reactor concentration, i.e.;

$$\frac{dV_1(t)}{dt} = F_0(t) - F_1(t) \tag{13}$$

$$V_1(t)\frac{dC_{A,1}(t)}{dt} = F_0(t)[C_{A,0}(t) - C_{A,1}(t)] - V_1(t)k_oC_{A,1}(t) \tag{14}$$

where $C_{A,1}(t)$ is the concentration of reactant A, $F_0(t)$ and $F_1(t)$ are the flow rates in and out of reactor 1, and k_o is the reaction rate constant. Analogous equations exist for reactors 2 and 3.

As shown, a high and low concentration stream mix to form, $C_{A,0}(t)$, the feed concentration to tank 1. The analyzer controller measures $C_{A,3}(t)$, the output concentration exiting the third reactor, and adjusts the high concentration feed flow rate, thus adjusting $C_{A,0}$. A flow controller adjusts the low concentration stream to maintain $F_0(t)$ at setpoint. Unmeasured disturbances are made by varying the setpoint on the flow controller, thus causing transients in tank volume to cascade through the three CSTR process.

QUALITATIVE MODELING OF PROCESS BEHAVIOR

Diagnosing process behavior on a timely basis is required both so the system can take preemptive corrective actions in adjusting estimator sensitivity, and so the system can generate an accurate gain schedule. The causal structure assumed for all processes is that if u(t) is changing significantly, then the process is experiencing excitation and a K_p fit will be meaningful. And if K_p is changing, then the process is experiencing a shift in operating level.

The gain schedule referred to in previous sections is essentially a table that relates u(t) to K_p. Schedule construction begins after the operator has input the minimum and maximum u(t). The span between these u(t) are first discretized into 500 equally spaced "slots," and each slot is associated with a discrete input value, u_i. After process identification, two slots are filled on the schedule. These are the u_i slots closest to the average u(t) used in the perturbation studies along with their associated K_p's. The empty slots between these two entries are then temporarily filled via linear interpolation.

The method of approach for process diagnosis, as illustrated in Figure 2, is through the construction of a "window of sufficient excitation." The horizontal axis of this window is sample number, and extends from 1 to 25 corresponding to the 25 most recent samples of input. The vertical axis is process input, and two methods are considered in determining its limits. The first method specifies the vertical axis as one half the span of u_i's used in the perturbation studies. The second specifies the axis as the span of u_i's which, based on the gain schedule, produce a 30% change in K_p. The actual method used during window construction is the one that results in the smallest vertical axis limits.

The logic for the above heuristic is as follows. If a process has a strong nonlinear component, then K_p will vary as u(t) varies. Any change in u(t) that is significant enough to cause K_p to vary by more than 15% within 25 samples qualifies as process excitation. Since a 15% change in either direction must be considered, a window must have a span of 30%. However, if a process is linear in its current range of operation, then large changes in u(t) will not cause K_p to vary significantly. The process would still be excited by these large changes, however. Thus, if the change in u(t) required to achieve a 30% change in K_p exceeds one half the input span used in the perturbation studies, then the process is considered to be in a linear region and 50% of the perturbation span is used as the vertical axis.

Figure 2 illustrates the window of sufficient excitation as it progresses with data from a simulation example. The window is located such that the most recently sampled u(t) is centered on the right vertical axis, and the previous 24 samples trail across the graph. In Figure 2b, future samples are depicted to show how the window shifts through successive sets of data. The reactor simulation is very nonlinear, so the vertical axis of the excitation window corresponds to the change in input concentration, $C_{A,0}$, required to produce a 30% change in K_p.

Every three samples, the most recent 25 u(t) samples are fit to a quadratic polynomial. The resulting curves are shown in Figure 3. A quadratic fit of 25 samples is used because this population contains enough data to substantially smooth noise, yet it is few enough samples such that new trends are rapidly reflected in the polynomial curvature. A quadratic fit also will discount data spikes that briefly diverge from an apparent trend.

As we have defined it, if any point of the quadratic polynomial when placed within the excitation window exits the window anywhere other than the vertical axes then the process is experiencing excitation. Such a situation occurs in Figure 3b. If a process is experiencing excitation, then a determination of K_p will be initiated and the resulting value will be used to fill or refine slots on the gain schedule.

ADJUSTING ESTIMATOR SENSITIVITY

With time, the gain schedule grows larger and more accurate. The computation of Q(t) from Eqn (11), using the scheduled gains, enables robust control. To maintain stability, however, decisions about when and how much to adjust estimator sensitivity must be made. The tools used in making this decision include the gain schedule and the ability to diagnose process excitation. Here, estimator sensitivity is adjusted only through covariance resetting, where the diagonal elements of the covariance matrix, P(t), are increased by a scaler multiple of the identity matrix, i.e.:

$$P(t) = P(t) + \gamma I \tag{15}$$

Because dynamic process information promotes estimator convergence, P(t) can be safely reset whenever the process is experiencing excitation. However, the result of resetting is a discounting of past dynamic information. Since an estimator is most reliable when it is based on a long process history, then P(t) should be reset only when the process has experienced a change in character sufficient to warrant such discounting. A change in character is determined in the simulation examples by a 15% change in K_p since the last covariance resetting. Hence, if a process has experienced such a change, and it is also experiencing excitation, then Eqn (15) is implemented. The examples determine γ by converting every 5% change in K_p to a increase of 1 in γ. A ceiling of 50 is imposed on the elements of P(t) to appropriately bound their growth.

SIMULATION EXAMPLES

The example process, as illustrated in Figure 1, is a nonlinear reactors in series simulation. Each reactor is initially run to steady state using a feed concentration of 5.0 gmol/liter. The associated steady state output, $C_{A,3}$, is 2.0 gmol/liter. Unmeasured disturbances are simulated by repeatedly manipulating the flow controller setpoint between 17 and 20 liter/sec in a slow cycle lasting 500 samples, thus causing the flow rate to the first tank to vary. Sample rate is determined to be 30 seconds based on the methods detailed earlier.

The first investigation studies the effect of gain scheduling on nonlinear self-tuning control. To test gain scheduling, the setpoint is initialized to 4.0 gmol/liter and then increased by steps of 1.0 gmol/liter every 500 samples until a final setpoint of 8.0 gmol/liter is reached. Throughout this and all other investigations, the process is experiencing significant unmeasured disturbances from the changing inlet flowrate. In Figure 4, the controlled output, $C_{A,3}$, is plotted along with the step setpoint changes to enable a visual appreciation of controller performance.

For this process, K_p decreases as u(t) increases. In Figure 4a, Q(t) is determined using the K_p associated with the setpoint of 4.0 gmol/liter, and this value is held constant throughout the step setpoint changes. As shown, the controller becomes increasingly sluggish as the penalty on incremental input change grows too large for the higher levels of operation. We note that in this investigation, a predetermined gain schedule is not being used. Rather, K_p is computed whenever the process is excited as previously defined. In Figure 4b, the controller using gain scheduling generation demonstrates superior performance.

Next, the effect of covariance resetting on controller performance is investigated. The gain schedule is again generated during the study and used to compute Q(t). To study the effectiveness of covariance resetting, the setpoint is initialized to 8.0 gmol/liter, and then decreased using a combination of steps and ramps to a final value of 4.0 gmol/liter. Figure 5 illustrates this setpoint trajectory along with the controlled output, $C_{A,3}$.

Figure 5a depicts controller performance with gain scheduling but without any covariance adjustments. In Figure 5b, the covariance is reset based on the change in operating level and degree of process excitation. Controller performance is good in Figure 5a, but controller error is visibly decreased in Figure 5b.

A final study demonstrates how important timing can be in covariance resetting. Specifically, a comparison is made between using estimation error as opposed to process behavior as indicators for covariance resetting. In this study, several small changes to the setpoint are made at the beginning of each run to cause the trace of the covariance matrix to fall. This results in an estimator that is very insensitive to changes in process character. A large and sudden decrease in setpoint is then made. Figure 6 shows the estimation error transient during this large and sudden change, and Figure 7 shows the response of $C_{A,3}$ versus setpoint when no resetting occurs.

Figure 6 shows the point in time where the large and sudden change in process behavior could reasonably be detected based on a diverging estimation error. Also shown is the point where the change is detected using excitation windows and gain scheduling. Figures 8a and 8b show the effect on controller performance using these different points in time. The case based on a changing process behavior ultimately produces a covariance matrix that has diagonal elements at the maximum value of 50. These maximum values are also used for the diverging estimation error case illustrated in Figure 8a.

The result of late resetting is catastrophic. Estimator parameters flip sign while attempting to converge on the free falling input/output sampled data and produce a diverging controller. The early resetting, shown in Figure 8b, permits the estimator to obtain enough dynamic input that it is able to converge on a reasonable process description and maintain control. We note that the results of this last example are rather extreme, and such catastrophic results are not the rule. In all cases studied, however, resetting based on changes in process behavior always produces results that are at least as good as resetting based on the detection of estimation error.

ACKNOWLEDGEMENTS

Acknowledgement is made to the Donors of the Petroleum Research Fund, administered by the American Chemical Society as Grant #18691-G7, and to the Connecticut Department of Higher Education through a High Technology Project Grant, for their support of this work.

NOMENCLATURE

a_i	Parameters corresponding to system output, y_i
b_i	Parameters corresponding to system input, u_i
C_A	Concentration of species A in reactor
c_i	Parameters corresponding to the system noise, v_i
d	Unmeasured disturbance variable
F	Flow in reactor
J	Objective function, defined in Eqn (5) and (8)
k	Dead time expressed as a multiple of sampling period, Δt
K_p	Steady state process gain
k_o	Reaction rate constant
m	Order of input in the ARMAX model in Eqn (1)
n	Order of output in the ARMAX model in Eqn (1)
P	Covariance matrix, defined in Eqn (7)
Q	penalty on incremental input change, defined in Eqn (11)
t	Integer number representing the sampling instant

t_d	Process dead time
Δt	Sampling period for estimator and controller
u	Process input variable
V	Reactor volume, Defined in Eqn (13)
y	Process output variable
y'	Estimate of output y
y_{sp}^f	Filtered setpoint, defined in Eqn (9)
y_{sp}	Controller setpoint

Greek Symbols

α	Filter rate parameter used in Eqn (9)
ϵ	Error of estimation in Eqn (4)
ψ	Vector of inputs and outputs, defined in Eqn (4)
τ_i	Process time constants in Eqn (12)
θ	Vector of the coefficient parameters, defined in Eqn (3)

REFERENCES

Astrom, K. J., and B. Wittenmark, "On Self-Tuning Regulators," *Automatica*, **9**, 185 (1973).

Clarke, D. W., and P. J. Gawthrop, "Self-Tuning Control," *Proc. of IEEE*, **126**, 633 (1979).

Clarke, D. W., "Introduction to Self-Tuning Controllers," in *Self-Tuning and Adaptive Control: Theory and Applications*, [Harris and Billings, Ed.], ch. 2, Peregrinus, London (1981).

Fortescue, T. R., L. S. Kershenbaum, and B. E. Ydstie, "Implementation of Self-tuning Regulators with Variable Forgetting Factors," *Automatica*, **17**, 831 (1981).

Goodwin, G. C., H. Elliot and E. K. Teoh, "Deterministic Convergence of a Self-Tuning Regulator With Covariance Resetting," *Proc. of IEEE*, **130**, 6 (1983).

Goodwin, G. C., and K. S. Sin, *Adaptive Filtering Prediction and Control*, Prentice-Hall, Englewood Cliffs, New Jersey (1984).

Goodwin, G. C., D. J. Hill and M. Palaniswami, "Towards a Robust Adaptive Controller," *Proc. of IFAC Conf. on Identification and System Parameter Estimation*, York, U. K. **997** (1985).

Ljung, L., and T. Soderstrom, *Theory and Practice of Recursive Identification*, MIT Press, Cambridge, Massachusetts (1983).

Ortega, R., L. Praly and I. D. Landau, "Robustness of Discrete-Time Adaptive Controllers," *IEEE Trans. on Automatic Control*, **30**, 1179 (1985).

Seborg, D. E., T. F. Edgar, and S. L. Shah, "Adaptive Control Strategies for Process Control: A Survey," *AIChE J.*, **32**, 881 (1986).

Shah, S. L. and W. R. Cluett, "RLS Based Estimation for Self-Tuning Control," *Proc. 1987 Annual AIChE Meeting*, New York, N. Y. (1987).

Stephanopoulos, G., *Chemical Process Control: An Introduction to Theory and Practice*, Prentice-Hall, Englewood Cliffs, New Jersey (1984).

Vogel, E. F., and T. F. Edgar, "Application of An Adaptive Pole-Zero Placement Controller to Chemical Processes with Variable Dead Time," *Proc. Amer. Control Conf.*, Arlington, VA, 536 (1982).

Ydstie, B. E., L. S. Kershenbaum and R. W. H. Sargent, "Theory and Application of an Extended Horizon Self-Tuning Controller," *AIChE J.*, **31**, 1771 (1985).

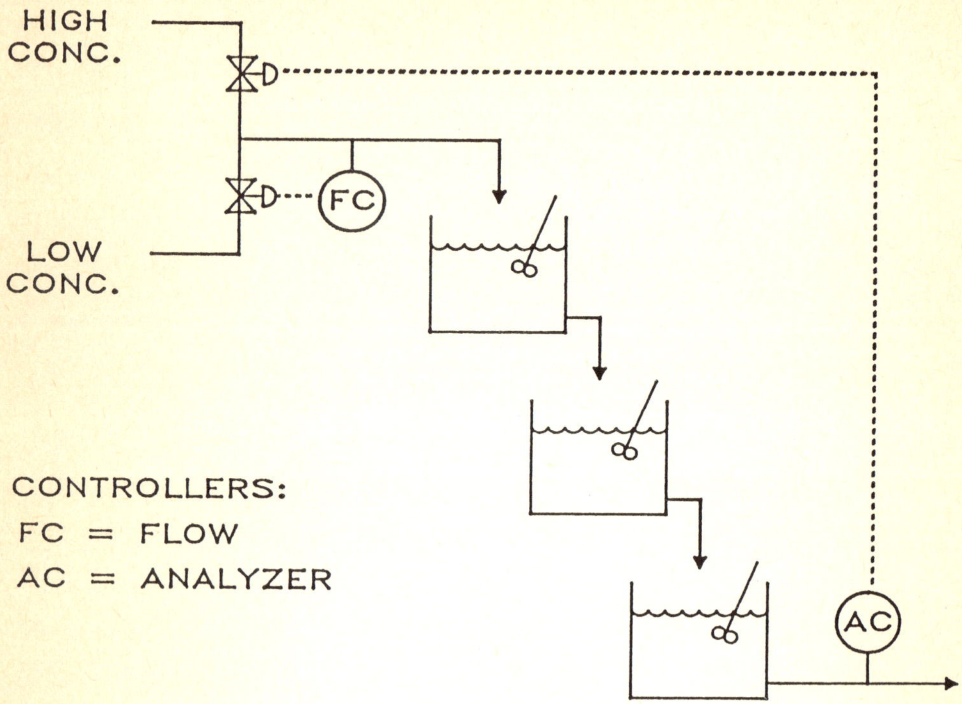

Figure 1 - Nonlinear reactor series simulation.

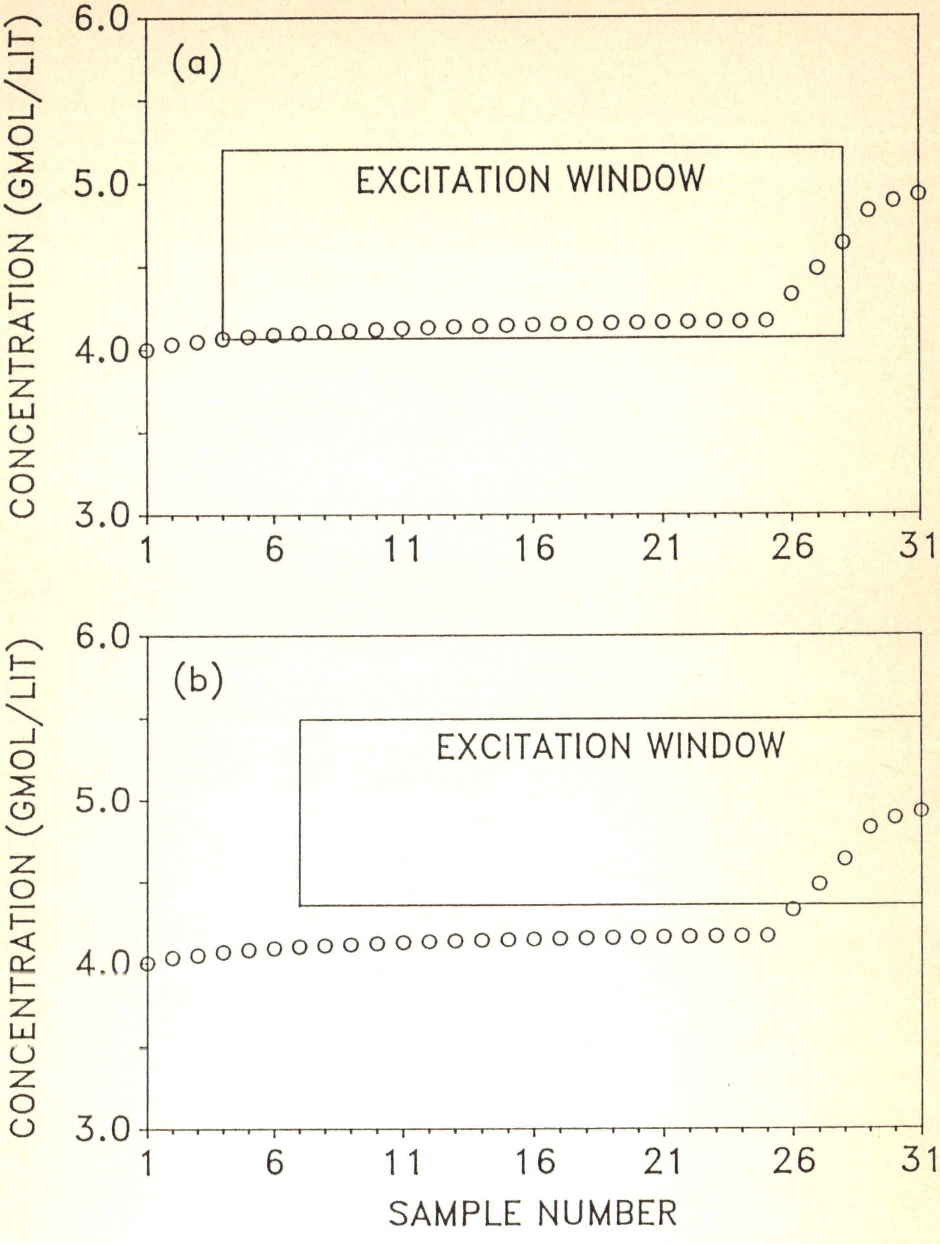

Figure 2 - Progression of excitation window.

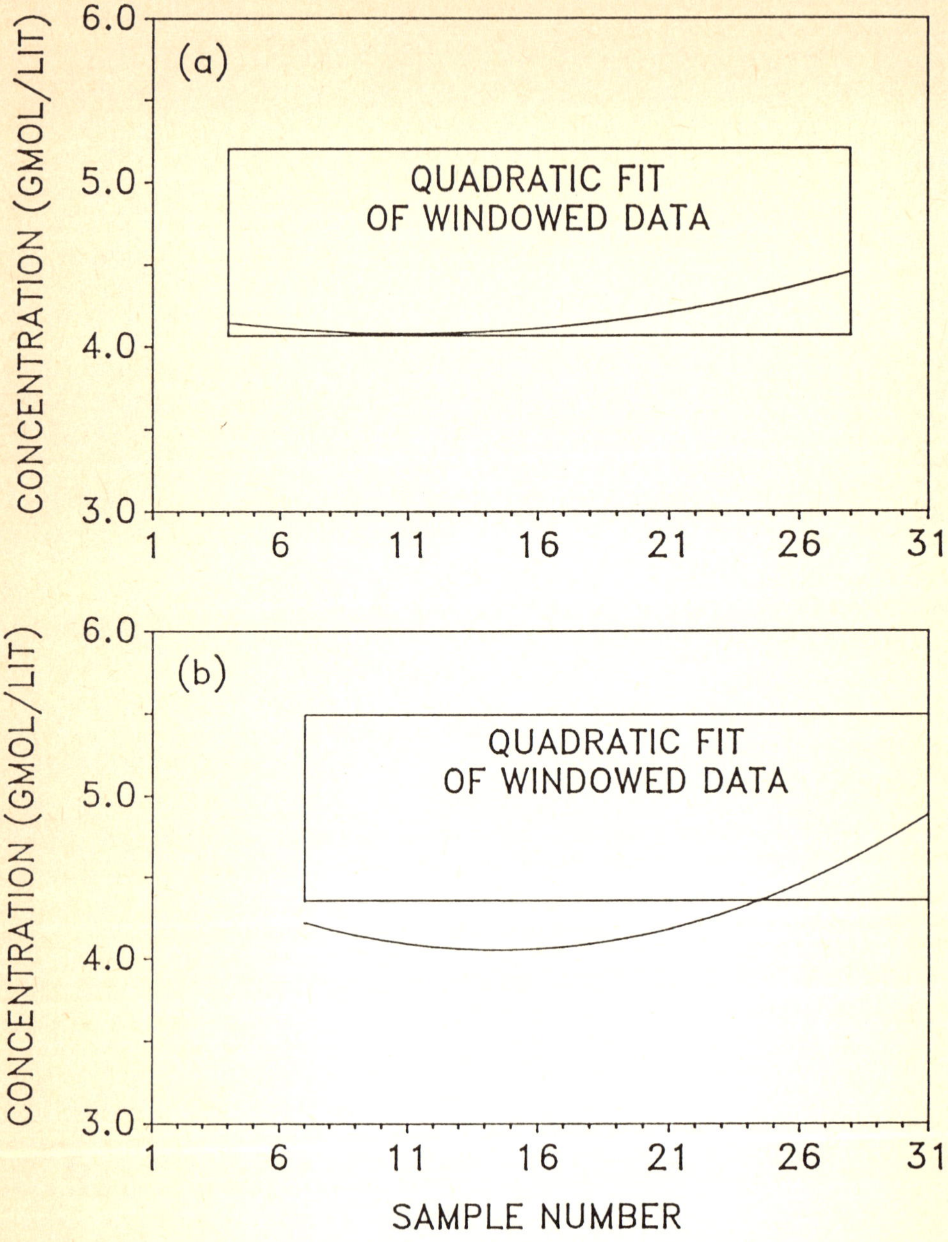

Figure 3 - Progression of process into excited state.

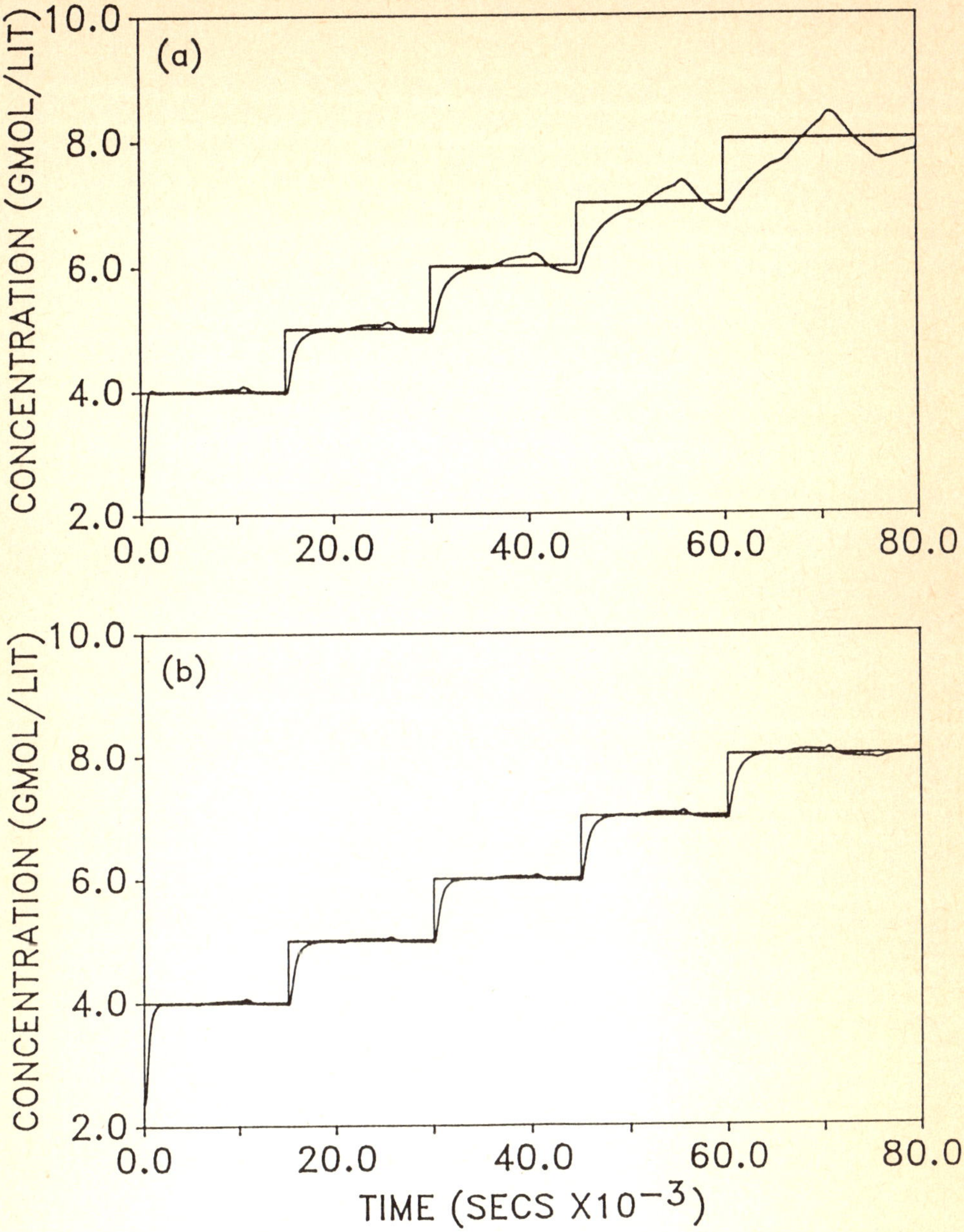

Figure 4 - Controller performance with a) constant gain
b) gain scheduling.

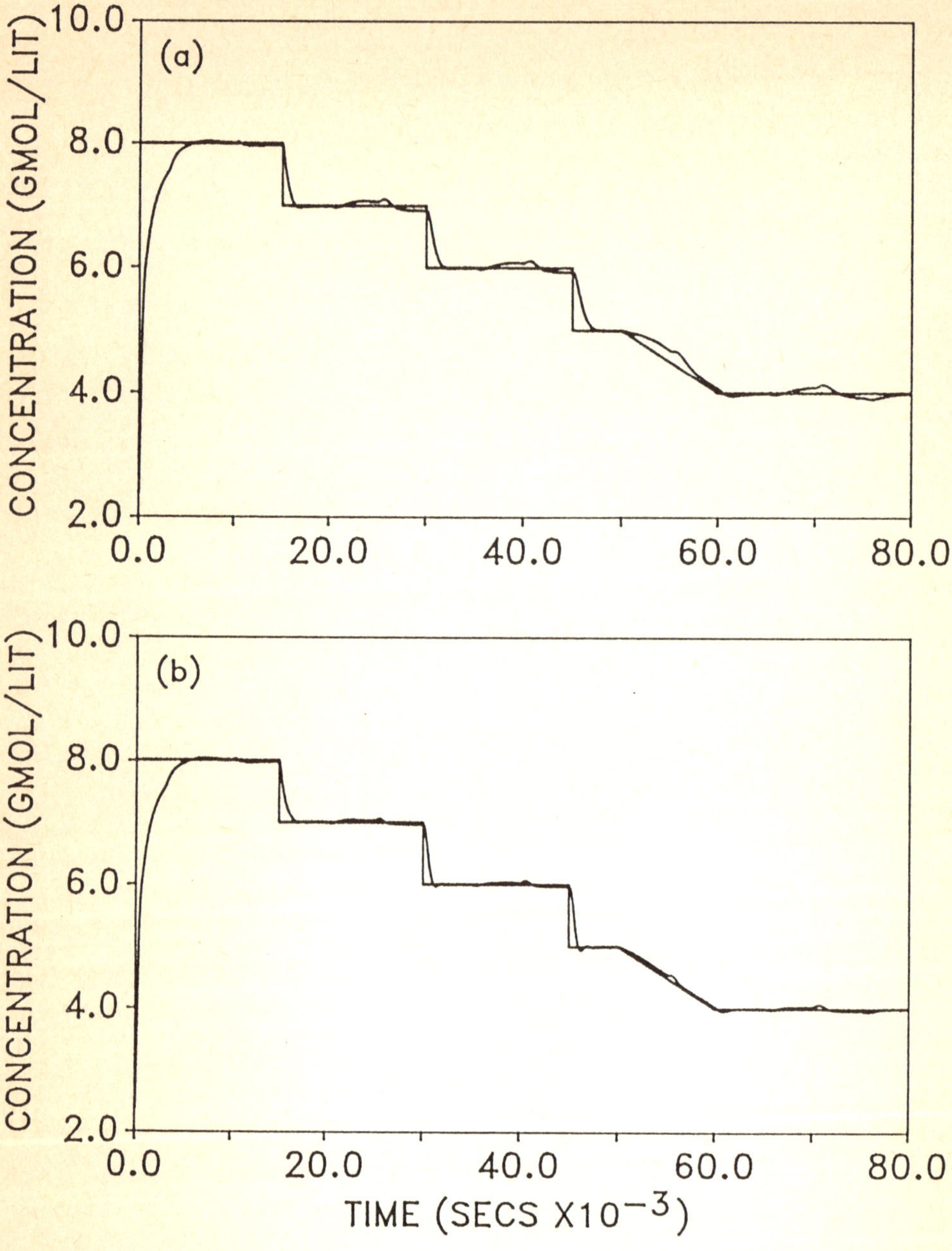

Figure 5 - Controller performance a) without and b) with covariance resetting.

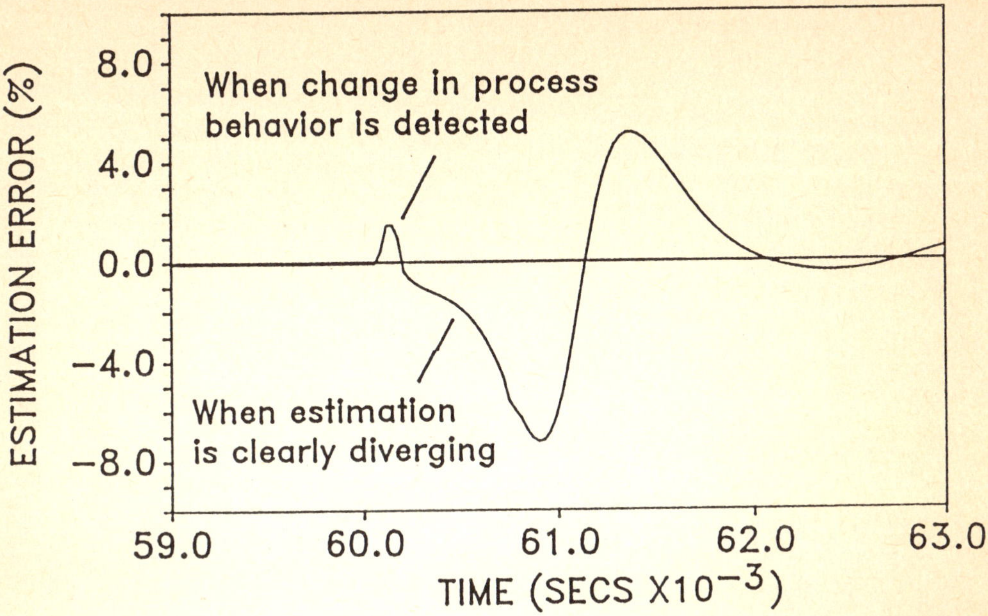

Figure 6 - Estimator error transient after a large, sudden change in setpoint.

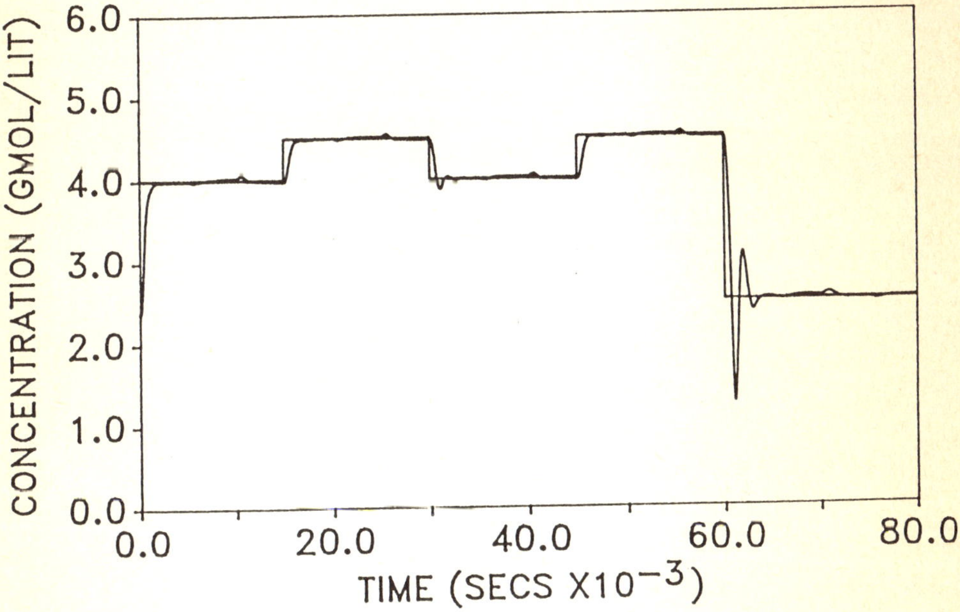

Figure 7 - Controller performance without resetting.

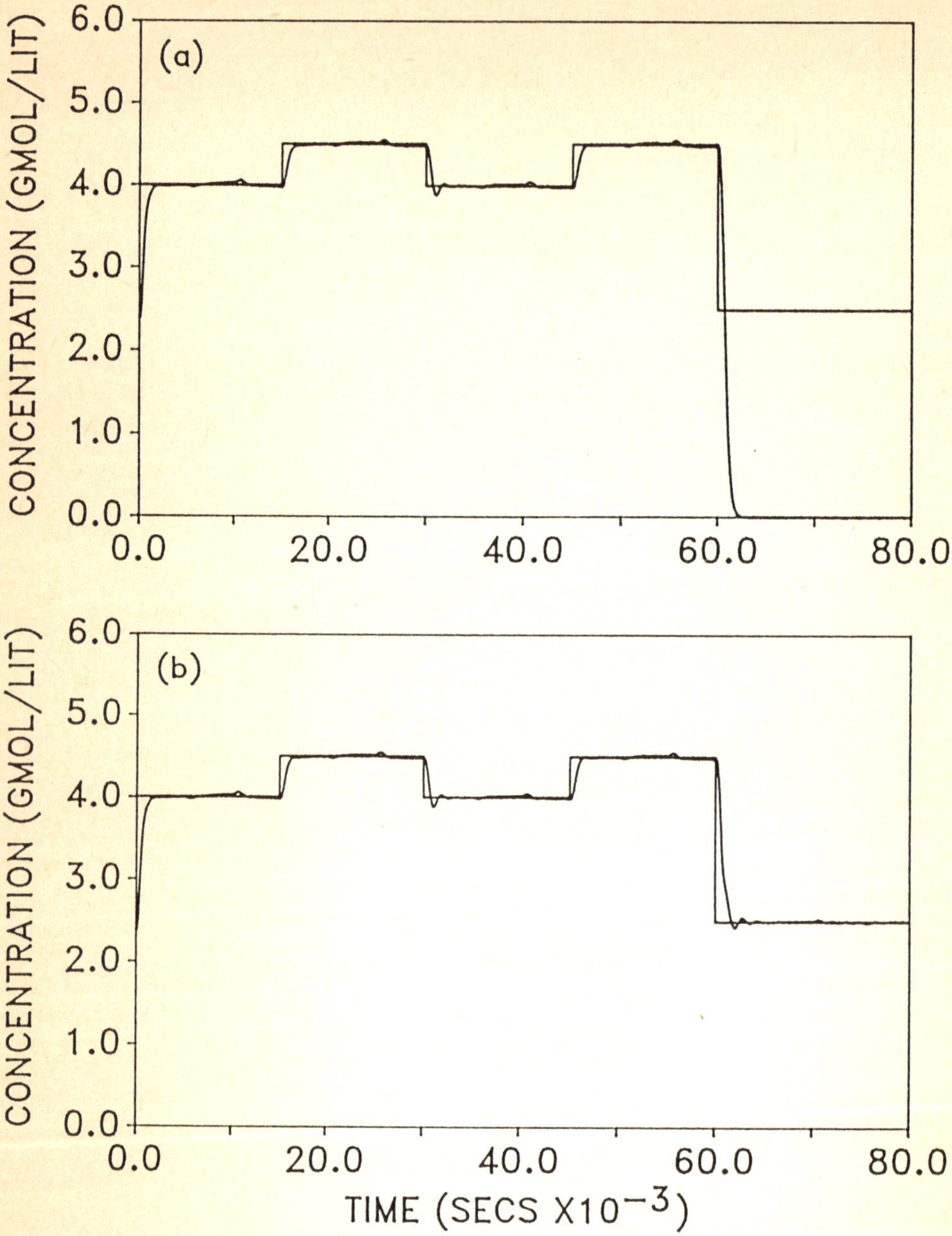

Figure 8 - Controller performance with resetting tied to a) estimation error and b) process diagnosis.

TWO DEGREES OF FREEDOM PID AUTO-TUNING CONTROLLER BASED ON FREQUENCY REGION METHODS

Takashi Shigemasa, Yutaka Iino and Masae Kanda*
Toshiba Research & Development Center, Toshiba Corporation
Kawasaki, Japan
*Instrument & Automation Development Dept., Fuchu Works,
Toshiba Corporation
Fuchu, Japan

ABSTRACT

Recently two degrees of freedom PID control systems have attracted attention, because of their better process control quality for both set point variable tracking and disturbance resistivity than that of conventional PID control systems. This paper describes a practical auto-tuning method for the two degrees of freedom PID control system based on frequency region methods; the well known recursive least square parameter estimation method, a frequency characteristics conversion method and a model matching method in frequency region. By combining frequency region method, the auto-tuning method can use for both minimum phase process and non-minimum phase process. This method was introduced into the two degrees of freedom PID auto-tuning controller. The superior features of the controller and the results of field tests are presented.

1. INTRODUCTION

From operational and economical points of view, every process control loop should be tuned optimally in accordance with the process dynamic characteristics and the operational conditions. Recently in order to save time for tuning work of field engineers and to improve control performance, auto-tuning PID (Proportional, Integral, Derivative) controllers (Shigemasa 1981, Nagakawa 1982, Åström 1984, Kraus 1984) were developed.

Conventional PID auto-tuning controllers have the following problems to be improved;

(a) weak disturbance resistivity when optimal PID control parameters are used for set point variable tracking,

(b) large overshoot response in set point variable tracking when optimal PID control parameters are used for disturbance rejection,

(c) restriction in application for particular process characteristics.

In order to overcome the above problems, the authors developed an auto-tuning method (Shigemasa 1987) for the two degrees of freedom PID control system (Araki 1985). The auto-tuning method is mainly based on a developed model matching method (Shigemasa 1986) using the process characteristics in the cut-off frequency region. The first model matching method was developed by T. Kitamori (1979) using the process characteristics in low frequency region and is applicable to the process including time delay and dead time. The model matching method described here, gives wide application to the process including not only time delay and dead time but also overshooting reverse-shooting and oscillating characteristics. Applying this auto-tuning method, a two degrees of freedom PID auto-tuning controller, named AdTune TOSDIC 211D8 (Picture 1) has been produced.

This paper describes the auto-tuning method for a two degrees of freedom PID control system. The basic scheme consists of

(1) a parameter estimation part to identify the dynamic characteristice of the process from the manipulated variable and the process variable.

(2) a perturbation signal generation part to satisfy the identifiability condition under closed loop control,

(3) a frequency characteristics calculation part from the identified transfer function in Z domain,

(4) a two degrees of freedom PID control parameter calculation part based on the model matching method in the cut-off frequency region.

The superior features of the controller and the results of field tests are presented later.

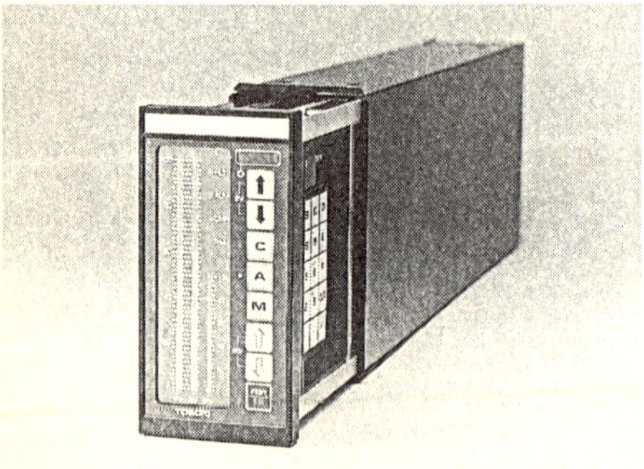

Picture 1 TOSDIC 211D8 two degrees of freedom PID auto-tuning controller

2. CONFIGURATION OF TWO DEGREES OF FREEDOM PID AUTO-TUNING CONTROLLER

The basic scheme of the two degrees of freedom PID auto-tuning control system consists of the following five functions, as shown in Figure 1;

(1) two degrees of freedom PID control function,

(2) two degrees of freedom PID control parameter calculation function,

(3) process parameter estimation function,

(4) perturbation signal generation function,

(5) Z-jω conversion function.

2.1 TWO DEGREES OF FREEDOM PID CONTROL FUNCTION

There are several control schemes in two degrees of freedom PID control systems. The most typical two degrees of freedom PID control system is shown in Figure 2. The manipulated variable u in these figures, are determined as the sum of integral operation for the deviation signal e, proportional operation and deviation operation for the process variable y and feed-forward operation for the set point variable r. The block diagram of the system in Figure 2(A) becomes the same as that of Figure 2(B) by the following equations;

$$K = K_P/T_I \quad , \tag{1-1}$$

$$f_0 = K_P \quad , \tag{1-2}$$

$$f_1 = K_P \cdot T_D \quad , \tag{1-3}$$

$$f_r = \alpha \cdot K_P \quad . \tag{1-4}$$

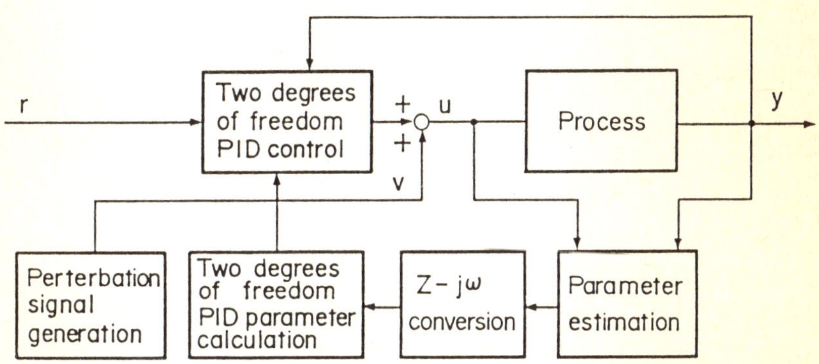

Figure 1 Scheme of the two degrees of freedom PID auto-tuning system

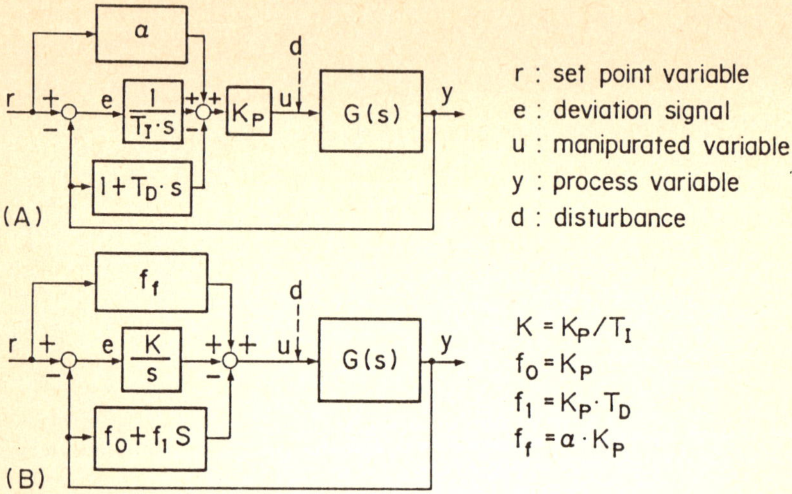

Figure 2 Two degrees of freedom PID (FF-I-PD) control system.

The control system works best, when the control parameters K, f_0, f_1 are tuned optimlly for disturbance rejection and the control parameter f_f is tuned optimally for set point value tracking.

2.2 TWO DEGREES OF FREEDOM PID PARAMETER CALCULATION FUNCTION

In order to calculate the control parameters, the authors developed a model matching method in the cut-off frequency region by extending the conventional model matching method in S domain.

The basic idea of this model matching method is to make the cut-off frequency characteristics of the control system equal to that of the reference model. The reference model representing the desired specification of the control system, plays an important role in the model matching method. Considering the structure of Kitamori's reference model (Kitamori 1979), the following open loop reference model $\tilde{G}_m(s)$ from e to y is used.

$$\tilde{G}_m(s) = [\sigma s + \alpha_2(\sigma s)^2 + \alpha_3(\sigma s)^3 + \alpha_4(\sigma s)^4]^{-1} \quad , \tag{2}$$

where σ is a time constant and α_i (i = 2,3,4) are dumping coefficients. Figure 3 shows the vector locus for $G_m(j\omega)$. The phase margin ϕ_M and the gain margin G_M of the reference model are the principal parameters representing the cut-off frequency characteristics and are determined by points P and G respectively, as shown in Figure 3. The dumping coefficients, the phase margins and the gain margins of the well-known reference models, for example, the butterworth model, the ITAE minimum model, the bessel model and the binominal model, are given in Table 1. The step response curves of these reference models are shown in Figure 4. A desired reference model is chosen from these response curves and the desired control system specification.

Next, the open loop transfer function Gey(s) from e to y in Figure 2(B) can be expressed as

$$Gey(s) = \frac{K}{s[f_0 + f_1 s + G(s)^{-1}]}, \tag{3}$$

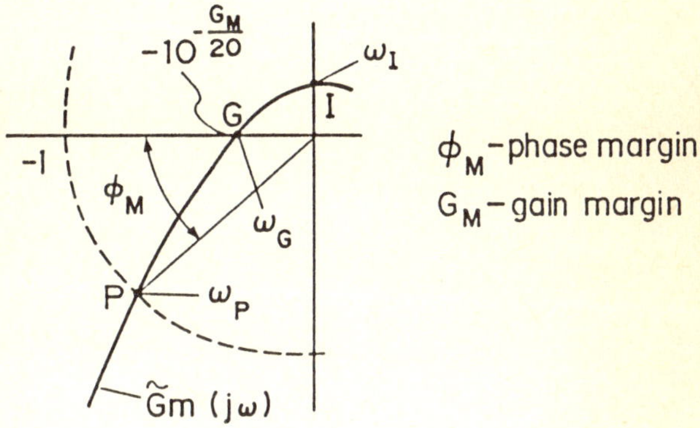

Figure 3 Vector locus of $\tilde{G}_m(j\omega)$.

Table 1 Phase Margin-Gain Margin of Reference Models.

NO.	Ref. Model	n	DUMPING COEFFICIENTS a_2	a_3	a_4	ϕ_M(deg)	G_M(dB)	$\lvert G(j\omega_I)\rvert$
1	Butter-worth	3	0.5	0.125	-	60.5	12.04	0.0
		4	0.5030	0.1479	0.02188	59.8	7.61	0.0869
2	ITAE minimum	3	0.3786	0.1006	-	66.5	11.51	0.0
		4	0.4664	0.1067	0.01882	63.4	8.69	0.1222
3	Bessel	3	0.4	0.0667	-	67.2	15.56	0.0
		4	0.4268	0.09524	0.009524	65.1	10.71	0.0457
4	Binomial	3	0.3333	0.03704	-	71.3	19.08	0.0
		4	0.3750	0.0625	0.003906	68.6	13.98	0.0204

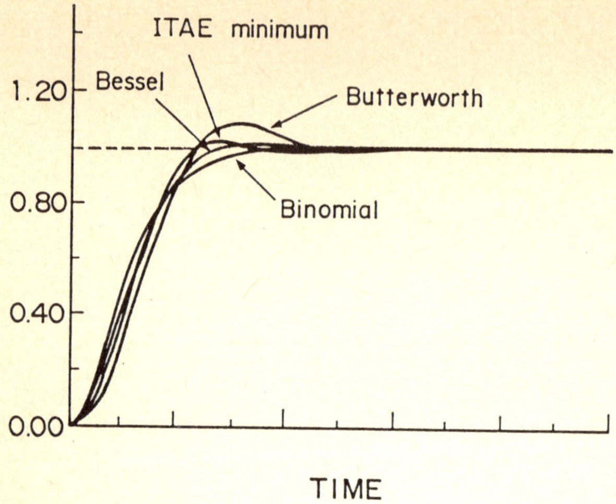

Figure 4 Response curves of the reference models.

where G(s) is the transfer function of the process, as shown in Figure 2. By matching the cut-off frequency characteristics of Gey(jω) with that of $\tilde{G}_m(j\omega)$, the following equations can be obtained;

[At the point P]

$$\tan(\phi_M) = \frac{f_0 + \cos(\omega_P)/|G(j\omega_P)|}{\omega_P \cdot f_1 + \sin(\omega_P)/|G(j\omega_P)|} , \tag{4-1}$$

$$K = \frac{\omega_P[\omega_P \cdot f_1 + \sin(\omega_P)/|G(j\omega_P)|]}{\cos(\omega_P)} , \tag{4-2}$$

[At the point G]

$$f_0 + \frac{\cos(\omega_G)}{|G(j\omega_G)|} = 0 , \tag{5-1}$$

$$\frac{K}{\omega_G[\omega_G \cdot f_1 + \sin(\omega_G)/|G(j\omega_G)|]} = 10^{-\frac{G_M}{20}} \tag{5-2}$$

[At the point I]

$$f_1 + \frac{\sin(\omega_I)}{\omega_I \cdot |G(j\omega_I)|} = 0 , \tag{6-1}$$

$$\frac{K}{\omega_I[f_0 + \cos(\omega_I)/|G(j\omega_I)|]} = |G_m(j\omega_I)| , \tag{6-2}$$

where ω_P, ω_G and ω_I are angular frequency at the point P, G and I, respectively as shown in Figure 3.

Based on only the cut-off frequency characteristics of the process $G(j\omega)$ and the desired specifications of the control system (ϕ_M, G_M and $|G_m(j\omega_I)|$), this design method is applicable for the process including not only time delay and dead time but also overshooting reverse-shooting and oscillating charactersitics. This implies the control parameter calculation method can use for both minimum phase process and non-minimum phase process. The control parameters K, f_0 and f_1 can be determined easily using Eq. (4-1) ∿ Eq. (6-2) recursively.

The closed loop transfer function Gry(s), from r to y in Figure 2(B), is shown by Eq. (7). The closed loop reference model $\hat{G}_m(s)$ from r to y is shown by Eq. (8).

$$Gry(s) = \frac{1 + f_f\, s/K}{1 + s(f_0 + f_1 s + G(s))/K} \tag{7}$$

$$\hat{G}_m(s) = \frac{1 + \beta\sigma s}{1 + \sigma s + \alpha_2(\sigma s)^2 + \alpha_3(\sigma s)^3 + \alpha_4(\sigma s)^4} \tag{8}$$

Finally, the feed-forward gain f_f can be determined by matching Eq. (7) with Eq. (8).

2.3 PROCESS PARAMETER ESTIMATION FUNCTION

The frequency characteristics of the process are necessary for designing the control system by the model matching method. In this auto-tuning system, the well-known recursive least squares (RLS) algorithm is used for the process parameter estimation function.

Generally, many processes can be expressed by the following Auto-Regressive Moving Average (ARMA) model;

$$A(z^{-1})y(k) = B(z^{-1})u(k) \quad , \tag{9-1}$$

$$A(z^{-1}) = 1 + a_1 z^{-1} + a_2 z^{-2} + \cdots\cdots + a_{na} z^{-na} \quad , \tag{9-2}$$

$$B(z^{-1}) = b_1 z^{-1} + b_2 z^{-2} + \cdots\cdots + b_{nb} z^{-nb} \quad . \tag{9-3}$$

Where y(k) is the process variable, u(k) is the manipulated variable, k is the discrete time with sampling period τ, z^{-1} is the backward shift operator, n_a and n_b are order of $A(z^{-1})$ and $B(z^{-1})$, respectively. The process model of Eq. (9-1) can be expressed as

$$y(k) = \phi(k)^T \cdot \hat{\theta} \quad . \tag{10}$$

Where T denotes transpose, $\phi(k)$ and $\hat{\theta}$ are the data vector and the process parameter vector defined by;

$$\phi(k) = [y(k-1), -y(k-2), \ldots, -y(k-n_a), u(k-1), \ldots, u(k-n_b)]^T , \tag{11-1}$$

$$\hat{\theta} = [\hat{a}_1, \hat{a}_2, \ldots, \hat{a}_{na}, \hat{b}_1, \hat{b}_2, \ldots, \hat{b}_{nb}]^T \quad . \tag{11-2}$$

The RLS algorithm is shown as follows;

$$\hat{\theta}(k) = \hat{\theta}(k-1) + P(k-1)\phi(k)\mu(k)(y(k) - \phi(k)^T\hat{\theta}(k-1)) \quad , \tag{12-1}$$

$$P(k) = [P(k-1) - P(k-1)\phi(k)\phi(k)^T P(k-1)\mu(k)]/\lambda(k) \quad , \tag{12-2}$$

$$\mu(k) = (\lambda(k) + \phi(k)P(k-1)\phi(k)^T)^{-1} \quad , \tag{12-3}$$

where $\lambda(k)$ is the forgetting factor and is chosen as

$$\lim_{k \to \infty} \lambda(k) = 1 \qquad (0 < \lambda(k) \leq 1) \quad .$$

The RLS algorithm can be started with the following initial values

$$\hat{\theta}(0) = \hat{\theta}_0 \quad , \qquad P(0) = \nu I \qquad (\nu >> 0) \quad .$$

2.4 PERTURBATION SIGNAL GENERATION FUNCTION

In order to satisfy the identifiability conditions for any process under closed loop PID operation, it is necessary to add a perturbation signal including many frequency components to the manipulated variable. The maximum period sequence signal is used as the perturbation signal because of its simple generation algorithm.

2.5 Z-jω CONVERSION FUNCTION

The frequency characteristics of the process $G(j\omega)$ is converted from the identified pulse transfer function $\hat{G}(z^{-1})$ using a precise conversion method shown by Eq. (13).

$$G(j\omega) = [\hat{G}(z^{-1})|z = \exp(j\omega\tau)] \cdot [\Delta(s)|s = j\omega] \quad , \tag{13}$$

where $\hat{G}(z^{-1})$ is the identified pulse transfer function and $\Delta(s)|s = j\omega$ is a correction term to reduce the frequency distortion near the nyquist frequency.

$$\hat{G}(z^{-1}) = \frac{\hat{b}_1 z^{-1} + \hat{b}_2 z^{-2} + \ldots\ldots + \hat{b}_{nb} z^{-nb}}{1 + \hat{a}_1 z^{-1} + \hat{a}_2 z^{-2} + \ldots\ldots + \hat{a}_{na} z^{-na}} \tag{14}$$

$$\Delta(s) = [1 + \tau s + (\tau s)^2/s][se^{-\tau s} - 3 + e^{\tau s} - e^{-2\tau s}]/(\tau s)^3 . \tag{15}$$

2.5 OTHER FUNCTIONS

In order to use easily and simply the auto-tuning controller, a pre-tuning function to set optimal tuning parameters: sampling period, amplitude of perturbation signal and initial control parameters, for the auto-tuning system and an automatic retuning function to start the auto-tuning function automatically by detecting the instability of the control system are provided.

3. FIELD TEST

Several field tests were done to verify the auto-tuning performance of the two degrees of freedom PID auto-tuning controller, AdTune TOSDIC 211D8. A typical example performed at CHIYODA CHEMICAL ENGINEERING & CONSTRUCTION CO., LTD. is presented here. Figure 5 shows the gas flow process of a pilot plant with a gas flow control loop (FIC) and the gas pressure control loop (PIC) which were used to test the controller.

Figure 6 shows an example of the process responses in a tuning test for the FIC. SV, PV and MV are the set point variable, the process variable and the manipulated variable, respectively. The process characteristics were identified within about 30 seconds using the perturbation signal of 3 percent. The two degrees of PI control parameters were calculated from the identified process characteristics, and were set into the control part, automatically. The set point value was suddenly increased and decreased by 5 percent at about 4 minutes 15 seconds and 5 minutes 20 seconds, respectively; fine gas flow rate responses was obtained, as shown in Figure 6.

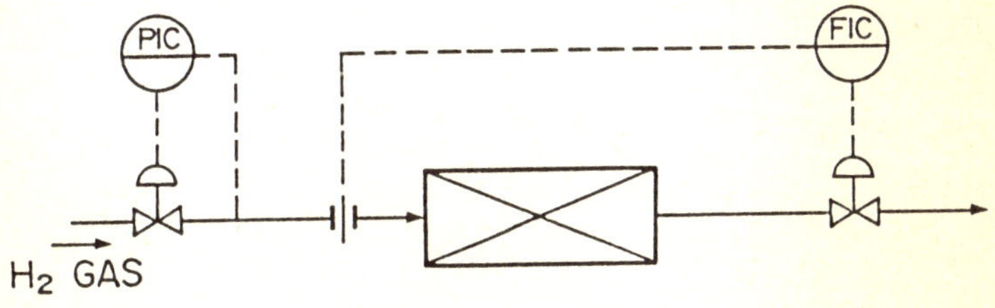

Figure 5 H_2 Gas flow process of pilot plant

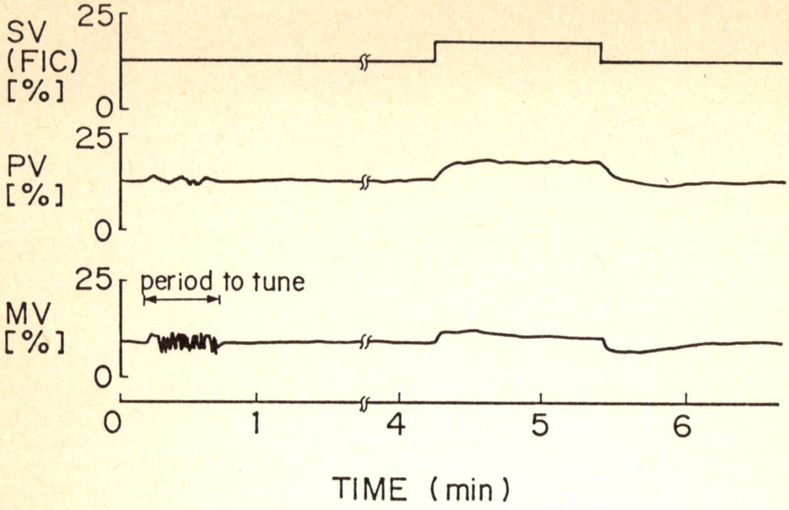

Figure 6 A tuning experiment for the FIC loop

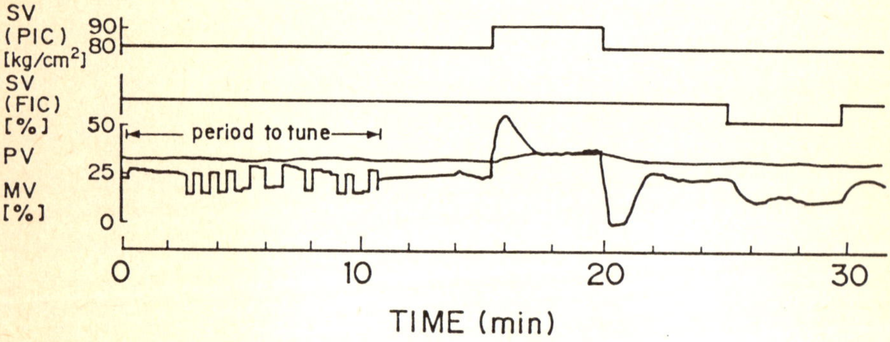

Figure 7 A tuning experiment for the PIC loop

Figure 7 also shows an example of the process response in a tuning test for the PIC loop. In the same manner, the process characteristics were identified within about 10.5 minutes using the perturbation signal of 3 percent. The two degrees of PI control parameter were calculated from the identified process characteristics, and were set into the control part, automatically. The set point value was suddenly increased and decreased by 10 kg/cm^2 at 15.7 minutes and 20 minutes respectively; fine gas pressure (PV) response was obtained. The PIC loop was intensively interacted from the FIC loop, however, the disturbance caused by the set point variable change in the FIC loop was well rejected from the PIC loop.

These control parameters were tuned optimally within a short period and were almost the same as the parameters manually tuned by expert engineers. These results show the effectiveness of the two degrees of freedom PID auto-tuning controller.

4. CONCLUSIONS

By combining the process parameter estimation method in z domain, the precise z-jω conversion method and the developed model matching method in cut-off frequency region, the authors developed auto-tuning method. The two degrees of freedom PID auto-tuning controller, AdTune TOSDIC 211D8, was developed using this method.

The superior features of the auto-tuning controller are summarized as follows;

(1) two degrees of freedom PID control system brings excellent process control qualities for both set point variable tracking and disturbance resistivity,

(2) auto-tuning method available both closed loop and open loop,

(3) wide application to process including not only time delay and dead time but also overshooting, reverse-shooting and oscillating characteristics, based on the developed model matching method in cut-off frequency region,

(4) chooseable function for the desired control system response,

(5) short tuning period with small perturbation signal,

(6) pre-tuning function for the auto-tuning system and

(7) automatic retuning function

Several tuning tests showed the effectiveness of the auto-tuning scheme for the two degrees of freedom PID control system.

REFERENCES

Araki, M. (1985). 2-degree of freedom controller part I, Trans. Japan Associ. of Auto. Cont. Eng. (Japan), 29, 649.

Åström, K. J., and T. Hagglund (1984). Automatic Tuning of Simple Regulators with Specifications on Phase and Amplitude Margins, Automatica, 20, 645.

Gustavsson, I., L. Ljung and T. Söderström (1977). Identification of processes in closed loop - identificability and Accuracy aspects -, Automatica, 13, 59.

Kitamori, T. (1979). A method of control system design based upon partial knowledge about controlled processes, Trans. Soc. Instrum. & Control Eng. (Japan), 15, 549.

Kraus, T. (1984). Self-tuning control: an expert system approach, Proceedings of ISA, 605.

Nagakawa, K., T. Shigemasa and Y. Hanakuma (1982). An auto-tuning PID controller. Proceedings of ISA, 1353.

Shigemasa, T. and K. Akizuki (1981). A closed loop auto-tuning method for digital PID controller, Proceedings of Applications of mini and microcomputers, 427.

Shigemasa, T. (1983). A design method for feed forward servo control system using reference model with zero, 6th Dynamical System Theory Symposium (Japan), preprint, 53.

Shigemasa, T. and Y. Ichikawa (1984). A closed loop auto-tuning method for a digital process control system, Trans. Soc. Instrum. & Control Eng. (Japan), 20, 592.

Shigemasa, T. and Y. Iino (1986). A design method for control system based on model matching in cut off frequency region, 29th Joint Conference of Automatic Control (Japan), proceedings, 1041.

Shigemasa, T., Y. Ichikawa and S. Sumida (1985). An Interactive online tuning system for process control — Applications to control system of BWR power plant, 7th IFAC Symposium on Identification and system parameter estimation, proceedings, 933.

Shigemasa, T., Y. Iino and M. Kanda (1987). Two degrees of freedom PID auto-tuning controller, Proceedings of ISA, 119.

Lecture Notes in Control and Information Sciences

Edited by M. Thoma and A. Wyner

Vol. 97: I. Lasiecka/R. Triggiani (Eds.)
Control Problems for Systems Described by Partial Differential Equations and Applications
Proceedings of the IFIP-WG 7.2 Working Conference
Gainesville, Florida, February 3-6, 1986
VIII, 400 pages, 1987.

Vol. 98: A. Aloneftis
Stochastic Adaptive Control
Results and Simulation
XII, 120 pages, 1987.

Vol. 99: S. P. Bhattacharyya
Robust Stabilization Against Structured Perturbations
IX, 172 pages, 1987.

Vol. 100: J. P. Zolésio (Editor)
Boundary Control and Boundary Variations
Proceedings of the IFIP WG 7.2 Conference
Nice, France, June 10-13, 1987
IV, 398 pages, 1988.

Vol. 101: P. E. Crouch, A. J. van der Schaft
Variational and Hamiltonian Control Systems
IV, 121 pages, 1987.

Vol. 102: F. Kappel, K. Kunisch, W. Schappacher (Eds.)
Distributed Parameter Systems
Proceedings of the 3rd International Conference
Vorau, Styria, July 6–12, 1986
VII, 343 pages, 1987.

Vol. 103: P. Varaiya, A. B. Kurzhanski (Eds.)
Discrete Event Systems: Models and Applications
IIASA Conference
Sopron, Hungary, August 3-7, 1987
IX, 282 pages, 1988.

Vol. 104: J. S. Freudenberg/D. P. Looze
Frequency Domain Properties of Scalar and Multivariable Feedback Systems
VIII, 281 pages, 1988.

Vol. 105: Ch. I. Byrnes/A. Kurzhanski (Eds.)
Modelling and Adaptive Control
Proceedings of the IIASA Conference
Sopron, Hungary, July 1986
V, 379 pages, 1988.

Vol. 106: R. R. Mohler (Editor)
Nonlinear Time Series and Signal Processing
V, 143 pages, 1988.

Vol. 107: Y. T. Tsay, L.-S. Shieh, St. Barnett
Structural Analysis and Design of Multivariable Systems
An Algebraic Approach
VIII, 208 pages, 1988.

Vol. 108: K. J. Reinschke
Multivariable Control
A Graph-theoretic Approach
274 pages, 1988.

Vol. 109: M. Vukobratović/R. Stojić
Modern Aircraft Flight Control
VI, 288 pages, 1988.

Vol. 110: B. J. Daiuto, T. T. Hartley, S. P. Chicatelli
The Hyperbolic Map and Applications to the Linear Quadratic Regulator
VI, 114 pages, 1989

Vol. 111: A. Bensoussan, J. L. Lions (Eds.)
Analysis and Optimization of Systems
XIV, 1175 pages, 1988.

Vol. 112: Vojislav Kecman
State-Space Models of Lumped and Distributed Systems
IX, 280 pages, 1988

Vol. 113: M. Iri, K. Yajima (Eds.)
System Modelling and Optimization
Proceedings of the 13th IFIP Conference
Tokyo, Japan, Aug. 31 – Sept. 4, 1987
IX, 787 pages, 1988.

Vol. 114: A. Bermúdez (Editor)
Control of Partial Differential Equations
Proceedings of the IFIP WG 7.2 Working Conference
Santiago de Compostela, Spain, July 6–9, 1987
IX, 318 pages, 1989

Vol. 115: H.J. Zwart
Geometric Theory for Infinite Dimensional Systems
VIII, 156 pages, 1989.

Vol. 116: M.D. Mesarovic, Y. Takahara
Abstract Systems Theory
VIII, 439 pages, 1989

Lecture Notes in Control and Information Sciences

Edited by M. Thoma and A. Wyner

Vol. 117: K.J. Hunt
Stochastic Optimal Control Theory with Application in Self-Tuning Control
X, 308 pages, 1989.

Vol. 118: L. Dai
Singular Control Systems
IX, 332 pages, 1989

Vol. 119: T. Başar, P. Bernhard
Differential Games and Applications
VII, 201 pages, 1989

Vol. 120: L. Trave, A. Titli, A. M. Tarras
Large Scale Systems: Decentralization, Structure Constraints and Fixed Modes
XIV, 384 pages, 1989

Vol. 121: A. Blaquière (Editor)
Modeling and Control of Systems in Engineering, Quantum Mechanics, Economics and Biosciences
Proceedings of the Bellman Continuum Workshop 1988, June 13–14, Sophia Antipolis, France
XXVI, 519 pages, 1989

Vol. 122: J. Descusse, M. Fliess, A. Isidori, D. Leborgne (Eds.)
New Trends in Nonlinear Control Theory
Proceedings of an International Conference on Nonlinear Systems, Nantes, France, June 13–17, 1988
VIII, 528 pages, 1989

Vol. 123: C. W. de Silva, A. G. J. MacFarlane
Knowledge-Based Control with Application to Robots
X, 196 pages, 1989

Vol. 124: A. A. Bahnasawi, M. S. Mahmoud
Control of Partially-Known Dynamical Systems
XI, 228 pages, 1989

Vol. 125: J. Simon (Ed.)
Control of Boundaries and Stabilization
Proceedings of the IFIP WG 7.2 Conference Clermont Ferrand, France, June 20–23, 1988
IX, 266 pages, 1989

Vol. 126: N. Christopeit, K. Helmes M. Kohlmann (Eds.)
Stochastic Differential Systems
Proceedings of the 4th Bad Honnef Conference June 20–24, 1988
IX, 342 pages, 1989

Vol.127: C. Heij
Deterministic Identification of Dynamical Systems
VI, 292 pages, 1989

Vol. 128: G. Einarsson, T. Ericson, I. Ingemarsson, R. Johannesson, K. Zigangirov, C.-E. Sundberg
Topics in Coding Theory
VII, 176 pages, 1989

Vol. 129: W. A.Porter, S. C. Kak (Eds.)
Advances in Communications and Signal Processing
VI, 376 pages, 1989.

Vol. 130: W. A. Porter, S. C. Kak, J. L. Aravena (Eds.)
Advances in Computing and Control
VI, 367 pages, 1989

Vol. 131: S. M. Joshi
Control of Large Flexible Space Structures
IX, 196 pages, 1989.

Vol. 132: W.-Y. Ng
Interactive Multi-Objective Programming as a Framework for Computer-Aided Control System Design
XV, 182 pages, 1989.

Vol. 133: R. P. Leland
Stochastic Models for Laser Propagation in Atmospheric Turbulence
VII, 145 pages, 1989.

Vol. 134: X. J. Zhang
Auxiliary Signal Design in Fault Detection and Diagnosis
XII, 213 pages, 1989

Vol. 135: H. Nijmeijer, J. M. Schumacher (Eds.)
Three Decades of Mathematical System Theory
A Collection of Surveys at the Occasion of the 50th Birthday of Jan C. Willems
VI, 562 pages, 1989

Vol. 136: J. Zabczyk (Ed.)
Stochastic Systems and Optimization
Proceedings of the 6th IFIP WG 7.1 Working Conference, Warsaw, Poland, September 12–16, 1988
VI, 374 pages. 1989